Advances in
MICROBIAL ECOLOGY

Volume 5

ADVANCES IN MICROBIAL ECOLOGY

Sponsored by International Commission on Microbial Ecology,
a unit of International Association of Microbiological Societies
and the Division of Environmental Biology of the
International Union of Biological Societies

A Continuation Order Plan is available for this series. A continuation order will bring
delivery of each new volume immediately upon publication. Volumes are billed only upon
actual shipment. For further information please contact the publisher.

Advances in
MICROBIAL ECOLOGY

Volume 5

Edited by
M. Alexander
Cornell University
Ithaca, New York

PLENUM PRESS · NEW YORK AND LONDON

The Library of Congress cataloged the first volume of this title as follows:

Advances in microbial ecology. v. 1—
  New York, Plenum Press c1977—
    v. ill. 24 cm.
  Key title: Advances in microbial ecology, ISSN 0147-4863

  1. Microbial ecology—Collected works.
QR100.A36   576'.15   77-649698

Library of Congress Catalog Card Number 77-649698
ISBN 0-306-40767-1

© 1981 Plenum Press, New York
A Division of Plenum Publishing Corporation
233 Spring Street, New York, N.Y. 10013

Printed in the United States of America

Contributors

Alexander B. Filonow, Department of Botany and Plant Pathology, Michigan State University, East Lansing, Michigan 48824

D. M. Griffin, Department of Forestry, Australian National University, Canberra, A.C.T., Australia

F. LeTacon, Institut National de la Recherche Agronomique CNRF, Champenoux 54280, Seichamps, France

John L. Lockwood, Department of Botany and Plant Pathology, Michigan State University, East Lansing, Michigan 48824

B. Mosse, Formerly in Department of Soil Microbiology, Rothamsted Experimental Station, Harpenden, Herts, England

Jeanne S. Poindexter, The Public Health Research Institute of New York, Inc., New York, New York 10016

D. P. Stribley, Department of Soils and Plant Nutrition, Rothamsted Experimental Station, Harpenden, Herts, England

J. G. Zeikus, Department of Bacteriology, University of Wisconsin, Madison, Wisconsin 53706

Preface

Advances in Microbial Ecology was established by the International Commission on Microbial Ecology to provide a vehicle for in-depth, critical, and, it is hoped, provocative reviews on aspects of both applied and basic microbial ecology. In the five years of its existence, *Advances* has achieved recognition as a major source of information and inspiration both for practicing and for prospective microbial ecologists. The majority of reviews published in *Advances* have been prepared by experts by invitation from the Editorial Board. Although the Board intends to continue its policy of soliciting reviews, some unsolicited review outlines have been approved and the authors invited to proceed with the preparation of manuscripts for publication in *Advances*. The Editorial Board continues to encourage microbial ecologists to submit unsolicited proposals on original topics, in outline form only, for consideration by the Board for subsequent publication.

Volume 5 of *Advances in Microbial Ecology* again covers a broad range of topics, with particular emphasis on the ecology of fungi and on the role that environmental extremes play in the overall behavior of microorganisms in natural habitats. J. G. Zeikus examines the biochemistry and ecological significance of lignins, with particular reference to the overall carbon cycle in nature. The responses of fungi to nutrient limitations and to inhibitory substances in natural habitats are considered in the review by J. L. Lockwood and A. B. Filonow. In a somewhat similar vein, J. S. Poindexter discusses the possible basis for oligotrophy, which describes the ability of certain bacteria to grow at very low nutrient concentrations. The contribution by D. M. Griffin looks at the stresses imposed upon microorganisms by water limitations in natural habitats. The mycorrhizal fungi and their beneficial contributions to plant growth are the subject of the review by B. Mosse, D. P. Stribley, and F. LeTacon.

The Editor and members of the Editorial Board of *Advances* are appointed by the International Commission on Microbial Ecology for fixed terms. With the publication of this volume, our colleague Hans Veldkamp completes his term as a Board member. The Editor and members of the Editorial Board offer to Hans their sincerest thanks for his outstanding help and guidance since the inception of the series.

M. Alexander, Editor
T. Rosswall
K. C. Marshall
H. Veldkamp

Contents

Chapter 3
Water and Microbial Stress
D. M. Griffin

Chapter 4
Ecology of Mycorrhizae and Mycorrhizal Fungi
B. Mosse, D. P. Stribley, and F. LeTacon

Chapter 5
Lignin Metabolism and the Carbon Cycle: Polymer Biosynthesis, Biodegradation, and Environmental Recalcitrance
J. G. Zeikus

Responses of Fungi to Nutrient-Limiting Conditions and to Inhibitory Substances in Natural Habitats

JOHN L. LOCKWOOD AND ALEXANDER B. FILONOW

1. Introduction

Fungi, as heterotrophic microorganisms, coexist with numerous other microorganisms, with whom they must compete for a share of nutrients. Since such nutrients, particularly energy substrate, are often in short supply, adaptive traits have evolved that enhance survival (Lockwood, 1977). Moreover, numerous organic metabolic products are produced as a result of microbial degradation of various substrates. These, together with various mineral components of a habitat, also may affect microorganisms, including fungi. In this review we attempt to discuss the responses of fungi to nutrient competition with other microorganisms and to the presence of inhibitory substances that occur in natural environments.

Because of the great potential breadth of the subject matter encompassed by this topic, it has been necessary to restrict the scope of our review. Thus, we have confined the bulk of our discussion to the soil, since this environment has been by far the most studied with respect to the effects of nutrient limitation and toxic substances on fungal ecology. Where evidence exists for the expression in other environments of phenomena related to those occurring in soil, e.g., mycostasis on plant leaves, this also is discussed. We have restricted the cov-

JOHN L. LOCKWOOD AND ALEXANDER B. FILONOW ● Department of Botany and Plant Pathology, Michigan State University, East Lansing, Michigan 48824.

erage of nutrient effects to those imposed by nutrient deficiencies, rather than include responses to an abundance of nutrients, though it is granted that ecologically important responses to the presence of nutrients (in addition to vegetative growth), e.g., zoospore attraction to roots, are known. Since effects owing to inhibitory substances are manifestly negative ones, restriction of the subject matter concerning nutrients to responses to shortages allows a comparison of the effects of nutrient deprivation and toxic substances. We have attempted to evaluate the relative significance of each for a particular phenomenon. We also have dealt only with naturally occurring inhibitory substances, and have avoided consideration of pesticide residues, toxic pollutants from the atmosphere, or phytoalexins involved in suppressing fungal infection of plants. These are major subjects in themselves, and have been reviewed elsewhere (respectively, Goring and Hamaker, 1972; Heagle, 1973; Kuć, 1976).

For purposes of this review, our use of the term *inhibitory* is broad. It includes inhibitory and toxic substances and lytic enzymes of microbial origin that may degrade fungal cell walls.

Our coverage of fungal responses to nutrient limitation and inhibitory substances will deal primarily with mycostasis and mycolysis. Other fungal activities that will be discussed are regermination of propagules, appressorium formation, and persistent structure formation. Unfortunately, little work has been done regarding the effect of inhibitory substances on fungal activities other than mycostasis and mycolysis.

2. The Soil Habitat

2.1. Energy Budgets

Fungi, and other heterotrophic microorganisms in soil, must subsist on (a) plant residues, such as root and leaf litter from trees, shrubs, herbs, and grasses; (b) animal bodies and excreta; (c) other microorganisms, in the case of mycoparasites; and (d) exudates from living roots. Humic materials are also present but probably do not serve as a major source of energy, since their turnover rate is very slow (Barber and Lynch, 1977).

There are few estimates of the amounts of these components in soils, but Lynch and Panting (1980) provide an estimate from an arable wheat field. Of the total annual substrate input of 3540 kg C/ha per year, 79% was straw residues, 11% decomposed roots, 7% root exudates, and 3% autotrophic microbes.

In soil, the supply of readily utilizable carbon is apparently severely limited (Clark, 1965; Lockwood, 1977) and hence is an object of intense competition. Recent studies relating microbial populations to substrate availability in soil verify this concept, and a consideration of these findings seems justified in terms of its importance to fungal ecology. A number of investigators have related measurements of microbial biomass to the amount of substrate available. Much use has been made of the Monod growth equation (Marr *et al.,* 1963) to calculate the amount of substrate required for cell maintenance (functions other than growth):

$$dx/dt + ax = Y(ds/dt) \qquad (1)$$

where a is the specific maintenance constant (per hour), x the concentration of cells in grams, Y the yield coefficient (efficiency of conversion of substrate to cells), and s the amount of substrate required for maintenance. Subtracting s from the total substrate input gives the amount available for growth. Values of a, or the maintenance coefficient m (g substrate/g dry weight per hr), have generally been obtained from laboratory studies. Values of a and m are interconvertible by $m = a/Y$. Estimates of a for aerobically grown bacteria in chemostats were 0.025–0.028/hr at 30°C and 0.005/hr at 15°C (Marr *et al.,* 1963), and 0.042/hr at 37°C (Pirt, 1965); Barber and Lynch (1977) advocate use of values of this order. However, when such values are used, maintenance energy requirements, in most cases, appear to consume far more energy than is available in the substrate (Barber and Lynch, 1977). Flanagan and Van Cleve (1977) obtained an m value of 0.00055 g/g substrate (equivalent to a = 0.00022). Most workers have used a values of the order of 0.001, and values of a obtained from experiments done in soil itself are of a similar magnitude (Shields *et al.,* 1973; Behera and Wagner, 1974).

The number of generations of microbial growth per year is given by the equation

$$Y(S + xR) = xR \qquad (2)$$

(Gray and Williams, 1971), where Y and x are as in equation (1), S is the substrate available for growth, and R is the number of generations per year. This equation allows for the cells formed to serve as secondary substrates (Gray and Williams, 1971).

Values for Y for individual bacteria in chemostats are usually in the range 0.35–0.60 (Marr *et al.,* 1963; Payne, 1970; Flanagan and Bunnell, 1976; Flanagan and Van Cleve, 1977), and those in soil from 0.39–0.60 (Shields *et al.,* 1973; Behera and Wagner, 1974).

In nine studies tabulated (Table I), the annual number of generations of microbial growth in soil was determined to be very low. These estimates range from less than 1 to 36 generations/year, and strongly indicate that substrate limitations place severe restrictions on the ability of fungi (and other microorganisms) to grow in soil.

Measurements needed to estimate the number of generations occurring in

Table I. Predicted Numbers of Generations of Microbial Cells per Year and Maintenance Energy Requirements from Estimates of Substrate Inputs, Yield Coefficients, Biomass, and Specific Maintenance Constants in Several Soils

Reference	Substrate input (S)	Yield coefficient (Y)	Biomass (x)	Specific maintenance constant (a)	Maintenance energy requirements (s)	Number of generations (R)
Clarholm and Rosswall (1980)	441 g/m²	0.5[a]	35 g/m²	0.001[a]	613 g/m²	<1
Lynch and Panting (1980)	3540 kg C/ha	0.5[a]	200 kg C/ha	0.002[a]	3504 kg C/ha	<1
Gray and Williams (1971)	764 g/m²	0.35	46 g/m²	0.001	1156 g/m²	<1
Babiuk and Paul (1970)	500 g/m²	0.35	55 g/m²[b]	0.001	1364 g/m²	<1
Shields et al. (1973)	937 μg C/g	0.60	165 μg C/g	0.002	300 μg C/g	5.8
Flanagan and Van Cleve (1977)	398 g/m²	0.40	5.7 g/m²	0.00055	22.6 g/m²	21
Flanagan and Bunnell (1976)	10 g/m²[c]	0.54	0.45 g/m²[d]	0.00032	1.2 g/m²	23
Behera and Wagner (1974)	1040 kg/ha	0.39	8.4 kg/ha	0.0036	265 kg/ha	36
Gray et al. (1974)	706 g/m²	0.35	3.7 g/m²[b]	0.001	92 g/m²	87
	706 g/m²	0.35	11.1 g/m²[e]	0.001	277 g/m²	20

[a]Value supplied, or substituted for that in reference.
[b]Bacteria only.
[c]Amount available for microbial growth of annual input of 70 g/m².
[d]Fungi only.
[e]Assuming fungal biomass = 2 × bacterial biomass.

soil are subject to error. It is not reasonable, for example, that maintenance energy requirements should consume all of the substrate available (or more!) without allowing for any growth (Table I). This could result from estimates of Y or S being too low, or from those of x or a being too high. For example, in the study of Clarholm and Rosswall (1980), the estimated number of generations is given in Table I as <1. The authors gave no estimate of a, and our calculations used $a = 0.001$, which is an order of magnitude less than many estimates in chemostats. Actual specific maintenance might have been still lower. Direct counts of bacterial production revealed that the population approximately doubled after each of 12 seasonal rain events. Assuming that the fungi contributed equally to soil respiration and $Y = 0.50$, the biomass produced by the 12 generations (220 g C/m^2) would have consumed all of the substrate available (220 g C/m^2).

In the work of Lynch and Panting (1980), the annual number of generations was also calculated to be <1. Of the total 3540 kg C substrate input, 2800 kg C/ha was straw; this figure was used in the calculations, but straw is normally burned and thus was not available to support the microbial population. Omitting the substrate contributed by straw would result in an even lower estimate of the number of generations.

Soil microbial biomass could be overestimated when direct counts of bacterial cells are made and lengths of fungal hyphae are measured microscopically (Babiuk and Paul, 1970; Gray and Williams, 1971; Shields *et al.,* 1973; Flanagan and Bunnell, 1976; Flanagan and Van Cleve, 1977), because some proportion of dead cells is probably included. In some studies (Behera and Wagner, 1974; Gray *et al.,* 1974), only soil bacterial or fungal biomass was included. Including the other component would decrease the number of generations. For example, Gray *et al.* (1974) estimated 87 bacterial generations per year; assuming that the fungal biomass was twice the bacterial biomass— a conservative estimate—the number of generations would be reduced to 20 (Table I). If the ratio of fungal to bacterial biomass is taken to be 5:1, the number of generations would be 3.6 per year.

Clark and Paul (1970) looked at the problem in more general terms. Assuming a yield coefficient of 0.7 (higher than any reported), each gram of cell dry weight synthesized would require 1.4 g glucose. Thus, 20 g microbial biomass/m^2 would require 280 g glucose, or ca 1100 kcal energy, to produce a new generation of microbial cells. Since net annual production rarely exceeds 5000 kcal/m^2, reproduction could only occur a few times per year—not taking account of maintanance energy requirements.

Conclusions corroborating the above have been derived from a comparison of respiratory data from soil and laboratory (Clark, 1967; Clark and Paul, 1970). In stationary populations of bacterial cells, CO_2 production per day is nearly equivalent to cell dry weight (Clark and Paul, 1970). However, in field soils a microbial biomass of 200 g/m^2 is 35–70 times the weight of CO_2 produced daily; moreover, the weight of CO_2 produced also includes that respired by fauna and roots, which may amount to 40% of the total soil respiration (Kucera and Kirkham, 1971). Therefore, growth and activity must be severely limited in soil. Similar conclusions were drawn by Gray and Williams (1971) in comparing soil biomass, CO_2 evolution in the field, and hypothetical growth rates in soil. Clarholm and Rosswall (1980) reported that fungi respired 110 g C/m^2 per year in a forest soil. Assuming a yield coefficient of 0.5, the bacterial production of 105 g C/m^2 per year would require 105 g C/m^2 per year. Thus, combined bacterial and fungal respiration was equal to 215 g C/m^2 per year, accounting for all of the above- and belowground litter production, which amounted to about 220 g C/m^2 per year.

In spite of the errors inherent in estimations of the data required for calculations such as those cited in Table I, it seems clear that the soil must be viewed as an impoverished medium insofar as carbon substrate is concerned. Inevitably then, a large proportion of the microbial population of the bulk soil must for the most part be confined to a state of enforced quiescence. Brief bursts of activity may occur in response to inputs of plant and animal litter (such as decomposing roots or leaf fall) or following physical disturbances (including wetting and drying, freezing and thawing, and addition of water) that kill microorganisms or cause redistribution of nutrients isolated from microbial contact with microsites. It may be argued that humic materials provide sources of energy for microbial growth (Griffin and Roth, 1979), but this seems unlikely in view of the average residence time of humus of over 1000 years (Barber and Lynch, 1977) and the fact that humus itself is ultimately derived from plant litter.

2.1.1. Root Exudates

Of the loci capable of supporting microbial growth in soil, probably that which offers the best opportunity for sustained growth of microorganisms is the rhizosphere, since it is here that energy sources are available more or less continuously for the life of the root. It is well established that roots release into soil a wide range of sugars, amino acids, proteins, organic acids, and other compounds which may be used by microorganisms. The rhizosphere is of particular significance to root-infecting fungi, since root exudates provide the nutrients which stimulate the fungal propagules to germinate and to grow toward the root. The exudates also provide the energy required for infection and may attract zoospores of Oomycetous fungi to the root (Schroth and Hildebrand, 1964). The rhizosphere has been the object of much study, and readers are referred to reviews for general information (Rovira and Davey, 1974; Bowen and Rovira, 1976; Hale *et al.*, 1978; Bowen, 1979). However, recent information on sites of exudation and on the quantities of photosynthate released from roots seem pertinent to this review.

Sites of exudation have been identified using plants grown with $^{14}CO_2$. Methods of detection include localizing the radioactivity of the roots by autoradiography, by collection of exudate on filter papers and scanning for radioactivity, and by hot ethanol extraction of roots cut into segments followed by counting the radioactivity by scintillation. Such studies have revealed, for wheat seedling roots, a very early (first 2 min) intense locus of exudation from the apices of emerging lateral roots (McDougall and Rovira, 1970). This represented about half of the exudate released during a 2-hr period. The discreteness of the spots in autoradiograms suggested that the labeled material is nondiffusible, and that it may consist of root cap cells and mucilaginous material. Later, a more diffuse zone of radioactivity appeared along the axis of the primary root and to a lesser extent along the axes of the lateral roots. These were thought to represent lower-molecular-weight diffusible substances. The area of major release was in the zone of elongation distal to the root tip.

Sloughed cells and tissues apparently constitute a major portion of the organic carbon and nitrogen in the rhizosphere. Of the carbon released from wheat roots into the soil, insoluble mucilaginous material including sloughed root cap cells accounted for about four times as much as water-soluble exudate (Bowen and Rovira, 1973; Martin, 1977a). Griffin *et al.* (1976) have attempted to determine directly the amount of sloughed cells and tissues released from the roots of solution-grown peanut seedlings. After 2–4 weeks, root caps, cortical tissue fragments and sheets, and individual cells were seen microscopically. Sloughed material ranged from 0.26 to 0.73 mg/plant per week and ca 1.5 mg/g root dry weight per week. Thus, about 0.15% of the root tissue was sloughed per week.

Total materials lost from roots, including exudates, lysates, mucigel, and cellular material, may constitute a surprisingly large amount of the carbon fixed. Martin (1975) grew wheat, clover, and ryegrass plants in nonsterile soil in an atmosphere continuously supplied with $^{14}CO_2$. The amount of ^{14}C lost from the roots after 8 weeks was 3.1–5.8% of the total ^{14}C fixed and 10.4–38.4% of that translocated to the roots. Roots of wheat and barley plants grown under nonsterile conditions released about twice as much material as did roots grown axenically for 3 weeks (Barber and Martin, 1976). Roots of wheat and barley grown in nonsterile soil released 18–21% and 25%, respectively, of the total dry matter production. Corresponding values as proportions of root dry matter production were 48–63% for wheat and 74% for barley. Most of the lost material was in the insoluble fraction. To account for the greater ^{14}C loss from roots grown in nonsterile soil, Martin (1977a,b) hypothesized that soil microorganisms induced autolysis of cortical root cells, and that this accounted for release of sugars and amino acids from the roots. The roots apparently were not invaded by microorganisms during autolysis, at least in the early stages. This hypothesis is supported by Holden's (1975) finding that $>70\%$ of the cortical cells in seminal roots of 3- to 4-week-old wheat seedlings were without nuclei and hence were moribund or dead, even though direct examination of unstained roots showed no overt evidence of cell death.

Bowen and Rovira (1973) have pointed out that it should be possible to relate the amount of exudate to potential microbial growth on or near the root. Such a determination was made by Barber and Lynch (1977), who grew barley plants in solution culture in the presence or absence of a mixed population of bacteria. For the first 4 days, there was a close correlation between actual bacterial production and that calculated from exudation data from plants grown axenically. Therefore, however, the actual bacterial yields were 6–12 times the calculated yields, reflecting increased exudation in the presence of microorganisms. From earlier data, the quantity of root exudate from barley seedlings grown in nonsterile soil over a 3-week period was ca 60% of the dry matter increment of the roots (Barber and Martin, 1976). Assuming a yield coefficient of 0.35, the expected weight of bacteria (omitting maintenance requirements) should be about 0.21 mg bacteria/mg root. The actual bacterial yield was in reasonable agreement, ranging from 0.1 to 0.26 mg/mg root during a 7- to 16-day period.

Exudate losses from roots of wheat plants grown at $10°$ or $18°C$ and 3–8 weeks old were reasonably consistent (Martin, 1977b). Expressed in terms of total ^{14}C recovered, these losses ranged from 14.3 to 22.6%, or, in terms of ^{14}C translocated to the roots, 29.2–44.4%. Thus, it seems likely that exudation losses of this magnitude are realistic in terms of field conditions.

Among the factors affecting the amount of material released from roots are plant age, species, temperature (especially extremes), mineral nutrition, O_2 concentration, and soil moisture (Hale et al., 1978). With respect to soil moisture, about twice as much material was released from portions of roots of wheat plants grown in soil at 8% water content as compared with that lost from portions of roots at 22% soil moisture; exudation was intermediate at 15% moisture (Martin, 1977c). In these experiments, the upper half of the root systems was maintained at the water contents indicated, but the lower half was held at saturation (26% moisture). Thus, exudation was increased even in conditions of localized water stress that had no differential effect on shoot growth. The material released consisted of both insoluble mucilaginous material and soluble material. It had been previously shown in experiments in which soil moisture was less well controlled that about four times more low-molecular-weight substances were released from roots in dry than in moist soil (Barber and Martin, 1976). The dry soil also had $< 1\%$ of the number of fluorescent pseudomonads that were present in moist soil (Martin, 1977c). Such a reduction in bacterial populations would reduce competition for the released materials and allow fungi, which are more tolerant of low soil moisture (Cook, 1973), to gain a greater share of resources. Increased exudation from roots at lower water potentials, coupled with decreased populations or decreased activity of bacteria, may account in part for increased severity of fungal root disease under conditions of soil water stress (Cook, 1973).

2.2. Nitrogen and Minerals

Many organic residues contain nitrogen and other minerals in sufficient quantity that the supply of these materials should not limit decomposition (Clark, 1967). For example, the C:N ratio of herbaceous plant residues may be as low as 10:1 (Burges, 1967; Levi and Cowling, 1969). However, that of woody stem tissue may be of the order of 50:1 or greater (Burges, 1967; Levi and Cowling, 1969; Frankland, 1974) and that of wood, greater than 100:1 (Levi and Cowling, 1969). Competition for nitrogen may be expected to occur during decomposition of tissues of higher C:N ratios, though Clark (1967) does not consider nitrogen to be limiting in plant residues if its content is not less than 1.5% (C:N ratio ca 28:1). Supplemental nitrogen is commonly applied to hasten the decomposition of plant residues during composting, and also will aid the decomposition of straw and other materials of high C:N ratio in or on soil (Alexander, 1977). An indirect consequence of enhanced decomposition of wheat stem tissue in nitrogen-supplemented soil is the more rapid competitive displacement of certain weakly saprophytic root- and foot-rotting pathogens of wheat that had previously colonized the stem pieces (Garrett, 1970, 1976).

The nitrogen content of wood is very low (ca 0.03–0.1%) and thus presents a very deficient substrate for microorganisms. Fungi adapted for this habitat have a remarkable ability to grow and produce cellulase at low nitrogen concentrations (Levi and Cowling, 1969), to scavenge efficiently for the small amount of nitrogen available, to concentrate it in sporophores and spores (Merrill and Cowling, 1966), and to recycle the nitrogenous constituents of their own mycelium through autolysis and reutilization (Levi et al., 1968).

The supply of nitrogen (or other minerals) does not appear to limit microbial activity in soils (Shields et al., 1974), except when the supply of carbon is artificially enhanced, thus rendering other materials unavailable. For example, in a natural sandy soil, increased CO_2 evolution in the presence of nitrogen, phosphorus, and sulfur could be shown only following the addition of supplemental glucose (Stotzky and Norman, 1961a,b). Similarly, growth suppression of Fusarium oxysporum f. sp. cubense by several non-antibiotic-producing bacteria in sterilized soil was overcome by the addition of sucrose but not nitrogen and phosphorus (Marshall and Alexander, 1960; Finstein and Alexander, 1962). These elements gave a slight growth response only after the carbon demand was satisfied.

In plant pathology, experimental control of Fusarium root rot of bean has been achieved with organic amendants of high C:N ratio, such as barley straw or cellulose, which presumably resulted in immobilization of nitrogen required by the pathogen (Snyder *et al.,* 1959; Maurer and Baker, 1965).

In water, the absolute amount of nitrogen may be very low. Park (1976), in a significant study, compared the ability of a pythiaceous fungus to decompose cellulose in a liquid medium containing the same total amount of nutrients but in different volumes of solution. Decomposition was manyfold faster in the larger volume than in the smaller, suggesting that low concentrations of nutrients may be more favorable than high concentrations for aquatic fungi. When the C:N ratio was varied from 40 to 4000, most decomposition occurred at the lower ratios, but the efficiency (amount decomposed/unit of N) was greatest at 400:1. There was some decomposition even at 1600:1.

In summary, it would seem that competition for nitrogen can occur in substrates in which nitrogen concentration is low, such as in woody stem tissues, wood itself, and water. Some fungi appear to be physiologically well adapted for growth in such habitats, in which they would be expected to have a competitive advantage.

2.3. Inhibitory Substances

Soil is a repository for numerous inorganic and organic chemical substances, both natural and man-made. It is a sink for numerous gaseous chemicals, such as ammonia (Malo and Purvis, 1964), ethylene (Abeles *et al.,* 1971), sulfur compounds (Bremner and Steele, 1978), and carbon monoxide (Inman *et al.,* 1971), to name a few. In addition to gaseous inputs, soil regularly receives organic and mineral inputs from living, dead, and decomposing plants and animals. These inputs are constantly, and in most cases, rapidly reworked to yield organic and inorganic substances that leave the soil or become part of its complex, recalcitrant humic phase. The net result is a scarcity of readily available nutrients in soil for microbial metabolism, as discussed in Section 2.1., and the formation of inhibitors. Many of the inhibitory substances that are formed in soil will not move beyond their foci of formation, whereas others will diffuse. In general, the extent to which such substances diffuse in soil depends on their chemical nature (volatility, solubility, and ability to react with metals), the soil pH, and their reaction with adsorptive matrices (Goring and Hamaker, 1972).

Adsorption to organic and mineral surfaces plays a major role in controlling the movement and degradation of volatile and nonvolatile soil substances. Adsorption may act as a focusing mechanism, increasing the probability of a fungal cell's encounter with an inhibitor. On the other hand, adsorption may also tenaciously immobilize many potential inhibitory compounds, thereby inactivating them or markedly attenuating their concentrations in the soil solution.

Presently, there appears to be no study of the overall composition of inhibitors and their effects on fungal development in a natural soil. Such an approach, although fraught with difficulty, would contribute valuable insights as to the inhibitory components of a soil and to the interactions of inhibitors with nutrients and environmental parameters that mediate the overall expression of fungal inhibition within the soil.

In one noteworthy study, volatiles from 12 soils were trapped on activated charcoal and later analyzed by gas chromatography (Pavlica *et al.*, 1978). Care was taken to minimize background contamination in the analyses. Classes of compounds detected included alkanes, alkenes, aldehydes, ketones, and inorganic compounds (ammonia and carbon dioxide). Several fungi were tested for germination responses to varying dosages of the most likely volatile inhibitors, the latter judged by frequency and level of detection. Ammonia, which was detected in five of the soils tested, was considered to be a likely candidate for an antifungal volatile. Formaldehyde was considered to be another possible inhibitor, although the frequency of its detection in soils was low. The possible presence of nonvolatile inhibitors in these soils was not explored, however.

There have been attempts (Vaartaja, 1974, 1977) to characterize nonvolatile inhibitory substances in soil leachates by molecular sieve chromatography. Fractionation according to molecular weight followed by bioassay appears to be a reasonable, first-line approach to identifying a nonvolatile substance in the soil solution. However, inhibition from soil leachates or saturated soil extracts should be viewed with great caution, as saturation liberates considerable organic and mineral substances which otherwise may not be available at inhibitory concentrations. In this regard, pressure plate extracts (King and Coley-Smith, 1969) of soil at water potentials less than saturation might be useful.

In lieu of a systematic study of inhibitory soil substances, researchers have chosen to demonstrate whether an inhibitor was volatile or nonvolatile. This seems to be a convenient approach to classifying toxic substances. More attention has been given to volatile compounds. This orientation may be due to the relative ease of analysis by gas chromatography of gaseous compounds compared to nongaseous componds, or it may indicate the great interest by researchers in the composition of the soil atmosphere as an indicator of microbial activity (Francis *et al.*, 1973).

Soil microorganisms produce the bulk of volatile compounds emitted from soil. Active, growing cultures of microorganisms also produce an array of volatile and nonvolatile metabolic products, some of which are inhibitory and others of which are stimulatory to fungal development (Fries, 1973; Hutchinson, 1973; Stotzky and Schenk, 1976). Staled cultures also produce a variety of substances exhibiting inhibitory and morphogenetic properties (Park, 1963; Park and Robinson, 1964; Robinson *et al.*, 1968). Although these studies have yielded valuable information concerning fungal physiology, great caution must be exercised when applying these results to natural soil.

Unlike a culture medium, natural soil is an impoverished medium in which the vast majority of microorganisms exist in a quiescent or dormant condition. Moreover, microorganisms in soil reside within pores of various diameters and structural complexities, which may place severe restrictions on the movement of a potential inhibitor. Finally, microorganisms within soil microhabitats are profoundly influenced by mineral and humic colloids of high surface area and varying physicochemical properties (Hattori and Hattori, 1976). Such a situation cannot be duplicated on an agar surface!

3. Mycostasis

Perhaps the most characteristic, and certainly the most studied, response of fungi to soil is the restriction imposed upon germination. The terms *mycostasis* and *fungistasis* have been applied to this phenomenon, which was reviewed in detail in 1977 (Lockwood, 1977). Specific aspects of mycostasis have been reviewed more recently (Balis and Kouyeas, 1979; Griffin and Roth, 1979; Lockwood, 1979). Aspects applicable to the current review will be discussed here.

Nearly all soils are mycostatic, with the exception of deep subsoils, in which microbial populations are low. Mycostatic expression appears to be maximal at high (ca −0.9 bars) but not saturated soil water potentials (Kouyeas and Balis, 1968) and over a broad range of soil pH values (Schüepp and Green, 1968; Schüepp and Frei, 1969). In assays that attenuated the mycostatic effect, greatest inhibition resulted at pH values near 6–7 and above. The lesser expression of mycostasis at low pH values could be due either to a direct effect of unfavorable pH or indirectly to suppression of bacterial and actinomycete activity. Sterilization of soil by heat, fumigants, or gamma radiation annuls mycostasis, but it can be restored through reinoculation with microorganisms, more or less nonspecifically. The addition of various antibacterial antibiotics to soil also diminished the mycostatic effect (Cook and Schroth, 1965; Adams *et al.*, 1968; Mircetich and Zentmyer, 1969).

Since mycostasis is mainly of microbial origin, it might be expected to occur in other habitats wherein microbial populations are high; in fact, mycostasis also occurs on the surfaces of living plant leaves when populations of leaf-surface microorganisms are sufficiently high (Blakeman and Brodie, 1977) and in seawater (Kirk, 1980). Considerable research has been done on the mycostatic mechanism in soil and on leaves, and this will be discussed in Section 3.4.1.

3.1. Sensitivities

Among the fungi, a range of sensitivity to mycostasis exists (Lockwood, 1977). Steiner and Lockwood (1969) ranked 29 different kinds of spores from 22 fungi, using germination data obtained from varying mixtures of natural and sterilized soils. The most sensitive spores tended to be small and to require exogenous nutrients for germination (nutrient-dependent), whereas the least sensitive spores were large and their germination did not require exogenous nutrients (nutrient-independent). The correlation coefficient for mycostasis sensitivity $\times$ spore volume was $r = -0.81$ and for mycostasis sensitivity $\times$ surface area, $r = -0.85$. The nutrient-dependent and nutrient-independent spore types also have been termed, respectively, *carbon-dependent* and *carbon-independent* (Griffin and Roth, 1979). Some large-spored, nutrient-independent types, including urediospores of rusts, conidia of powdery mildews, activated ascospores of *Neurospora tetrasperma* (Ko and Lockwood, 1967), and conidia and ascospores of *Calonectria crotalariae* (Hwang and Ko, 1974), germinated freely on soil. The constitutively dormant ascospores of carbonicolous Ascomycetes, like those of *N. tetrasperma,* once activated, are mycostasis insensitive (Wicklow and Zak, 1979). These will be discussed in Section 4. It should be emphasized that except for the mycostasis-insensitive group, ranking of sensitivity depends on attenuating the mycostatic effect of soil; as assayed directly in soil, most of even the least sensitive types had low germination. Different species within a genus (Griffiths, 1966; Steiner and Lockwood, 1969), different isolates of the same species (Chinn and Ledingham, 1967), or different spore types within a species (Papavizas and Adams, 1969; Steiner and Lockwood, 1969) may also differ in mycostasis sensitivity. Hyphae appear to be less sensitive than spores of the same species (Steiner and Lockwood, 1969).

An attempt to understand the basis for differences in sensitivity to mycostasis was made by Steiner and Lockwood (1969), who found that sensitivity of 17 of 19 spores tested was directly correlated with germination time as determined on sterilized soil ($r = 0.80$), and was inversely correlated with spore volume ($r = -0.82$). The relation between germination time and mycostasis sensitivity was explained to be the result of a race between the fungal spores and competing microflora for utilization of transitory nutrients. Thus, spores with long germination times would show greater sensitivity and presumably would be thereby disadvantaged. Artificial reduction in germination time by preincubating spores on sterilized soil correspondingly reduced mycostasis sensitivity ($r = 0.72$). The lesser sensitivity of hyphae than of spores was attributed to the absence of a germination period for hyphae (Steiner and Lockwood, 1969). Sensitivity of hyphae also differed among fungi. The mycostatic response of 20 fungi was inversely correlated with growth rate in nonmycostatic conditions ($r = -0.91$) and with hyphal diameter ($r = -0.82$), and was directly correlated with the mycostasis sensitivity of the spores of the of the same species ($r = 0.74$) (Hsu and Lockwood, 1971). Hyphal diameter was directly correlated with growth rate ($r = 0.88$) and with spore volume ($r = 0.84$), Thus, species with large spores tended to germinate rapidly and to exhibit low sensitivity to mycostasis; they gave rise to large hyphae whose early growth was rapid and relatively uninhibited in soil. The opposite was observed for species with small spores (Lockwood, 1977).

An inverse relationship between speed of germination and mycostasis sensitivity was also found for species of *Penicillium* (Dix, 1972) and *Trichoderma* (Mitchell and Dix, 1975). However, Dix (1972) found a tendency for germination of 11 species of *Penicillium* to be inversely correlated with germ tube growth rate ($r = -0.75$), the opposite of the relationship found by Hsu and Lockwood (1971). This discrepancy was attributed to Dix's measuring hyphal growth much later (10–40 hr) than Hsu and Lockwood (0–4 hr). Germ tube growth is probably dependent upon spore reserves, whereas later mycelial growth is obviously independent of the spore. Very early growth rates may differ from later growth rates (Trinci, 1971).

Though most research has been directed toward *mycostasis* in soils, a more appropriate designation might be *microbiostasis* (Ko and Chow, 1977), since there is sufficient information to establish that the activity of bacteria and actinomycetes also is suppressed in soil. Stevenson (1978) also has made a case for bacterial dormancy of exogenous origin in natural aquatic systems. Lloyd (1968) found that spores of 8 isolates of streptomycetes germinated only 3–7% on three soils as compared with 79% on sterilized soil and 89% on water agar. The inhibitory effect was annulled by soil sterilization or upon addition of glucose (Mayfield *et al.*, 1972). Bacteria also show similar suppression (Brown, 1973; Davis, 1976). Perhaps the most convincing work was that of Ko and Chow (1977) showing that streptomycin-resistant mutants of 20 unidentified soil bacteria and 6 identified bacteria failed to increase in numbers when added to natural soil. Like mycostasis, the effect was annulled by soil sterilization or by addition of nutrients. These findings suggest that mycostasis and the general microbiostasis probably have a common origin.

3.2. Annulment

Mycostasis is frequently annulled by the addition of energy-yielding nutrients to soil, with complex sources usually being more effective than single compounds (Lockwood, 1977). Mineral salts have been shown to partially annul bacteriostasis in soil (Brown, 1973), mycostasis in sea water (Kirk, 1980), and mycostasis of leached spores of *Botrytis cinerea in vitro* (Sztejnberg and Blakeman, 1973). However, in most instances mineral salts were ineffective. In soil, recognized sites for germination and growth of fungi include rhizospheres and spermospheres (Schroth and Hildebrand, 1964) and the proximity of undecomposed plant residues (Schroth and Hendrix, 1962; Griffiths and Dobbs, 1963), all of which are relatively rich in soluble nutrients.

It seems likely that carbon sources, in annulling mycostasis, are serving as energy sources rather than as nonnutritive "activators," but there has been some doubt with volatile stimulants of germination. An interesting example is the stimulation of germination of sclerotia of *Sclerotium rolfsii* (Linderman and Gilbert, 1969) and *Verticillium albo-atrum* (Gilbert and Griebel, 1969) effected with only short exposures to the vapors of a distillate of alfalfa hay. Contrary to expectations, when ^{14}C-labeled ethanol was applied to soil or to conidial suspensions of *Fusarium solani* f. sp. *phaseoli,* almost all of the respired $^{14}CO_2$ was derived from oxidation of the ethanol (Griebel and Owens, 1972). Thus, the stimulatory effect of the ethanol and, by implication, of other volatiles was attributed to their serving as energy sources rather than as "activators."

Many nutrient-dependent propagules appear to require the presence of an energy source throughout the germination period (Steiner and Lockwood, 1969; Yoder and Lockwood, 1973). That is, withdrawal of the carbon source before emergence of the germ tube has resulted in cessation of progress toward germination. The nutrient-independent conidia of *Cochliobolus sativus,* however, germinated freely in soil after an initial exposure to water (Bristow and Lockwood, 1975b). Perhaps such large spores are able to mobilize endogenous reserves for germination during exposure to noncompetitive conditions.

Although supplementing soil with carbon sources such as cellulose, glucose, and plant residues may temporarily relieve mycostasis, after a longer period mycostasis may be enhanced. For example, germination of conidia and chlamydospores of *Thielaviopsis basicola* was enhanced 4 days after incorporation of 1% alfalfa hay, but thereafter more alfalfa was required to promote a given level of germination in supplemented soil than in nonsupplemented soil (Adams and Papavizas, 1969). Prior amendment of soil with carbon sources may also reduce germination in response to plant exudates. Germination of *T. basicola* chlamydospores in the spermospheres of bean was reduced from 38% before amendment to zero by alfalfa hay or corn stover added 7 or more days previously (Papavizas and Adams, 1969).

3.3. Ecological Significance

The repression of germination in the absence of a potentially exploitable substrate has obvious survival value for fungal propagules. This concept was experimentally confirmed by Chinn and Tinline (1964), who showed that isolates of *C. sativus* capable of spontaneous germination in soil disappeared rapidly, whereas isolates whose germination was restricted persisted. The apparent general ability of rust urediospores to germinate freely in soil (Ko and Lockwood, 1967) may contribute to their well-known inability to survive in soil from season to season.

Few attempts have been made to relate differences in sensitivity to fungistasis to fungal ecology. Dix (1967) found a relationship between mycostasis sensitivity of 12 fungi colonizing bean roots and their colonization sites. The least sensitive were pioneer colonizers of the surfaces of root tips; slightly more sensitive were secondary colonizers of the root tissues farther back; still more sensitive were fungi invading moribund tissues. The most sensitive fungi were restricted to the rhizosphere. This colonization pattern was related to increasing amino acid exudation from the root tip back, with the greatest exudation occurring from invaded portions of the root.

It would be of interest to know the ecological significance, if any, of the relationship demonstrated among mycostasis sensitivity, spore size, and germination time (Steiner and Lockwood, 1969). The association of large spore size, rapid germination, and independence from exogenous nutrients in many foliar pathogens may reflect selection pressure for these traits. Gregory (1973) has suggested that the large spores characteristic of pathogens of cereal leaves may be a result of selection for high impaction efficiency, an important mechanism of spore deposition on such leaves. Rapid germination may be advantageous since the duration of adequate moisture for germination (e.g., from dews) on foliage is short as compared with soil. In addition, the ability to germinate without exogenous nutrients would permit germination to proceed whenever moisture was available.

The fact that the mycostatic level of soil is manipulatable suggests that enhanced mycostasis might result in reduced severity of root diseases via suppressed germination and infection. This possibility has not been evaluated to the extent merited, but a reduction in the severity of *Verticillium dahliae* wilt of strawberries by soil amendment with 0.2% chitin or laminarin was associated with increased soil mycostasis (Jordan *et al.*, 1972). Amendments that were not effective in reducing wilt severity had a lesser effect on mycostasis. Certain root pathogens have been controlled in the laboratory by stimulating germination by organic amendments, thus rendering the pathogen germ tube vulnerable to lysis. Examples, among others (Lockwood, 1977), are reduction in the population density of *T. basicola* by alfalfa meal and other dried green plant material (Adams and Papavizas, 1969; Sneh *et al.*, 1976) and of *C. sativus* by soybean meal (Chinn and Ledingham, 1961). This approach is effective only for pathogens that do not readily form new persistent structures. Whether sufficient organic material can be practically incorporated to reduce pathogen populations in the field by this mechanism is not known.

As soils differ in mycostasis capacity, so they differ in responsiveness to nutrient additions to annul mycostasis. Filonow and Lockwood (1979) found that mycostasis of conidia of *Cochliobolus victoriae* was more readily annulled by nutrient "titration" in light-textured soils than in heavier-textured soils. By this method of assessing mycostasis, heavier-textured soils were more mycostatic. However, more exudate was lost from the spores in light-textured than in heavier-textured soils. These findings raise the question of which aspect has the greater significance for the fungus—the ease with which it can germinate in response to exogenous nutrients or its ability to retain endogenous reserves.

Since soils differ in their ability to inhibit fungal germination, one might expect consequent ecological differences. Soils "suppressive" or "conducive" to certain root pathogens are known (Arjunarao, 1971; Smith and Snyder, 1972;

Alabouvette *et al.*, 1979; Hancock, 1979; Hornby, 1979), and these properties may be related to the germinability of the pathogen in the soils (Arjunarao, 1971; Smith and Snyder, 1972). However, the relation of soil suppressiveness to the general mycostasis has not been investigated.

3.4. Mechanisms of Mycostasis

3.4.1. Nutrient Depletion

3.4.1a. Soil. The information presented in Section 2.1 on the energy budgets of soils provides a solid basis for the concept that fungi and other microorganisms in soil exist in an environment characterized by insufficient resources for sustained growth. More direct evidence that nutrient deprivation can restrict germination of fungal propagules is based on the similarity of the behavior of propagules incubated in soil and in model systems that impose nutrient stress upon them. For example, propagules that germinated poorly in distilled water also germinated poorly in soil, whereas those of several other fungi which germinated in water also germinated in soil (Ko and Lockwood, 1967). A third group of propagules—which includes conidia, chlamydospores, and sclerotia—was strongly inhibited when incubated in model systems designed to create a steepened diffusion gradient away from the propagules, such as is thought to exist in soil due to microbial activity. In these systems, propagules borne on membrane filters are usually exposed to either dripping water or salt solutions in a filter funnel, or are incubated on a bed of washed, sterile sand through which the solutions percolate. Nutrient solutions used as leaching media supported germination, and activated ascospores of *N. tetrasperma,* which germinate freely on soil and in water, were not inhibited in the leaching system (Ko and Lockwood, 1967). Thus, the leaching systems, qualitatively, appeared to mimic the mycostatic effect of soil.

The imposition of mycostasis by nutrient deprivation appears to be related to increased exudation losses from the propagules. Leached sand induced greater losses of exudate from conidia of *C. victoriae* and *Curvularia lunata* and from sclerotia of *Sclerotium cepivorum* than occurred on nonleached sand; intermittent leaching induced greater losses than occurred during static periods (Bristow and Lockwood, 1975a; Sneh and Lockwood, 1975). Passage of exudates from the same species through the leaching system stimulated germination nonspecifically, and the exudate was taken up by propagules made nutrient-dependent by prolonged leaching (Bristow and Lockwood, 1975a).

The leaching system has been calibrated by comparing the amount of exudate lost from ^{14}C-labeled conidia of *C. victoriae* on soil and in the leaching system operated at various flow rates. The faster the rate of flow, the greater the amount of exudate lost and the lower the germination response (Sneh and Lockwood, 1975). Conidia incubated on soil always lost larger amounts of exudate than they did during incubation on the leaching apparatus at flow rates sufficient to impose mycostasis. This result was confirmed using the same fungus with eight different soils (Filonow and Lockwood, 1979). At flow rates sufficient to reduce germination to levels of $<10\%$, exudation in the leaching system was 20–70% of that from conidia incubated on seven of the eight soils and was equal to that on the other soil. At least for *C. victoriae*, the leaching system did not appear to impose an excessive stress on the propagules, as compared with that of soil. A corollary of these results is that each of the eight soils apparently imposed a sufficient nutrient sink to account for mycostasis without the obligatory participation of other factors.

The rapid metabolism of exudates from fungal spores and sclerotia by microbes in soil indicates that the exudates can readily serve as substrates for microbial growth (Lingappa and Lockwood, 1964; Bristow and Lockwood, 1975a). For example, $^{14}CO_2$ was detected within a few minutes of adding ^{14}C-labeled glucose or exudate from conidia of *C. victoriae* to soil (Bristow and Lockwood, 1975b). Other evidence for the existence in soil of an efficient microbial sink for nutrients is that loss of glucose from paper disks was more rapid on natural soil than on sterile soil, with the difference easily detectable within 8 min; the mean glucose half-life on several natural soils was 80 min and for autoclaved soils, 240 min (Lockwood, 1975).

The soil's microbial nutrient sink can effect losses from fungal propagules not only of preformed materials but also of substrate taken up during the course of germination. Spores were first exposed to sterile soil for an interval just short of that required for emergence of the germ tube, and they were then transferred to natural soil or the leaching system (Yoder and Lockwood, 1973). After several hours or days, they were transferred again to sterilized soil to measure any change in the time required for germination, as compared with control spores incubated continually on sterilized soil. A few hours' exposure to soil or the leaching system caused no change in the germination status of the spores, but incubation for several days caused regression proportional to the time of incubation, i.e., the germination time was proportionately prolonged. When spores of *Penicillium frequentans* were allowed to take up ^{14}C-labeled glucose and then were incubated on the leaching system, regressions as measured by prolonged germination time and by loss of radioactivity were closely parallel. The label lost was about equally divided between respiration and exudation.

The role of nutrient deprivation in soil mycostasis is also indicated by experiments on the restoration of mycostasis by inoculating sterilized subsoil, which was fortified with glucose, with various microorganisms (Steiner and Lockwood, 1970). Subsoil was used because it released no detectable carbohydrates on autoclaving and thus allowed uncomplicated detection of glucose. The incidence of germination of conidia of *C. sativus* was inversely correlated with the amount of glucose remaining in the soil at different times. By 55–70 hr after inoculation, the glucose had disappeared; aqueous extracts of the soils did not support germination of nutrient-dependent spores, whereas nutrient-independent spores germinated freely. Similarly, aqueous leaching made autoclaved soil fungistatic to the nutrient-dependent conidia of *T. basicola,* but not to the nutrient-dependent conidia of *F. solani* f. sp. *phaseoli* (Adams *et al.,* 1968).

3.4.1b. *Leaves.* A similar mycostasis occurs on plant leaves, if the population density of foliar microorganisms is high enough. This has been studied in detail by Blakeman and his co-workers at Aberdeen, Scotland, and in many respects their findings are similar to those reported above concerning mycostasis in soil. Bacterial populations increased to large numbers when conidia were applied as droplets to leaf surfaces, and this increase was accompanied by substantially decreased germination of conidia of *B. cinerea* (Fraser, 1971). The increase in bacteria was a result, in part at least, of amino acids exuded from the spores. The mycostasis was annulled by nutrients, and no antifungal substances were found in the spore droplets. Exudate from ^{14}C-labeled spores was rapidly taken up by the ambient bacteria, whereas in the absence of bacteria, some of the released exudate was taken up by the spores during germination (Brodie and Blakeman, 1975). Germination of the spores also was prevented by leaching (Sztejnberg and Blakeman, 1973), and ^{14}C-labeled conidia of *B. cinerea* incubated on a leaching apparatus lost exudate in proportion to the rate of flow (Brodie and Blakeman, 1977). Conidia undergoing leaching for 24 hr lost about twice as much exudate as did unleached control spores. These results are similar to those of Bristow and Lockwood (1975a) for conidia of *C. sativus, C. lunata,* and *T. basicola* and suggest that microbial competition for nutrients might be the causal mechanism for the foliar mycostasis (Sztejnberg and Blakeman, 1973).

Mycostasis of *B. cinerea* conidia was closely related to microbial competition for amino acids. A common leaf saprophytic bacterium, *Pseudomonas* sp. isolate 14, competed very actively for amino acids in glucose–amino acid mixtures and removed 80% from solutions in 5 hr (Brodie and Blakeman, 1976). Inhibition of conidial germination was closely correlated with amino acid uptake by this bacterium. The fungal spores, on the other hand, took up amino acids relatively slowly. Their germination was stimulated more by amino acids than by glucose, for which the bacterium competed less well. Suppression of germination was also proportional to the ability of different epiphytic bacteria and yeasts to take up amino acids ($r = 0.95$), and also to amino acid uptake by a series of cell densities of different leaf microorganisms ($r = 0.94$) (Blakeman and Brodie, 1977). Wetting beet leaves for 24 hr allowed build-up of a natural microflora to numbers sufficient to strongly suppress germination and to take up 80% of the applied amino acids. A similar relationship between amino acid uptake and germination suppression by *Pseudomonas* sp. isolate 14 was shown for two other pathogens of beet, *Phoma betae* and *Cladosporium herbarum* (Blakeman and Brodie, 1977).

It is significant that about one-tenth as many cells of competing microbes was required to bring about an equivalent reduction in germination on leaves as on glass (Blakeman and Brodie, 1977). The difference could be due to (a) more favorable nutritional conditions on leaves, (b) the concentrations of major nutrients being less on leaves than on glass, or (c) chemical inhibitors of germination associated with leaf wax (Blakeman and Sztejnberg, 1973; Blakeman and Atkinson, 1976).

3.4.1c. Interactions Involving Pollen. A complex interaction involving certain plant parasites, saprophytic fungi, and pollen on plant leaves has been revealed by Fokkema and co-workers in the Netherlands. Development of lesions by the pathogen, *C. sativus,* on rye leaves was greatly enhanced in the presence of rye pollen (Fokkema, 1971a,b). However, prior or co-inoculation of the pathogen with each of several common foliar saprophytes largely cancelled the increase (Fokkema, 1973). The fungal groups used were *Cladosporium* spp., *Cryptococcus* spp., *Sporobolomyces* spp., and *Aureobasidium pullulans.* The degree of suppression of lesion development was related to the amount of reduction in development of superficial mycelium of the pathogen; germination itself was not inhibited. Pollen-stimulated lesion development by *P. betae* on beet leaves also was suppressed by the saprophytic fungal flora (Warren, 1972).

To test whether the foliar saprophytes could suppress pollen-induced infection under field conditions, field-grown rye plants enclosed in plastic chambers were sprayed with the fungicide benomyl to reduce the populations of the saprophytic fungi (Fokkema *et al.,* 1975). *Cochliobolus sativus* was unaffected by benomyl at the concentrations used. When inoculated after flowering so that pollen was deposited on the leaves, benomyl-sprayed leaves had more than twice the necrotic leaf area as water-sprayed leaves, providing evidence for biological control of the disease under conditions of high relative humidity. Counts of the saprophytic mycoflora on benomyl-treated leaves were $400-1200/cm^2$, as compared with $3000-10,000$ cells/cm^2 on water-sprayed controls. The benomyl sprays eliminated *A. pullulans,* reduced *Sporobolomyces roseus* and *Cladosporium* spp. to $<6\%$ of controls, but did not affect *Cryptococcus* spp.

Natural populations of *S. roseus* on wheat leaves exposed to high humidity ($>85\%$ relative humidity) increased to a steady state of $5 \times 10^3-10^4$ cells/cm^2 of leaf surface (Bashi and Fokkema, 1977). Application of a nutrient solution in the presence of high humidity gave a further increase in populations to ca 5×10^5 cells/cm^2. Infection by *C. sativus* on wheat leaves at these high populations of saprophytes was reduced to 15% of that on control leaves.

Some of the same saprophytes also reduced superficial mycelial growth and infection of leaves of wheat by *Septoria nodorum* (Fokkema and Van der Meulen, 1976) and of onion by *Alternaria porri* and *B. cinerea* (Fokkema and Lorbeer, 1974). However, the presence of rye pollen negated the suppressive effect of the foliar saprophytes against *S. nodorum* (Fokkema and Van der

Meulen, 1976). This effect, which is the converse of that with *C. sativus,* was thought to be due to the ability of *S. nodorum* to cause high infection in the absence of pollen.

The mechanism of the inhibition of growth and lesion development by foliar saprophytes has not been thoroughly investigated. It apparently is not due to the induction of phytoalexins by the host plant, since similar growth inhibition was obtained on agar-coated slides (Fokkema, 1978). Fokkema (1978) believes that suppression of pathogen development is most likely due to competition for nutrients among foliar microorganisms.

3.4.1d. Cellular Mechanism. The cellular mechanism by which nutrient deprivation suppresses germination is not known. The enhanced exudation of energy-rich materials from propagules incubated in the presence of microorganisms in soil or on leaves, or in leaching systems, may represent soluble energy reserves which are lost at a rate sufficient to prevent their being used to synthesize a germ tube (Lockwood, 1977). The large initial flushes of exudation (Bristow and Lockwood, 1975a; Brodie and Blakeman, 1975; Sneh and Lockwood, 1975) may be related to reorganization of the spore membrane on hydration (Blakeman and Brodie, 1977), but the extent to which these drain insoluble reserves is not known. The smaller losses incurred thereafter are very likely derived from insoluble reserves.

Propagules may differ in their ability to conserve reserve materials under nutrient stress. Nutrient-independent conidia of *C. sativus* and *C. victoriae* (Bristow and Lockwood, 1975a), *C. oryzae* (Bhattacharya and Samaddar, 1977), and *B. cinerae* (Sztejnberg and Blakeman, 1973) became nutrient-dependent after several days or weeks of incubation on soil or leached sand; this was followed upon further incubation by progressive loss of viability (Bristow and Lockwood, 1975a; Bhattacharya and Samaddar, 1977). However, many types of propagules are known to survive for months or years in soil. Exhaustion of reserves in these fungi could be minimized by a feedback mechanism, based on the propagules "sensing" the competitive status of the environment. If exudates are removed from the immediate vicinity of the propagule beyond a specific rate, enzymes hydrolyzing insoluble reserves are repressed, conserving the reserves. However, under conditions of reduced competition, the enzymes are derepressed and reserve materials, made available via hydrolysis, are used to synthesize the germ tube. Such a mechanism would require a slow drain on the propagule's storage reserves to produce the minimum quantity of exudate necessary for "sensing" the environment. Differences in longevity or in ability to maintain nutrient independence could be functions of differences in amount of exudation inherent in the species or of the efficiency of enzyme repression. The mechanism of exogenous dormancy at the cellular level merits investigation. The following aspects are suggested for study: (a) the nature of the reserve materials, (b) the nature of the exuded materials and whether a specific chemical might be involved as a "sensor," and (c) whether under nutrient stress, exudation can be minimized or arrested completely to conserve endogenous resources.

3.4.2. Inhibitory Substances

Since Dobbs and Hinson (1953) first reported the presence of mycostasis in soil, considerable study has been directed toward substances inhibitory to fungal germination in soil. Much of this literature must be viewed with caution, as the assay techniques employed do not rule out microbial competition for nutrients as an inhibitory mechanism (Lockwood, 1977). In addition, many of these reports describe the phenomenon of inhibition but do not attempt to identify the inhibitory substances. Knowledge of the chemical nature of an inhibition is crucial to evaluating the role of an inhibitor in fungal ecology. Some progress, however, has been made in this direction.

Volatile inhibitors from soil which have been implicated in mycostasis are CO_2, ammonia, ethylene, and sulfur compounds. A comprehensive review of the effects of O_2 and CO_2 on the growth and metabolism of fungi has been published (Tabak and Cooke, 1968). In addition, Griffin (1972) discusses many aspects of CO_2 chemistry in soil in relation to fungal growth.

Nonvolatile substances exhibiting antifungal effects which have been identified in soil include metals and salts, nitrous acid, molecular-sieve fractions of soil leachates, and aromatic compounds from leaf litter and detritus.

3.4.2a. Nonvolatile. Nonvolatile inhibitors in soil derive both from the inorganic component of the soil matrix (i.e., metals and salts of metals) and from the organic components.

Fungi show wide latitude in their tolerance to naturally occurring metals. A study of microfungi in a copper swamp (Kendrick, 1962) showed that several fungi had developed a physiological tolerance to high copper concentrations. Thirty-one species of fungi were isolated from soil and peat samples of the swamp, and these were divided into three groups according to copper concentration of the samples. Nine species occurred only in samples with less than 5 ppm copper, suggesting sensitivity to copper. Thirteen others occurred in samples containing over 7500 ppm. Indeed, four species of this group were isolated from peat samples containing up to 68,000 ppm copper. It appeared that this second group might have developed a dependence on high copper habitats. A third group of fungi had developed some tolerance to copper but did not appear to be dependent on its presence.

In acid soils, soluble aluminum may exhibit inhibitory properties (Ko and Hora, 1971, 1972a). Heat-activated ascospores of *N. tetrasperma,* which are not sensitive to the common soil mycostasis, were completely inhibited by an aqueous extract from a soil showing inhibitory properties (Ko and Hora, 1971). The inhibitor was removed from the extract by both cationic and anionic exchange resins, and activity was retained after autoclaving. Inhibition by the soil extract was attenuated by diluting the extract. The inhibitor was effective at pH 4.5 but not at 7.0. Of the several metals in the soil extract which were evaluated as possible inhibitory agents, only aluminum was found to be active. Furthermore, a 0.65 ppm solution of aluminum salts gave complete inhibition of ascospore germination. Autoclaving solutions of aluminum salts did not destroy their inhibitory nature. Dormant ascospores (not heat activated) were not affected by the fungitoxin.

Calcium carbonate in some limestone soils also inhibits fungal germination (Dobbs and Gash, 1965). In certain marine sands exhibiting mycostasis, iron compounds were considered to be involved in the inhibition. The inhibitor could be removed from these sands with dilute HCl but not with hot water. Several cations were obtained in the acid leachate, but only iron compounds were inhibitory when added to acid-leached sand at concentrations found in sands prior to leaching.

Nitrous acid from nitrite in solution is an intermediate product of nitrification in soil, and it has been shown to inhibit fungi. In urea-amended soils, both ammonia and nitrous acid were inhibitory to *Phytophthora parasitica* and *Phytophthora cinnamomi* (Tsao and Zentmyer, 1979). At alkaline pH, ammonia was the inhibitory agent, whereas at pH 6 or less, nitrous acid was the inhibitor. Zentmyer and Bingham (1965) observed that the inhibition of *P. cinnamomi* by nitrite increased as the solution pH decreased. At pH 6.5, about 10 times as much nitrite was required for inhibition of mycelial growth and zoospore germination as at pH 4.5. Sequeira (1963) has also suggested that nitrite in urea-amended soil may be toxic to *Fusarium* spp.

Application of anhydrous ammonia to soil has yielded nitrite concentrations toxic to *Fusarium* populations (Smiley *et al.,* 1970). Near the center of the injection zone, ammonia appeared to be the toxic compound; however, nitrite accumulated with time in the peripheral regions of the injection zone and was also considered to be involved in reducing *Fusarium* populations. Nitrite accumulations of 30–35 ppm were detected in a silt loam incubated at 24°C, and up to 10 ppm nitrite was detected when the soil was incubated at 6°C. Additions of 35 ppm nitrite (as KNO_2) to the silt loam reduced *Fusarium* populations to undetectable levels in one week. A high pH in the absence of ammonia and a nitrate concentration of 100 ppm in the soil, however, were not inhibitory to *Fusarium* populations.

Nonvolatile fungal inhibitors have been demonstrated in fractions of aqueous soil extracts (Vaartaja, 1974, 1977). Membrane filter-sterilized soil extracts were fractionated by use of molecular-sieve columns, and fractions in agar were then assayed for effects on hyphal growth of *Pythium ultimum*. In an early study, 66 crude soil extracts, of which 54 were inhibitory and 12 were noninhibitory, were analyzed (Vaartaja, 1974). Most of the extracts yielded one or more fractions inhibitory to growth of *P. ultimum,* and a few yielded only stimulatory fractions, whereas others gave both stimulatory and inhibitory fractions. Molecular weights of the inhibitory fractions ranged from 100 to 55,000, but 74% of these were in the range of 3000–55,000. Of 36 inhibitory fractions, the effects of all except one were counteracted by nutrients.

Vaartaja (1977) has detected a nonvolatile, stable, low-molecular-weight inhibitor (mol. wt. = 150) in a rainwater extract of a clay loam soil. The *in vitro* effects of this extract were tested against 42 soil fungi. The descending order of sensitivity was: (a) soil Basidiomycetes, (b) *Pythium* spp., (c) miscellaneous soil fungi, (d) *Fusarium* and *Cylindrocarpon* spp., and (e) *Penicillium* and *Gliocladium* spp. The inhibitory effects were reduced significantly by adding nutrients to the extract.

Soil that was treated with varied combinations of nutrients and/or antifungal antibiotics has been used to assess the importance of bacteria as producers of an inhibitor in soil extracts (Vaartaja, 1978). After treatment, soil samples were watered for 5 days, and a membrane-filtered extract was added to an agar medium on which radial growth of *P. ultimum* was measured. The inhibitory effect increased with increasing levels of sucrose plus antifungal antibiotics. Additional experiments, however, showed that soil extracts could be either stimulatory or inhibitory.

Lignin-related aromatic compounds may play a role in the ecology of soil fungi. A study by Lingappa and Lockwood (1962) showed that aromatic model substances and decomposition products of lignin at 10^{-2} M and 10^{-3} M inhibited spore germination. Mycostasis by some compounds was annulled by addition of nutrients; however, autoclaving increased or decreased the inhibition in an inconsistent manner. Consequently, the authors concluded that these aromatic compounds are not likely to be a significant factor in soil mycostasis. Nevertheless, aromatic compounds such as tannin and ferulic acid, which are normal constituents of soil or plant litter, have been shown to inhibit fungal growth in agar media.

The slow microbial breakdown of oak leaf litter as compared to other litters, such as that of ash, birch, and elm, has been attributed to the inhibition of fungal growth by the greater amounts of tannins in oak leaves (Harrison, 1971). Of 19 fungal species assayed in aqueous extracts of oak leaf litter or in aqueous solutions of prepared tannins, 14 species were inhibited. Low concentrations of contaminating lead in the tannin preparations may have adversely affected fungal growth, however. An earlier study by Cowley and Whittingham (1961) suggested that tannin in an agar medium inhibited the growth of several fungi collected from prairie soils. In contrast, the fungal species prevalent in forest soils, or in both forest and prairie soils, showed mixed reactions toward tannin. Some of these species not only grew when tannin was present, but they utilized it as a carbon source.

Ferulic acid in water agar disks inhibited or stimulated germination and hyphal growth of several fungi obtained from litter and soil (Black and Dix, 1976). Of the dozen or so fungi tested, the germination of several was inhibited by 200 μg ferulic acid per water agar disc after 16–18 hr. Soil-inhabiting fungi, such as some species of *Penicillium,* showed no change or increased in germination. At 600 μg ferulic acid per disc, the germination of all fungi except one was significantly reduced. The exception, *Thysanophora penicillioides,* is a saprophyte of conifer needles. For the most part, the trend in hyphal growth paralleled germination. Thus, ferulic acid and perhaps other phenylpropane constituents of lignin and humic materials found in litter and in soil may play a role in the ecology of fungi associated with these habitats.

In the discussion so far, we have examined examples of heteroinhibitors, which are inhibitory substances produced by one organism but active against another. Self-inhibitors of fungi also exist, and considerable literature concerning them has accumulated. Allen (1976) and Macko *et al.* (1976) offer admirable discussions of this subject.

Cook (1977) has proposed that soil mycostasis is caused by fungal self-inhibitors. According to this hypothesis, the biotic or abiotic environment of the fungus provides signals that regulate the response of a self-inhibitor in propagules. The signals may be activators or deactivators of germination.

The following lines of evidence militate against acceptance of this hypothesis: (a) it is not certain whether all fungi contain self-inhibitors; (b) urediospores of rust fungi, some of which contain chemically identified and potent self-inhibitors, are not sensitive to soil mycostasis (Ko and Lockwood, 1967); (c) it is difficult to explain the occurrence of mycostasis on the leaching apparatus (see Section 3.4.1) by this mechanism since water flow would tend to remove or dilute self-inhibitors; and (d) the high density of spores in soil required for expression of a self-inhibitory effect does not occur in normal populations (Griffin and Ford, 1974).

Numerous soil microorganisms growing in rich media produce substances which are toxic to other microorganisms. Many of these substances are diffusible and highly potent, and have traditionally been termed antibiotics. Since many antibiotic producers were obtained from soil, might they play a role in soil ecology, especially in mycostasis?

Despite several attempts, specific antibiotics have not been detected in natural, unamended, and unaltered soil (Griffin, 1972; Gottlieb, 1976). Nor has any direct mycostatic effect yet been associated with a specific antibiotic in natural, unamended soil. Antibiotics have been detected, however, in sterilized soils which were inoculated with antibiotic producers. They have also been found in organic materials incorporated into soil, such as seeds (Wright, 1956) or buried straw (Bruehl *et al.,* 1969) which had been inoculated with antibiotic producers.

Within nutrient-rich loci in soil, antibiotics may very well play a role in microbial colonization of a substrate, and this may be a rewarding area of research. In natural, unamended soil, however, the paucity of nutrients places severe restrictions on microbial growth and presumably on antibiotic production. Adsorption to colloidal surfaces and degradation by nonsensitive microorganisms would also attenuate the concentrations of these substances in soil (Gottlieb, 1976).

3.4.2b. Volatile. Numerous workers have demonstrated that volatile toxic substances are released from soil (Lockwood, 1977). In most of the reports, the toxic substance was not identified; however, some volatile toxicants have been identified and their importance to soil biology assessed.

Ammonia. Ammonia has been shown to be toxic to fungal cells (Leach and Davey, 1935; Neal and Collins, 1936; McCallan and Weedon, 1940; Gilpatrick, 1969a; Pavlica *et al.,* 1978). It enters fungal cells primarily by free diffusion of the undissociated molecule (MacMillan, 1956). It is a volatile waste product of microbial metabolism, and in the immediate vicinity of a nitrogen-rich substrate, considerable quantities can be released. Ammonia has a high affinity for water, and in solution an equilibrium is established: $NH_3 + H_2O \rightleftarrows NH_4^+ + OH^-$, K_b [dissociation constant] $= 1.86 \times 10^{-5}$. It is apparent from this equilibrium reaction that pH is an important factor controlling the concentration of ammonia in a given habitat. It is also an important factor in determining the level of ammonia taken up by the fungal cell, with greater accumulation against a concentration gradient occurring at high external pH (MacMillan, 1956).

Ammonia concentration in the soil solution is greatly influenced by factors other than soil pH. Reactions of ammonia in soil cannot be completely separated from reactions of the ammonium ion, although ammonia does exhibit its own chemistry in soil (Mortland, 1958). The ammonia concentration in the soil solution is constantly regulated by adsorption and fixation reactions of ammonia and the ammonium ion with clay minerals and soil organic matter, immobilization of microbial nitrogen, volatilization of ammonia, and losses related to nitrification and subsequent removal of nitrate. Given these mechanisms of removal and considering the dearth of nutrients in soil, ammonia most likely is not a primary antifungal substance operative in the majority of soils. Nevertheless, there are special situations where ammonia is present in soil at levels that are inhibitory to fungal development, namely in alkaline soils and soils treated with anhydrous ammonia or organic amendments.

Ammonia has been shown to be a mycostatic agent in alkaline soils (Ko and Hora, 1972b; Ko *et al.,* 1974; Pavlica *et al.,* 1978) and in soils made alkaline by the addition of lime (Hora and Baker, 1974; Pavlica *et al.,* 1978). The influence of elevated soil pH on ammonia availability and fungal cell uptake is most evident in these cases. For instance, liming a Colorado loamy sand (pH 6.4) raised the soil pH to 8.5 and increased the ammonia trapped in soil headspaces from 20 $\mu g/g$ air to about 55 $\mu g/g$ air (Pavlica *et al.,* 1978). Conidia of *Gonatobotrys simplex* on agar disks in jars containing ammonia at 20 $\mu g/g$ air showed considerably more germination than conidia incubated in the higher ammonia concentration.

Fungistatic levels of ammonia evolved from an alkaline soil were derived from ammonium salts (Ko *et al.*, 1974). The volatile inhibition was generated from soil extracts at pH 7.6 but not at pH 2.0. A 100 ppm solution of NH_4Cl and $(NH_4)_2SO_4$ released a volatile inhibitor at pH 7.6 but not at pH 2.0.

Anhydrous ammonia injected into some Florida soils drastically reduced the numbers of nematodes and fungi in the soil (Eno *et al.*, 1955); only 4.9% of the fungi survived when 608 ppm of ammoniacal nitrogen was present in the injection zone. Within the injection zone, toxic levels of ammonia may persist for several weeks. In a laboratory study, anhydrous ammonia injected into three soils was fungicidal to *Fusarium* spp. within the 0–5 cm ammonia retention zone but not outside the zone (Smiley *et al.*, 1970). Fusaria outside the zone were not significantly affected, whereas *Fusarium* could not be recovered from within the zone 225 days after ammonia injection. Applying results obtained in the laboratory to the field, however, can be frustrating. Further research (Smiley *et al.*, 1972) has shown a limited potential for use of anhydrous ammonia in the field for controlling *Fusarium* foot and root rot of wheat and peas. Although exposures of 400 or 600 μg NH_3–N/g dry soil for 5 or 1 min, respectively, were inhibitory to chlamydospore germination in the laboratory, the use of ammonia in the field at comparable concentrations failed to control *Fusarium* infections of winter wheat.

The toxic effect of ammonia released from decomposing plant tissues in soil depends on the rate and duration of ammonia production in soil (Lewis, 1976) and on the state of the target fungus in the soil during ammonia production (Lewis and Papavizas, 1974). In this regard, surviving mycelia of *Rhizoctonia solani* within precolonized buckwheat stems were less sensitive to volatiles from decomposing plant tissue in soil than was the active saprophytic phase of the fungus (Lewis and Papavizas, 1974). These same authors suggested that ammonia, a probable volatile inhibitor detected in this study, may alter the ability of *R. solani* to competitively colonize plant debris in soil.

Ammonia production in soil will depend on many interacting biotic and abiotic factors, including the physical and chemical composition of the debris, the nature of the decomposer populations, soil pH, soil temperature, and soil moisture. It is beyond the scope of this review to elaborate on all the factors determining the level of ammonia production in soil. Discussions on this subject have been offered elsewhere (Allison, 1973; Terman, 1979). Nevertheless, all things being equal, the C:N ratio of the amendment is a very important factor in determining the overall level of ammonia production in soil and its effectiveness as an inhibitory compound. In general, immature plant tissue with a low C:N ratio or other materials high in nitrogen, such as soybean seed meal, will yield, upon decomposition in soil, more ammonia than materials with high C:N ratios (Allison, 1973; Lewis and Papavizas, 1974; Lewis, 1976).

Ammonia was produced in high amounts (60–197 ppm) in soil one week after amendment (5% w/w) with alfalfa meal (Gilpatrick, 1969a). *Phyto-*

phthora cinnamomi was not recovered from infected avocado roots placed in amended soil for 3–4 days during high ammonia production. Zoospore germination was completely inhibited and mycelium killed by 17 ppm ammonia in buffered solution after 24 hr. However, other mechanisms may have contributed to *P. cinnamomi* decline in the alfalfa-meal-amended soil because, in a parallel study, several materials in addition to alfalfa meal stimulated microbial growth on avocado roots, offering some biological control against *P. cinnamomi* (Gilpatrick, 1969b).

Germination of *F. solani* f. sp. *cucurbitae* and *Aspergillus flavus* conidia was inhibited by ammonia from chitin-amended soil (Schippers and Palm, 1973). Ammonia also has been implicated in the suppression of *Fusarium* spp. populations in a sandy soil amended with 1% (w/w) oilseed meals (Zakaria *et al.*, 1980). The effectiveness of the meals in reducing *Fusarium* populations was directly related to ammonia levels released by the meal-amended soils in the following descending order: soybean seed > linseed > cottonseed meal. In the presence of the oilseed meals, *Fusarium* chlamydospores were formed, but these were soon killed. Alfalfa-meal-amended soil produced negligible ammonia and likewise did not significantly reduce *Fusarium* populations.

Urea at 0.1% (w/w) concentration in sandy loam or loamy sand soils suppressed *P. cinnamomi* and *P. parasitica* populations (Tsao and Zentmyer, 1979). Sporangium germination was reduced to 0–9% by undiluted aqueous extracts from amended soils, whereas extracts from unamended soils allowed 36–70% germination of sporangia. In this study, both ammonia and nitrous acid were considered to be the likely inhibitors.

Ammonia is an uncoupling agent of oxidative phosphorylation and most likely interferes with other metabolic processes in the cell. Several investigators have suggested that ammonia expresses its inhibitory effect on fungal cells through a direct, internal interference with metabolic processes (Gilpatrick, 1969a; Smiley *et al.*, 1970; Tsao and Zentmyer, 1979; Zakaria *et al.*, 1980). However, in a study of the mode of action of ammonia on *S. rolfsii*, Henis and Chet (1967) concluded that the direct effect of ammonia on the fungus was a function of high external pH and time. Sclerotia did not germinate when incubated for 24 hr in buffered solutions at pH 10.4 either with ammonia (at 300 ppm) or without ammonia. However, the data presented do not support the claim that there was no effect of ammonia at pH values of 9.2 and below. For instance, sclerotia incubated for 24 hr in buffered solutions of pH 7.0, 7.6, 8.7, and 9.2 showed complete germination when removed and placed on an agar medium for one or two days. However, germination in these same buffered solutions with added ammonia (300 ppm) at 24 hr was significantly reduced. Moreover, as noted above, increased solution pH favors the movement of ammonia into fungal cells against a concentration gradient (MacMillan, 1956). Thus, high external pH and direct effects of ammonia are interactive, and both mechanisms are probably responsible for the overall inhibitory effect exerted by this compound.

Ethylene. Ethylene is an unsaturated hydrocarbon gas that has been reported to cause soil mycostasis (A. M. Smith, 1973, 1976a). In addition, Smith and Cook (1974) have assigned to ethylene a broad role in regulating microbial activity in soil, via an "oxygen–ethylene cycle." The role of ethylene in soil biology has recently been reviewed (A. M. Smith, 1976a; Primrose, 1979).

In pure culture, ethylene is produced by fungi (Ilag and Curtis, 1968; Lynch, 1972; Bird and Lynch, 1974; Lynch, 1975) and bacteria (Freebairn and Buddenhagen, 1964; Primrose and Dilworth, 1976). Methionine is required as a precursor to ethylene production in fungi (Lynch, 1974) and bacteria (Primrose and Dilworth, 1976). Ethylene production from phenolic acids by *Penicillium* sp. has also been reported (Considine and Patching, 1975).

Although it is generally accepted that ethylene is formed in soil by microbial activity, there remains a controversy as to whether the process is primarily anaerobic (K. A. Smith and Russell, 1969; K. A. Smith and Restall, 1971; A. M. Smith and Cook, 1974; A. M. Smith, 1976a,b) or aerobic (Lynch and Harper, 1974a; Lynch, 1975). Spore-forming bacteria in anaerobic microsites have been implicated as the main producers of ethylene (A. M. Smith and Cook, 1974; A. M. Smith, 1976a,b; Sutherland and Cook, 1980). On the other hand, Lynch and Harper (1974a) and Lynch (1975) have demonstrated that fungi, particularly *Mucor hiemalis,* also may be important producers.

Anaerobiosis appears necessary for the mobilization of methionine, which is required for ethylene biosynthesis (Lynch, 1972; Lynch and Harper, 1974a). Products of carbohydrate fermentation in anaerobic soil have been shown to stimulate ethylene production (Goodlass and K. A. Smith, 1978b). Certainly, anaerobic conditions favor the accumulation of ethylene, as diffusion and volatilization of ethylene would be restricted to water-filled pores. Moreover, ethylene production by *M. hiemalis* seems to be greatest at low growth rates (Lynch and Harper, 1974b), which would occur in soil under anaerobic conditions.

Ethylene utilization by soil microbes is an aerobic process (Abeles, 1973; Cornforth, 1975; deBont, 1975). Cornforth (1975) has shown that ethylene utilization proceeds approximately 50 times faster than its anaerobic production. Thus, more than 90% of a given volume of soil would have to be anaerobic and producing ethylene before the gas could accumulate in the remaining aerobic volume. Volatilization and adsorption to organic matter in soil (Witt and Weber, 1975) would further attenuate ethylene concentrations in the soil atmosphere. Ethylene concentrations in most natural mineral soils usually range from <1 ppm to 10 ppm (K. A. Smith and Restall, 1971; Rovira and Vendrell, 1972; A. M. Smith and Cook, 1974; Primrose and Dilworth, 1976). In a recent survey of 17 soils from arable and grassland locations, Goodlass and K. A. Smith (1978a) reported concentrations ranging from <0.01 to 7.9 ng/g soil in surface layer samples of 0–10 cm. Air-dried soils that had been rewetted produced ethylene in amounts ranging from 1.9 to 30.0 ng/g after 10 days. Air-dried soils produced more ethylene than oven-dried (105°C) soils, but both treatments resulted in far greater yields of ethylene than were present in fresh soils. It appears, therefore, that ethylene concentrations emitted from air-dried or oven-dried soils may be an over-estimation of those levels normally present in soil in the field.

Colonizable substrate is required for any appreciable ethylene production in soil. A study of ethylene formation in forest soils in Sweden (Lindberg et al., 1979) reported 17 ppm ethylene in the headspace above humus and 153 ppm above pine-needle litter held in sealed bottles. Mineral soils with pine roots removed evolved ethylene at a very slow rate, whereas in the presence of pine roots ethylene was formed at a considerably faster rate. Athough ethylene formation in soil requires the presence of soil organic matter (Goodlass and Smith, 1978a), there appears to be no difference between the amount produced in grassland soils, which are high in organic matter, and in arable soils, which contain less organic matter. Addition of wheat and barley straw to soil has been shown to increase ethylene formation under anaerobic conditions (Goodlass and Smith, 1978b). Fermentative degradation of carbohydrates released from plant debris decomposition stimulated the production of ethylene and other hydrocarbon gases. Ethanol was found to stimulate ethylene formation under both aerobic and anaerobic conditions. In another study, plots of a loam soil augmented with shredded tomato vines and flooded for varying periods were monitored for the changes in soil gas composition and effects on V. dahliae populations in the soil and debris (Ioannou et al., 1977b). Concentrations of ethylene ranged from trace quantities to 20 ppm, with an overall average of 6.5 ppm. Flooded or nearly saturated soil enhanced ethylene formation in the presence of decomposable plant debris. Ethylene at trace levels was even detected in very dry soils (-15 to -19 bars soil matric potential).

Apart from the work of A. M. Smith (1973), who showed sensitivity of S. rolfsii to 1 ppm ethylene, most fungi tested to date do not appear to be sensitive to ethylene. Germination of B. cinerea conidia was not reduced in a 10 ppm concentration of ethylene in the headspace of flasks which did or did not contain soil (Schippers et al., 1978). Conidia of B. cinerea that had endogenous nutrients removed by leaching were similarly not sensitive to ethylene. Pretreating conidia in culture with 5–10 ppm ethylene for 1–3 days prior to germination did not increase the sensitivity of conidia to ethylene in the germination assay, nor did pretreating the soil with ethylene for several days prior to the assay. Archer (1976) has reported that ethylene concentrations of 10 or 100 ppm did not reduce radial growth of 22 species of fungi cultured on various agar media, nor did ethylene significantly affect the dry weight of fungi in liquid shake cultures. Other colony characteristics including frequency of hyphal branching, hyphal dimorphism, and sporulation also were not significantly affected by ethylene. Although these studies were performed with pure cultures, the concentrations of ethylene found to be ineffective suggest that ethylene does not inhibit fungal growth in soils where its concentration rarely rises above a few parts per million. Likewise, Ioannou *et al.,* (1977a) found that ethylene concentrations of up to 35 ppm did not effect any change in growth, sporulation, microsclerotia production, and microsclerotia germination of *V. dahliae.* Several other workers also have shown that very high levels of ethylene (>500 ppm) are needed to inhibit the germination of some fungi (Cornforth, 1975; Lynch 1975; Balis, 1976; Pavlica *et al.,* 1978).

Evidence accumulated to date, therefore, suggests that ethylene is not a direct mycostatic agent in soil. However, ethylene in soil may induce the formation of other volatile toxic compounds (Balis, 1976). A 1% ethylene-in-air mixture in the headspace over soil produced another unsaturated hydrocarbon that was soluble in water condensates collected above the soil. The compound was identified as allyl alcohol. It was not found in water condensates from soil incubated with air alone. In bioassays, 4 ppm of allyl alcohol greatly inhibited germination of *Arthrobotrys oligospora* conidia. However, it is not known whether enough ethylene could accumulate in soil to produce allyl alcohol.

Does ethylene play a role in regulating microbial activity in soil? According to a proposed "oxygen–ethylene cycle" (A. M. Smith and Cook, 1974; A. M. Smith, 1976a), microbial utilization of O_2 near organic matter produces anaerobic microsites, allowing ethylene producers to proliferate. Ethylene diffuses from these sites, inactivating aerobic microbes and reducing O_2 demand. Oxygen diffuses back into the microsite, inactivating the ethylene producers. Aerobes are activated, and the cycle continues. Regulation of several soil processes by the oxygen–ethylene cycle has been postulated, including the turnover of organic matter, the availability of plant nutrients, and the incidence and control of soil-borne plant diseases. Recent evidence (K. A. Smith, 1978) casts doubt on the sensitivity of soil respiration to ethylene. Headspace concentrations of 50 ppm ethylene over arable, grassland, and woodland soils in flasks, which were incubated for 7 days, did not significantly depress respiration in these soils. In the oxygen–ethylene cycle, nitrate is assigned an important role in regulating the redox potential of ethylene-producing microsites in the soil. Concentrations of 20–200 μg nitrate/g soil reduced or stopped ethylene production until denitrification was completed (A. M. Smith and Cook, 1974; A. M. Smith, 1976a). However, Goodlass and K. A. Smith (1978b) found no correlation, in most cases, between the soil nitrate level and ethylene production in the soils they studied. Only in air-dried arable soils was there a corresponding decrease in ethylene production with increasing nitrate concentration. Addition of 100 μg nitrate/g soil had a negligible effect on ethylene formation, whereas up to 2000 μg nitrate/g soil reduced ethylene production by 85%. However, ethylene production was not completely arrested, even at this level of nitrate addition. Therefore, normal soil concentrations of nitrate and even those following fertilization would not appear to be inhibitory to ethylene formation.

Sulfur Compounds. Volatile sulfur compounds have been shown to be toxic to fungi (Walker *et al.,* 1937; Lewis and Papavizas, 1971). Several sulfur volatiles have been identified as emanating from soil (Lewis and Papavizas, 1970; Banwart and Bremner, 1976; Farwell *et al.,* 1979). Sorption of sulfur gases by soils has also been demonstrated (Bremner and Banwart, 1976).

Most of the sulfur found in soils is associated with soil organic matter, and microorganisms are the major agents of its mineralization and transformation (Freney, 1967). Greater quantities of sulfur volatiles are emitted from soil under anaerobic than aerobic conditions; consequently, the moisture content of soil greatly influences the nature and the quantity of these volatiles emitted from soil (Banwart and Bremner, 1976; Farwell *et al.,* 1979). The only significant sulfur emissions from an agricultural soil were dimethyl sulfide and carbon disulfide from water-saturated samples (Farwell *et al.,* 1979). No volatiles were detected from a forest soil. The largest emissions of sulfur from the soils tested came from an anaerobic tidal marsh soil, which emitted principally hydrogen sulfide and carbon disulfide with smaller amounts of methyl mercaptan and other sulfides. Soils low in organic matter produced very little or no detectable sulfur volatiles. There appears to be no work regarding the effects of sulfur volatiles on fungal development in unamended soils. Some studies have been reported, however, concerning the effects on fungi of cruciferous amendments to soil.

Crucifers such as cabbage, kale, and mustard contain an abundance of sulfur-containing compounds. When these materials were incorporated into a greenhouse soil, *Aphanomyces euteiches* root rot of pea was appreciably reduced (Papavizas, 1966; 1967). In contrast, corn-amended soil was not effective in reducing disease severity. Decomposition products of cabbage and other crucifers in soil included methane thiol, dimethyl sulfide, and dimethyl disulfide (Lewis and Papavizas, 1970). None of these volatiles was detected from decomposing corn tissue in soil.

Several volatile sulfur-containing compounds were tested on the various stages in the life cycle of *A. euteiches* (Lewis and Papavizas, 1970). Mycelial growth, zoospore formation, and germination were prevented by allyl isothiocyanate vapor concentrations of 0.04, 0.10, and 0.30 ppm, respectively. Dimethylsulfide concentrations needed for the same effects were about 100 times greater. Vapors resulting from decomposing cabbage in soil were not as inhibitory to *A. euteiches* as were authentic compounds *in vitro.* Vapors from cabbage-amended soil, however, adversely affected oospore development and resulted in deformed or abnormal hyphae. It was suggested that deformed hyphae may be more prone to lysis in soil.

4. An Alternative Strategy — Carbonicolous Fungi

A very different means of coping with an energy-deficient environment has been evolved in the carbonicolous (postfire) fungi—Ascomycetes whose ascospores are constitutively dormant until activated by a brief exposure to temperatures in the range of 50–60°C, which in nature occur as a result of heat from a prairie or forest fire. Production of fruiting structures that bear a new generation of ascospores is normally completed in 2–10 weeks, depending on the species (Wicklow, 1975). One of the principle benefits of fire appears to be the reduction of competitors to which these fungi are intolerant. Artificially activated ascopores of 11 Ascomycetes from burned soil could not develop and produce ascocarps in natural, unburned soil, but all were able to do so in sterilized soil (Wicklow, 1975), indicating that activation alone cannot account for colonization of substrate and production of ascospores. Species of *Penicillium* and *Trichoderma,* which are common antagonistic fungi, were greatly reduced in numbers in burned *Pinus contortus* plots (Widden and Parkinson, 1975). When carbonicolous fungi were inoculated into sterile soil in growth tubes together with common soil saprophytes or with nonsterile soil, growth and apothecia formation were strongly suppressed (El-Abyad and Webster, 1968b).

Not surprisingly, a general lack of mycostasis sensitivity was shown by activated ascospores of four species of carbonicolous Ascomycetes (Wicklow and Zak, 1979). These results corroborate those of Ko and Lockwood (1967) for activated ascospores of *N. tetrasperma,* which also germinate freely on soil. By contrast, conidia of *Trichoderma viride,* like most other types of spores, were strongly inhibited in soil (Wicklow and Zak, 1979).

Carbonicolous fungi may be favored in the exploitation of burned soil by their relatively high growth rates. Eleven carbonicolous discomycetes had a mean growth rate in sterile soil of 22.5 mm/day, whereas seven other soil fungi had a mean growth rate of 10.4 mm/day (El-Abyad and Webster, 1968b).

Burning of soil also increases alkalinity and produces ash, which appear to favor carbonicolous Ascomycetes. Mycelial growth of carbonicolous Ascomycetes showed greater tolerance to alkaline conditions (El-Abyad and Webster, 1968a) and to extracts of burned litter than did mycelia of noncarbonicolous fungi (Widden and Parkinson, 1975). Fruiting in carbonicolous fungi also was favored by alkaline substrates. Soil temperature and moisture, however, did not differentially influence the two groups. Deposition of ash from burned dead prairie vegetation over a layer of steamed soil underlain with a layer of unheated soil (simulated burn) closely matched naturally burned soil in the number of species of carbonicolous Ascomycetes and in the order of appearance of sporocarps on the soil (Zak and Wicklow, 1978a). Steamed soil overlying natural soil, or ash over steamed soil, had either fewer species or disturbed the normal successional order.

Thus, through a number of adaptations, the carbonicolous fungi are able to exploit the postfire environment. Fire activates the constitutively dormant ascospores and reduces populations of many competing species that normally suppress the poorly competitive Ascomycetes, a sequence that eliminates the need for an exogenous mechanism of dormancy. The alkaline conditions produced are favorable for vegetative development and fructification of the rapidly growing carbonicolous fungi, and the ash probably provides mineral nutrients. By contrast, alkaline conditions and ash are not well tolerated by competing fungi. It is likely, also, that heating the soil releases energy-yielding nutrients usable by the carbonicolous species.

Among the carbonicolous Ascomycetes, differences have been demonstrated in activation temperatures and times (Zak and Wicklow, 1978b) and in growth rates (Wicklow and Hirschfield, 1979), and these differences have implications for competition. Those fungi whose sporocarps appeared late had slow growth rates. The late species were also strongly inhibitory in culture to the earlier species, but the early species showed no antagonism toward any of the carbonicolous Ascomycetes. Wicklow and Hirschfield (1979) view the successional pattern as contributing to increasing species richness by maximizing the distribution of limited resources among numerous postfire colonists.

5. Mycolysis

Another characteristic response of fungi to soils is the lysis of their hyphae. Spores, especially hyaline conidia, are often lysed in soil also, but less readily than are hyphae. The hyphae of 20 different fungi were extensively lysed in 7–14 days in the presence of natural soil (Lockwood, 1960). Some fungi were completely lysed in 4–8 days (Lloyd and Lockwood, 1966). Fungi differ in sensitivity to lysis (Lockwood, 1960; Lloyd and Lockwood, 1966; Bumbieris and Lloyd, 1967a), and those with darkened hyphae, such as *C. sativus, Alternaria solani,* and *R. solani,* tend to be more resistant than those with hyaline hyphae (Lloyd and Lockwood, 1966). Different isolates of the same species may also differ in susceptibility to mycolysis. For example, Naiki and Ui (1972) tested hyphae of 12 isolates of *R. solani* in soils. After 3 weeks, the mean percentage of hyphae lysed ranged from 20% to 100%. The capacity of soil to lyse hyphae is removed by sterilization, but it may be restored by inoculation of sterilized soils, especially with actinomycetes but also with bacteria (Lloyd and Lockwood, 1966). Bacteria (Mitchell and Alexander, 1963) and actinomycetes (Carter and Lockwood, 1957; Ko and Lockwood, 1970) also have been shown to lyse fungal hyphae *in vitro.* Actinomycete hyphae apparently are also subject to lysis in soil. Streptomycete hyphae added to soil grew briefly then sporulated and lysed, or formed arthrospores without growth (Lloyd, 1968).

The rate of lysis may differ with different soils and soil moisture contents. Lysis of living mycelium of 13 fungi was more rapid in a fertile garden soil than in an impoverished wheat field soil, both of the same soil type (Bumbieris and Lloyd, 1967b). Total bacterial and actinomycete counts in the two soils did not differ greatly, but the pH of the garden soil was 7.1 and that of the wheat field was 5.0. Lysis was more rapid at high than at low soil moisture contents, and the characteristics of lysis also differed (Bumbieris and Lloyd, 1967b). Concurrent disappearance of protoplasm and cell walls occurred at high and intermediate soil moisture contents, whereas at lower soil moistures the protoplasm disappeared first, leaving an empty cell wall, which eventually lysed (sequential lysis). In wet soil, increases in bacterial numbers were associated with the early stages of concurrent lysis, and actinomycete numbers increased later. At low soil moisture, actinomycetes increased to a greater extent than bacteria.

The onset of lysis in soil may be delayed by supplementing soil with energy sources, such as glucose or peptone (Lloyd and Lockwood, 1966; Bhattacharya and Samaddar, 1977). Increments of an aqueous extract of alfalfa meal mixed with soil delayed the lysis of three different fungi in proportion to the concentration applied (Hsu and Lockwood, 1971). Carbon sources of high C:N ratio may be more effective in delaying lysis than carbon sources of low C:N ratio. For example, Bumbieris and Lloyd (1967b) found that glucose delayed the onset of lysis, whereas peptone hastened it. Cook and Snyder (1965) observed that germ tubes of *F. solani* f. sp. *phaseoli* lysed readily in soil treated with amino acids alone or together with glucose or sucrose, but not with glucose or sucrose alone. Germ tubes also lysed readily in the vicinity of bean seeds that exuded amino acids and sugars, but in the presence of bean hypocotyls that exuded mainly sugars, hyphae persisted and thalli were formed on the hypocotyls.

When soil was supplemented with glucose, peptone, or bone or blood meal several days prior to the application of fungal hyphae, lysis in the amended soils was more rapid than in unsupplemented soil (Bumbieris and Lloyd, 1967b). The enhanced mycolytic capability of soil previously amended with nutrients is probably due to the increase in microbial populations, including lytic organisms. However, the cause of enhanced lysis in the presence of low C:N sources and its suppression in the presence of high C:N sources has not been elucidated. The effect could be due either to nutritional requirements of the affected populations, rendering the fungus more or less susceptible, or to stimulatory or suppressive effects on the lytic microflora. It would clarify the matter to test the proneness to lysis of hyphae grown on media of different C:N ratios on the leached sand model system, which induces autolysis without the intervention of other microorganisms.

Cultural conditions under which the fungus is grown can affect susceptibility to lysis. Living mycelia of *A. solani* and *C. sativus* lysed more readily on peptone agar than on Czapek's agar (Lockwood, 1960). The resistance was related to the production of dark pigmentation on Czapek's agar and the absence of pigmentation on peptone agar. However, the opposite susceptibility was found for the mycelium of *V. albo-atrum*. For 17 other fungi, mycelium produced on either medium was equally prone to lysis. Cell wall composition can be affected by cultural conditions. Walls of *Cryptococcus albidus* grown on a medium supporting good growth contained α- and β-glucans and chitin as major components and were lysed by two *Streptomyces* species but not by a myxobacterium, *Cytophaga johnsonii* (Jones *et al.,* 1969). However, when the yeast was grown in a medium unfavorable for growth, the content of the α-glucan component was reduced, and all three organisms were capable of lysing the walls.

5.1. Resistance to Mycolysis

Biochemical factors may confer a measure of resistance to lysis. One such factor is melanin. This substance comprises about 8.5% of the weight of the cell wall of *R. solani* (Potgieter and Alexander, 1966), a fungus which may survive in soil as dark mycelial fragments that resist lysis. Cell walls of *R. solani* were resistant to attack by a variety of organisms capable of producing enzymes that were capable of degrading cell wall components of other fungi. More direct evidence for a protective role of melanin was provided in studies with *Aspergillus nidulans* (Kuo and Alexander, 1967). Isolated cell walls of the wild type were resistant to enzymatic degradation, but walls of a melanin-less mutant were not resistant. Moreover, increasing resistance to degradation as the wild-type culture aged was associated with increased melanin content. The relative resistance of three varieties of *Gaeumannomyces graminis* also was correlated with melanin content (Tschudi and Kern, 1979).

Many persistant structures of fungi are characteristically dark, and their persistance may be related to melanin content. The conidia of *Aspergillus phoenicis* were resistant to degradation by chitinase and β-1,3-glucanase, which will lyse hyphal walls of this and many other fungi (Bloomfield and Alexander, 1967). Removal of the dark spicules on the surface of the conidia rendered them vulnerable to attack by lytic enzymes, and analysis revealed that the spicules contained melanin. The rind of the sclerotia of *S. rolfsii* was found to be rich in melanin, which apparently protects the underlying hyaline hyphae (Bloomfield and Alexander, 1967). Conidia of two hyaline isolates of *C. sativus* were readily lysed when incubated in soil or with lytic enzymes, whereas those of two wild-type pigmented isolates were highly resistant (Old and Robertson, 1970). The resistance of the pigmented spores of *C. sativus* (Old and Robertson, 1970) and *C. spiciferum* (Berthe *et al.,* 1976) to lysis was associated with an electron-dense outer cell wall layer that was presumed to be melanin. Melanin has the capacity to inhibit polysaccharases, including chitinase and β-1,3-glucanase (Bull, 1970), and melanin extracted from mycelium of *G. graminis* strongly inhibited the lysis of living mycelium of the same fungus by a crude lytic enzyme preparation from *Streptomyces lavendulae* (Tschudi and Kern, 1979). The resistance to decomposition of melanin itself was shown by incubating ^{14}C-labeled melanin in soil and finding little evolution of $^{14}CO_2$ (Kuo and Alexander, 1967). The relative rates of decomposition of ^{14}C-labeled lyophilized mycelium of four hyaline and five melanic fungi were determined by Hurst and Wagner (1969). After 3 months, the hyaline fungi had a mean decomposition of 64%, whereas the mean for the melanic fungi was 45%.

Resistance to lysis in some fungi may be due to structural or physiological features. Hyphae of *Mortierella parvispora* (Pengra *et al.*, 1969) and *Zygorrhynchus vuilleminii* (Ballesta and Alexander, 1971) were resistant to lysis by chitinase and β-1,3-glucanase. Resistance was ascribed to cell wall heteropolysaccharides that required enzymes other than these two for hydrolysis. The outer cell wall layer of chlamydospores of *F. solani* f. sp. *cucurbitae* was resistant to degradation by the same two enzymes, but whether melanin or a resistant polysaccharide was responsible was not determined (Van Eck, 1978). It was suggested that the resistant outer layer contributed to resistance of the chlamydospores to microbial heterolysis in soil.

5.2. Mechanisms of Mycolysis

5.2.1. Heterolysis

Mycolysis *in vitro* has been much studied by Alexander and co-workers. The model used by this group has been to isolate from soils microorganisms capable of growing on dead mycelium or cell walls of fungi susceptible to lysis (Skujins *et al.*, 1965). Most of the fungi investigated contained chitin and β-1,3-glucan in their walls, and evidence has accumulated that the corresponding enzymes can account for the major part of lysis of isolated cell walls (Skujins *et al.*, 1965; Ballesta and Alexander, 1972). Lysis, to varying extents, of killed conidia by wall-degrading enzyme preparations and by soil also has been observed (Chu and Alexander, 1972).

This approach had led quite naturally to a concept that the lysis of intact living mycelium cells in soil takes place through direct heterolytic attack by microorganisms capable of producing the appropriate enzymes (Skujins *et al.*, 1965; Bloomfield and Alexander, 1967). Unfortunately, the question has not been addressed experimentally to the extent that is merited. Co-culture of living mycelia together with lytic microorganisms has resulted in lysis of the fungi (Horikoshi and Iida, 1958; Mitchell and Alexander, 1963; Lloyd *et al.*, 1965), but this cannot be taken as evidence for heterolysis, since induced autolysis is not ruled out. More directly, however, protoplasts have been released from hyphae (Mitchell and Alexander, 1963; Aguirre *et al.*, 1963; Sietsma *et al.*, 1967; Eveleigh *et al.*, 1968; Musílková and Fencl, 1969) or conidia (Bachmann and Bonner, 1959; Garcia Acha and Villanueva, 1963; Garcia Acha *et al.*, 1964) of several filamentous fungi, and from yeast cells (Ochoa *et al.*, 1963; Garcia Mendoza and Villanueva, 1962, 1964) incubated in crude enzyme preparations from microorganisms or snail digestive juice. Some preparations, in addition, caused extensive or complete wall degradation (Garcia Acha and Villanueva, 1963; Garcia Acha et al., 1964; Ochoa et al., 1963; Garcia Mendoza and Villanueva, 1964; Sietsma et al., 1967; Eveleigh et al., 1968). All of this points to a capability for exogenously originating enzymes to lyse living fungal cells.

The concept that mycolysis in soil is due to heterolysis has led to attempts to control soil-borne fungal plant pathogens by additions of chitin to soil to increase the populations of chitin-decomposing microorganisms. This has met with mixed success (Baker and Cook, 1974), but the pathogen reduction achieved might have been due to the production of volatile toxic products (Sneh et al., 1971) such as ammonia (Schippers and Palm, 1973), rather than to enhanced heterolytic attack on the fungus.

The lysis of cell wall preparations of a number of filamentous fungi and yeasts by fungi, actinomycetes, and myxobacteria also has been studied by Jones and co-workers in Scotland (Jones and Webley, 1967; Bacon et al., 1968, 1970; Jones, 1970; Jones et al., 1968, 1969, 1974). These studies have been done to improve the understanding of biological transformation of fungal residues in soil and have provided much information on the composition and structure of fungal cell walls and the nature of the enzymes produced by the lytic microorganisms. However, the orientation of the work has limited relevance to the ecology of living fungi.

5.2.2. Autolysis

Perhaps significant to the question of the mechanism of mycolysis in soil is the common finding that living mycelium is more prone than heat-killed mycelium to lysis in the presence of crude enzymes (Mitchell and Alexander, 1963), in mixed cultures with lytic microorganisms (Ko and Lockwood, 1970), and in soil (Lloyd and Lockwood, 1966; Ko and Lockwood, 1970). Such findings would not be expected if heterolytic activity were the primary cause of lysis. It was also found that enzyme preparations that partially lysed heat-killed mycelium of *Fusarium solani* f. sp. *pisi* had no effect on living mycelium or actually supported their growth (Lloyd et al., 1965). In earlier work on lysis of fungi in soil, no association of soil microorganisms with the lysing hyphae was seen until degradation was nearly complete (Lockwood, unpublished data). Lysis of walls of living conidia of *Fusarium oxysporum* f. sp. *raphani* (Takeuchi, 1976) and of *F. solani* f. sp. *cucurbitae* (Old and Schippers, 1973), also was not visibly associated with microorganisms.

These findings led one of us (Lockwood) to investigate the possibility that hyphal mycolysis in soil could be due to autolysis. Several experimental approaches were taken. Germinated conidia of *Glomerella cingulata, C. victoriae*, and *F. solani* f. sp. *pisi* were applied to Pellicon membrane filters with a pore size small enough to exclude the passage of compounds of molecular weight greater than 1000, and these membranes were incubated on soil (Ko and Lockwood, 1970). The membranes also were highly resistant to microbial degradation in soil. Living hyphae were lysed in a manner similar in character to that occurring directly in soil. Heat-killed hyphae, on the other hand, did not lyse. Any lytic enzymes present in the soil should have been excluded by the membrane or, if not, should have lysed the killed hyphae. Therefore, lysis of living mycelia must have been due to autolysis. The rate of lysis on membranes lagged somewhat behind that of hyphae incubated directly on soil. After 4 days, it was estimated that 30–70%, 70–90%, and 10–30% of the hyphae of *G. cingulata, C. victoriae,* and *F. solani* f. sp. *pisi,* respectively, had lysed.

To determine whether nutrient deprivation and/or inhibitors would induce autolysis, hyphae on membrane filters were incubated on dialysis tubing through which a salt solution slowly circulated, a procedure designed to impose nutrient stress (Lloyd and Lockwood, 1966). Complete autolysis of the hyphae of *G. cingulata* and *C. victoriae* and partial lysis of *F. solani* f. sp. *pisi* resulted within 2 days when each of 4 antifungal antibiotics was applied to the hyphae, whereas exposure to dialysis tubing alone, or antibiotic treatment alone, gave partial or no lysis. The synergistic action of dialysis-imposed stress and antibiosis suggested that microorganisms colonizing fungal hyphae might produce antibiotics from hyphal exudates. When mycelium of *G. cingulata* was mixed with natural soil in an amount equivalent to 1% dry weight, extracts of the soil made 2 and 4 days later showed antifungal activity in three of five tests (Lloyd and Lockwood, 1966).

This model for mycolysis was reexamined (Ko and Lockwood, 1970), and extensive lysis was obtained under axenic starvation conditions alone with a more efficient method of imposing nutrient stress. The leached sand system used has been described in Section 3.4.1. Lysis of living hyphae in this model system was similar in extent and character to that of hyphae incubated on the Pellicon membranes on soil; dead hyphae were not lysed. It was further found that soil to which the equivalent of 0.1% mycelium of *G. cingulata* had been added did not yield an inhibitor of germination of *G. cingulata* or *C. victoriae* upon extraction, nor did 0.2 g mycelium incubated on the surface of a 20 g bed of soil and then scraped off and extracted in various ways. Chitinase and β-D-glucanase activities in hyphae undergoing lysis on soil or in the leached sand system increased very rapidly, whereas much less was produced by hyphae left to grow in a nutrient medium. Decreased activities of chitinase and β-D-glucanase were correlated with a decreased rate of lysis as hyphae aged. A similar relationship was shown between activities of these enzymes and the relative susceptibilities to lysis of the three test fungi. After 24 hr of incubation of hyphae of *C. victoriae* directly on soil, 65% of the hyphal cells were dead, and chitinase and β-D-glucanase activities in the lysing hyphae had increased severalfold. With heat-killed mycelium a much lower level of β-D-glucanase activity was detected after 2 days, and chitinase was not detected until after 4 days. Activities of these heterolytic enzymes remained low.

These results indicate that autolytic enzymes in mycelia are rapidly activated in hyphae exposed to a nutrient-deprived environment such as soil. The activated enzymes degrade cell constituents, and cell death occurs rapidly. Cell-wall-degrading enzymes may or may not be responsible for death of hyphal cells. Death may be due to exposure of the protoplasts to soil during cell wall lysis, or it may be due to other autolytic enzymes. The earlier model involving nutrient deprivation and antibiotics, proposed by Lloyd and Lockwood (1966), does not seem necessary to account for lysis of a major part of fungal hyphae or cell death. However, the intimate conditions required could conceivably exist in instances of "hyphosphere" colonization by other microorganisms during lysis (Lockwood, 1968), and may have prevailed during the lysis of conidia of *Rhynchosporium secalis* by a pseudomonad in lesions on barley leaves (Rotem *et al.*, 1976).

Autolysis of batch fungal cultures occurs coincident with the exhaustion of the carbon source (Lahoz and Ibeas, 1968; Lahoz *et al.*, 1966, 1967, 1970; Trinci and Righelato, 1970; Reyes *et al.*, 1977). Autolysis also was induced by transferring mycelia from growing cultures to energy-deficient media (Arima *et al.*, 1965) or by reducing the energy sources in continuous-flow cultures to near-maintenance rations or below (Trinci and Righelato, 1970; Bainbridge *et al.*, 1971). Autolysis of *Aspergillus niger* cultures transferred to distilled water was prevented by addition of glucose (Arima *et al.*, 1965). The loss in dry weight during autolysis may be quite rapid and very great. For example, the "half-life" of mycelial dry weight for *Penicillium chrysogenum* during autolysis in batch culture was 91 hr, and in steady-state, continuous-flow cultures, 62 hr (Trinci and Righelato, 1970). In depleted batch cultures, *Aspergillus flavus* lost 85% of its dry weight during 16 days of autolysis (Lahoz and Ibeas, 1968), and *Neurospora crassa* lost 70–80% of its dry weight after 30 days (Reyes *et al.*, 1977).

During autolysis, polymeric cell components break down, and the monomers are apparently used to maintain the fungus (Trinci and Righelato, 1970; Zevenhuizen and Bartnicki-Garcia, 1970; Fencl, 1978). In some studies, cell walls appear to resist autolysis (Trinci and Righelato, 1970; Zevenhuizen and Bartnicki-Garcia, 1970; Bainbridge *et al.*, 1971), whereas in others the walls are degraded (Mitchell and Sabar, 1966; Polacheck and Rosenberger, 1975). It would be interesting to know whether cell wall autolysis would be enhanced by incubation of hyphae under conditions of greater nutrient stress—such as in the leached sand apparatus or in soil—than occurs in autolysing spent cultures.

Autolysis resulting in extinction of the propagule would appear to be a character without fitness value, but in fact it may have redeeming aspects. In autolysing cultures at least, the hyphal morphology is frequently heterogeneous, with some compartments empty and others retaining the appearance of normally growing cells (Trinci and Righelato, 1970; Bainbridge *et al.*, 1971). It is likely that the autolysis of some hyphal compartments provides substrates for maintenance and growth of others, thus maintaining viability of the culture. It is interesting that in carbon-limited cultures undergoing autolysis, soluble "melanin" increased in mycelia of *A. nidulans* (Bainbridge *et al.*, 1971), and that cell wall glucan was synthesized from soluble carbohydrate derived from insoluble cytoplasmic glucan in *P. cinnamomi* (Zevenhuizen and Bartnicki-Garcia, 1970). It is tempting to speculate that such metabolic changes represent adaptations to insure survival of a few cells. If similar adaptations occur in nature, this could explain the survival of hyphal fragments of fungi such as *R. solani* in soil. Autolysis of vegetative hyphae and of spores may also be involved in the production of more specialized persistent structures, such as chlamydospores or sclerotia (see Section 8).

A volatile factor in membrane-filtered aqueous extracts of an Australian soil caused lysis of *Gaeumannomyces graminis* var. *tritici* (Sivasithamparam *et al.*, 1975). The soil had a history of take-all disease of wheat, which is caused by *G. graminis* var. *tritici*. The lytic factor appeared to be of chemical rather than of biological origin. It was not heat-sensitive, was active after passing through a 25-nm filter, and was not sensitive to oxytetracycline, as are mycoplasmas, or to nucleases. No virus particles were detected in persistent hyphae taken from lysed areas. The lytic factor appeared to be extremely potent, as a 15-sec exposure of *G. graminis* var. *tritici* over soil filtrates containing the factor induced lysis. To our knowledge, this work has not been confirmed or continued.

6. Regermination

In some fungi, the original spore or sclerotium may survive after destruction of the germ tube. Sclerotia of *V. albo-atrum* were shown to survive as many as nine cycles of germination in moist, nutrient-amended soil, followed by death of germ tubes upon exposure to dry soil (Farley *et al.*, 1971). Some of 17 individual sclerotia of the same fungus regerminated up to 13 times when incubated in moist conditions followed by air-drying (Ben-Yephet and Pinkas, 1977). The number germinating declined with each transfer, with the larger sclerotia remaining germinable longer and producing the most germ tubes. Sclerotia of *Macrophomina phaseolina* (Locke and Green, 1977) and of *R. solani* (Pitt, 1964) also were able to regerminate several times during exposure to favorable conditions after air drying. The carpogenic sclerotia of *Sclerotium trifoliorum* germinated at least 3 times in successive seasons in the field (Williams and Western, 1965), and those of *Claviceps purpurea* were able to produce new stromata after three excisions of their stipes (Cooke and Mitchell,

1967). In both fungi, the number of sclerotia forming apothecia and stromata, respectively, decreased after each cycle.

Some of a population of the multicellular conidia of *C. sativus* and *C. victoriae* remained viable after several cycles of germination on acidified potato-dextrose agar following lysis of germ tubes on soil (Hsu and Lockwood, 1971). All germinated conidia of *C. sativus* regerminated at least twice and 56% regerminated 5 times. Seventy-five percent of the conidia of *C. victoriae* regerminated twice, and 18% regerminated 5 times. A general relationship between the number of cells in the conidia and the number of germinations was observed, but whether each new germination originated from a different cell was not determined. The single-celled sporangia of *P. ultimum* germinated at least three times in soil following pulses of low levels of energy sources (Stanghellini and Hancock, 1971). Upon exhaustion of nutrients, the protoplasm in the germ tube was observed to retract into the sporangium.

The ability of a propagule to survive lysis or desiccation and thus be able to regerminate when exogenous nutrients again present themselves is of obvious survival value. It would be of much interest to know whether multicellular conidia other than those of *Cochliobolus (Helminthosporium)* possess this property. It would seem likely that other genera of dematiaceous hyphomycetes with large, multicellular conidia, such as *Curvularia, Alternaria,* and *Stemphylium,* might behave similarly. The capacity of the single-celled sporangia of *P. ultimum* to regerminate suggests the possibility that sporangia of other fungi, or perhaps other types of single-celled spores, might do the same.

7. Appressorium Formation

Appressoria are specialized swellings on germ tubes of fungal pathogens that attach tightly to plant surfaces, particularly leaves, and give rise to a thin infection hypha that penetrates the plant cuticle and cell wall. The formation of appressoria is a normal prerequisite for infection in the fungi that produce them. Their formation is often enhanced in the presence of microorganisms (Lenne and Parbery, 1976; Blakeman and Parbery, 1977). For example, a strain of *Bacillus* induced lysis of germ tubes of germinating conidia of *Colletotrichum gloeosporioides* on citrus leaves, but appressorium numbers were greatly increased and they resisted lysis (Lenne and Parbery, 1976).

Nutrient deprivation imposed by bacteria can stimulate appressorium formation (Blakeman and Parbery, 1977). Two isolates of *Pseudomonas* known to compete strongly for amino acids on leaf surfaces were more stimulatory to appressorium formation by *Colletotrichum acutatum* on beet leaves or on glass slides than were two less competitive bacteria. Addition of glucose or glutamine nullified the formation of appressoria. Leaching the conidia between membrane filters stimulated the formation of appressoria, but leaching with nutrient solution depressed their formation. Since more appressoria were formed when leaching was followed by a static period than when it was continuous, the degree of nutrient stress required for appressorium formation may be less than that required for inhibition of germination or for lysis.

Chemical factors on the surface of leaves may also contribute to the induction of appressorium formation. Anthranilic acid, an active component of aqueous leachates from the surfaces of banana fruits, stimulated germination and appressorium formation by *Colletotrichum musae* (Swinburne, 1976). Six different species of *Colletotrichum* formed appressoria more readily on leaves than on glass, and a chloroform extract of the surface of sugar beet leaves increased the numbers of proto-appressoria formed on glass (Parbery and Blakeman, 1978). However, the presence of the bacterium, *Pseudomonas* isolate 14, stimulated much greater numbers of mature appressoria. Possibly, nutrient-deprived conditions and the presence of materials in leaf waxes together promote the formation of appressoria. The environmental cue signaling the change from vegetative growth to cessation of growth and the production of appressoria is most likely given via the competitive bacteria. This is not to say that all fungi that penetrate plant leaves by way of appressoria require the presence and activity of competitors to induce their formation (Emmett and Parbery, 1975). Much more work is required to establish the role of competition in inducing or enhancing appressorium formation as a general phenomenon.

8. Persistent Structure Formation

The formation of some asexually derived persistent structures, such as chlamydospores or sclerotia, often occurs upon transfer of mycelia or, in some cases, conidia to soil (Nash *et al.*, 1961; Christias and Lockwood, 1973) or to soil extracts (Alexander *et al.*, 1966; Ford *et al.*, 1970a,c). In the case of *Fusarium* spp., such results have been interpreted as indicating the dependence of the fungus on specific chlamydospore-inducing factors of bacterial origin in soil (Ford *et al.*, 1970a,b,c). Maximum chlamydospore formation in *F. solani* f. sp. *phaseoli* was usually obtained in the anionic fraction of soil extracts, and evidence was found for clonal specificity of extracts from different soils and from different fractions of the same soil (Ford *et al.*, 1970a). Chlamydospore formation was induced by filtrates from bacterial cultures, but the culture filtrates were less effective than the bacteria themselves. Since soil extracts are low in energy sources and chlamydospore formation closely resembling that which occurs in soil can be obtained in other energy-deprived media such as salt solutions or even distilled water (Qureshi and Page, 1970; Meyers and

Cook, 1972; Hsu and Lockwood, 1973), the chlamydospore-promoting effect of soil extracts may be due to the presence of a weak salt solution rather than to specific morphogenetic factors. The formation of sclerotia of *Sclerotinia sclerotiorum, R. solani, S. rolfsii,* and *S. cepivorum* was rapidly induced by transferring mycelia from a culture medium to a water-saturated glass bead substratum (Christias and Lockwood, 1973). The rate and characteristics of the transformation in the model system and in soil were similar.

The degree of nutrient stress required for formation of persistent structures may be less than that required for mycostasis and for complete mycolysis. The steepened gradient imposed by the leached sand system was not necessary for their formation (Christias and Lockwood, 1973; Hsu and Lockwood, 1973). Leaching conditions were not detrimental to sclerotium formation (Christias and Lockwood, 1973) but were either favorable or unfavorable for chlamydospore formation by *Fusarium,* depending on the salt solution used (Hsu and Lockwood, 1973). Vegetative growth was prolonged and sclerotium production delayed in a medium continually supplemented with nutrients (Christias and Lockwood, 1973).

The formation of persistent structures, both on soil and in substrate-deficient conditions *in vitro,* is often associated with mycelial lysis, sometimes of massive proportions (Christias and Lockwood, 1973). This suggests that lysis may be a mechanism for recycling mycelial constituents for synthesis of persistent survival structures when substrate becomes limiting, either through resource competition by other microorganisms or through the fungus's exploitation of its substrate. The large size of sclerotia made possible their collection and an estimation of the efficiency of the conversion of mycelium of sclerotia. Sclerotia of *S. rolfsii, S. sclerotiorum,* and *S. cepivorum* formed on leached glass beads converted 38, 40, and 52%, respectively, of their mycelial constituents into sclerotia. These values approached, but were less than, the dry matter of sclerotia formed on soil—i.e., 51, 55, and 68%, respectively, of that in the original mycelium. In the leaching system, there was no other possible source of materials for sclerotium formation than the mycelium, and the same was doubtless true for soil also. *Rhizoctonia solani* was exceptional in that greater dry weight and amounts of carbohydrate and nitrogen were present in the sclerotia formed on soil than were present originally in the mycelium, apparently due to the unique ability of this fungus to utilize humic substances in natural soil (Lloyd and Lockwood, 1966). The leaching system provided a means of determining a balance sheet for mycelial conversion during sclerotium formation. After 7 days' incubation of mycelial mats of *S. rolfsii,* distribution of carbohydrate was 42% in the sclerotia, 25% in the leachate, and 7% in the residual mycelium. The 26% unaccounted for was assumed to represent loss as CO_2. Of the total nitrogen content in the original mycelium, 44% was recovered in the sclerotia, 43% in the leachate, and 13% remained in the residual mycelium. Under conditions of nutrient deprivation, constituents of established mycelia apparently become translocated to sites of sclerotial synthesis. In view of the high rates of conversion and extensive lysis occurring, it is likely that some of the materials utilized in sclerotial formation were derived from cell wall polysaccharides made available through autolysis.

Utilization of mycelium in fruit body formation also is documented in the Basidiomycetes *Agaricus campestris* and *Coprinus lagopus*. As the mushroom of *A. campestris* expanded, its wet weight:dry weight ratio remained the same, and it was concluded that growth was dependent upon materials derived from the vegetative mycelium (Bonner *et al.*, 1956). Mycelial weight of *C. lagopus* was shown to decrease as fruit bodies enlarged and matured (Madelin, 1956). More direct evidence for the utilization of mycelial cell wall materials in fruit body formation has been derived from studies of *Schizophyllum commune* (Niederpruem and Wessels, 1969). Use of replacement cultures showed that initiation of fruit body primordia requires exogenous carbon and nitrogen sources, but that pileus expansion proceeds at the expense of carbon, including cell wall polysaccharides, and nitrogen in the mycelium. In fact, high concentrations of exogenous glucose completely inhibited pileus expansion. The decrease in cellular carbohydrate, concomitant with pileus formation, was due largely to the disappearance of R glucan (alkali-insoluble glucan fraction). R glucanase activity in mycelial extracts and in the culture fluid increased when glucose was exhausted from the medium, and its appearance was repressed by adding glucose. Decreased susceptibility of pileus cell walls, as compared with mycelial cell walls, to enzymic hydrolysis was shown, and this may allow for construction of pilei while R glucan is being broken down in cell walls of non-developing primordia and of the mycelia.

The formation of chlamydospores from macroconidia of *F. solani* f. sp. *cucurbitae* also is associated with autolysis of the conidia walls, according to work of Schippers and co-workers in the Netherlands. Treatments that suppressed the formation of chlamydospores, such as addition of chitin or other nitrogen sources to soil or *in vitro*, also suppressed lysis of the macroconidia (Schippers, 1972; Schippers and deWeyer, 1972; Schippers and Old, 1974). Ultrastructural studies of chlamydospore formation *in vitro* (Van Eck and Schippers, 1976) and in soil (Van Eck, 1976) showed that lysis of the conidial wall accompanied synthesis of the chlamydospore wall. Since lysis *in vitro* could only have been autolytic, autolysis was considered to contribute to the synthesis of new wall material in soil as well (Van Eck and Schippers, 1976). Moreover, lysis of the conidial wall occurring in soil was not visibly associated with microbial activity (Old and Schippers, 1973).

Ultrastructural studies of chlamydospore formation from macroconidia of *F. oxysporum* (Stevenson and Becker, 1972) and *F. sulphureum* (Schneider and Seaman, 1974) *in vitro* and of *F. solani* f. sp. *phaseoli* and *F. solani* f. sp.

pisi in soil (Van Eck, 1976) also showed lysis of macroconidial walls accompanying formation of chlamydospores, suggesting that this mechanism of chlamydospore formation is widespread in the fusaria.

9. Perspective

The question of the relative participation of nutrient depletion versus inhibitory substances in mycostasis is not yet resolved. However, from our perspective, nutrient depletion appears to be an underlying mechanism of major importance. This is indicated by the similarity of fungal behavior in soils and in leaching model systems and the fact that exudate losses in the leaching systems operated at inhibitory flow rates were never greater, and usually were much less, than those imposed by soils (Section 3.4.1). Supporting these results is a substantial body of information indicating that because of substrate limitations, the microbial community in soil must be operating at energy levels only slightly above maintenance (Section 2.1). In addition, when inhibitory factors, such as metals (Dobbs and Gash, 1965; Ko and Hora, 1971) or volatiles (Bristow and Lockwood, 1975b; Ko and Chow, 1978), have been removed from soils, mycostasis has persisted. These results would tend to suggest that inhibitory compounds, where they exist, are superimposed upon the more pervasive mycostasis imposed by nutrient depletion. Arguments for this point of view have been developed further by Lockwood (1977).

Given the paucity of substrate in soil in relation to the microbial community that must be supported, it should not be surprising that fungi have evolved various adaptations in order to cope with resource depletion. We have discussed mycostasis, mycolysis, regermination, appressorium formation, and persistent structure formation. The soil is not the only habitat wherein nutrients are in short supply. Mycostasis can also occur in the presence of a sufficient density of microorganisms on plant leaves (Section 3.4.1b), and appressorium formation has been studied only with respect to leaves (Section 7). It is likely that any of the five phenomena would be induced anywhere that substrate is depleted by microbial competition or, in the case of persistent structure formation, by exhaustive colonization of substrate by the fungus itself. Since exposure of propagules to microbial nutrient sinks in soil or on leaves results in greater exudation losses than occur during incubation in the absence of microorganisms, long-term consequences other than germination restriction might be expected. It would be of great interest to know whether soils with intense nutrient sinks would correspondingly decrease virulence or longevity of fungal propagules.

The extent to which nutrient depletion is involved in morphogenesis of appressoria and persistent structures needs further investigation, in terms both of the range of fungi affected and of alternative mechanisms. Other aspects of fungal development for which induction by nutrient depletion seems at least possible include the production of persistent hyphal fragments, as has been observed in autolysing cultures (Section 3.4.1), and of sexually derived persistent structures, but no research on these possibilities has been done.

Further research to evaluate the relative contributions of nutrient depletion and inhibitory substances to mycostasis would be aided by some standardization in methodology. Carriers of fungal propagules should be biologically inert materials such as polycarbonate membranes. If agar is used, it should be highly purified and essentially nutrient-free. Nutrient-independent propagules should be included in any work in which inhibitory substances are being evaluated, to avoid misinterpretations due to lack of substrate (Lockwood, 1977). Test fungi should be exchanged so researchers in different laboratories can evaluate common biological material. Finally, to avoid self inhibition, propagule densities in bioassays should not be excessive.

With modern analytical methodology, it should be possible to evaluate better the role of inhibitory substances in soils. The inhibitory effect of soluble aluminum in an acid Hawaiian soil has been mentioned (Ko and Hora, 1972a). However, copper, iron, manganese, zinc, and cobalt also may reach inhibitory or toxic levels in naturally acid soils or in soils receiving repetitive inputs of acid-forming fertilizers (Section 3.4.2a). For such soils, it may be possible to profile chemically the array and relative quantities of metals and correlate the chemical profiles with ecological behavior of fungi. A similar approach also may be applied to volatile substances in soils.

Nutrient depletion and mycostatic chemicals could conceivably augment each other. The mycostatic effects of inhibitory compounds might be expressed at lower concentrations against propagules that have been under intense nutrient stress, as suggested by Balis and Kouyeas (1979). Conversely, inhibitory compounds might induce greater leakage from propagules under conditions of nutrient stress. Neither possibility has yet been evaluated, but the leached sand model system offers a means for such evaluation.

A better understanding of the relative roles of nutrient stress and of fungistatic compounds in mycostasis is crucial to any attempts to manipulate soil to suppress root pathogens. Practices to enhance competitive stress would likely differ from those that increase the concentration of inhibitory substances. The possible relationship of mycostasis to the natural "suppressiveness" of some soils to certain root-infecting pathogens should also be investigated. While differences between "conducive" and "suppressive" soils might depend upon more specific interactions, e.g., mycoparasitism, the mechanisms of suppressiveness are not yet well understood.

The relative importance of autolysis and heterolysis in the destruction of fungal hyphae in soil and elsewhere in nature is not yet fully elucidated. This would seem to be a central question in fungal ecology, and it is surprising that it has not been studied to a greater extent. Its resolution is not only important for a better understanding of lytic mechanisms; like mycostasis, the information may also have practical importance in suppressing soil-borne plant pathogens. The management of soil to increase nutrient stress on pathogens might differ from practices designed to increase populations of microbes with a high capability for degrading cell walls. Our tentative view is that for young hyphae, autolysis from starvation is primarily responsible for death and at least partial degradation of the cell walls, with the residues degraded by heterolytic enzymes. With older hyphae, which autolyse less readily, the situation is less clear. Chemical agents which induce autolysis do not appear to be widespread in nature, but this may reflect lack of study.

ACKNOWLEDGMENT. This contribution is journal series article No. 9539 from the Michigan Agricultural Experiment Station.

References

Abeles, F. B., 1973, *Ethylene in Plant Biology,* Academic Press, New York.
Abeles, F. B., Craker, L. E., Forrence, L. E., and Leather, G. R., 1971, Fate of air pollutants: Removal of ethylene, sulfur dioxide, and nitrogen dioxide by soil, *Science* **173:**914–916.
Adams, P. B., and Papavizas, G. C., 1969, Survival of root-infecting fungi in soil. X. Sensitivity of propagules of *Thielaviopsis basicola* to soil fungistasis in natural and alfalfa-amended soil, *Phytopathology* **59:**135–138.
Adams, P. B., Lewis, J. A., and Papavizas, G. C., 1968, Survival of root-infecting fungi in soil. IV. The nature of fungistasis in natural and cellulose-amended soil on chlamydospores of *Fusarium solani* f. sp. *phaseoli, Phytopathology* **58:**378–383.
Aguirre, M. J. R., Garcia-Acha, I., and Villanueva, J. R., 1963, An enzyme(s) from a *Streptomyces* sp. to prepare mould 'protoplasts,' *Experientia* **19:**1–4.
Alabouvette, C., Rouxel, F., and Louvet, J., 1979, Characteristics of *Fusarium* wilt-suppressive soils and prospects for their utilization in biological control, in: *Soil-Borne Plant Pathogens* (B. Schippers and W. Gams, eds.), pp. 165–182, Academic Press, London.
Alexander, M., 1977, *Introduction to Soil Microbiology,* John Wiley and Sons, New York.
Alexander, J. V., Bourret, J. A., Gold, A. H., and Snyder, W. C., 1966, Induction of chlamydospore formation by *Fusarium solani* in soil extracts, *Phytopathology* **56:**353–354.
Allen, P. J., 1976, Spore germination and its regulation, in: *Physiological Plant Pathology* (R. Heitefuss and P. H. Williams, eds.), pp. 51–85, Vol. IV of *Encyclopedia of Plant Physiology* (A. Pirson and M. H. Zimmermann, eds.), Springer-Verlag, New York.
Allison, F. E., 1973, *Soil Organic Matter and Its Role in Crop Production,* Elsevier, New York.
Archer, S. A., 1976, Ethylene and fungal growth, *Trans. Br. Mycol. Soc.* **67:**325–326.
Arima, K., Uozumi, T., and Takahashi, M., 1965, Studies on the autolysis of *Aspergillus oryzae.* Part I. Conditions of autolysis, *Agric. Biol. Chem.* **29:**1033–1041.
Arjunarao, V., 1971, Biological control of cotton wilt. I. Soil fungistasis and antibiosis in cotton fields, *Proc., Indian Acad. Sci., Sect. B* **73:**265–272.
Babiuk, L. A., and Paul, E. A., 1970, The use of fluorescein isothiocyanate in the determination of the bacterial biomass of grassland soil. *Can. J. Microbiol.* **16:**57–62.
Bachmann, B. J., and Bonner, D. M., 1959, Protoplasts from *Neurospora crassa, J. Bacteriol.* **78:**550–556.

Bacon, J. S. D., Jones, D., Farmer, V. C., and Webley, D. M., 1968, The occurrence of α (1–3) glucan in *Cryptococcus, Schizosaccharomyces,* and *Polyporus* species, and its hydrolysis by a *Streptomyces* culture filtrate lysing cell walls of *Cryptococcus, Biochim. Biophys. Acta* **158:**313–315.

Bacon, J. S. D., Gordon, S. H., Jones, D., Taylor, I. F., and Webley, D. M., 1970, The separation of β-gluconases produced by *Cytophaga johnsonii* and their role in the lysis of yeast cell walls, *Biochem. J.* **120:**67–78.

Bainbridge, B. W., Bull, A. T., Pirt, S. J., Rowley, B. I., and Trinci, A. P. J., 1971, Biochemical and structural changes in non-growing maintained and autolysing cultures of *Aspergillus nidulans, Trans. Br. Mycol. Soc.* **56:**371–385.

Baker, K. F., and Cook, R. J., 1974, *Biological Control of Plant Pathogens,* W. H. Freeman and Co., San Francisco.

Balis, C., 1976, Ethylene-induced volatile inhibitors causing soil fungistasis, *Nature* **259:**112–114.

Balis, C., and Kouyeas, V., 1979, Contribution of chemical inhibitors to soil mycostasis, in: *Soil-Borne Plant Pathogens* (B. Schippers and W. Gams, eds.), pp. 97–106, Academic Press, London.

Ballesta, J.-P. G., and Alexander, M., 1971, Resistance of *Zygorhynchus* species to lysis, *J. Bacteriol.,* **106:**938–945.

Ballesta, J.-P. G., and Alexander, M., 1972, Susceptibility of several Basidiomycetes to microbial lysis, *Trans. Br. Mycol. Soc.* **58:**481–487.

Banwart, W. L., and Bremner, J. M., 1976, Evolution of volatile sulfur compounds from soils treated with sulfur-containing organic materials, *Soil Biol. Biochem.* **8:**439–443.

Barber, D. A., and Lynch, J. M., 1977, Microbial growth in the rhizosphere, *Soil Biol. Biochem.* **9:**305–308.

Barber, D. A., and Martin, J. K., 1976, Release of organic substances by cereal roots into soil, *New Phytol.* **76:**69–80.

Bashi, E., and Fokkema, N. J., 1977, Environmental factors limiting growth of *Sporobolomyces roseus,* an antagonist of *Cochliobolus sativus,* on wheat leaves, *Trans. Br. Mycol. Soc.* **68:**17–25.

Behera, B., and Wagner, G. H., 1974, Microbial growth rate in glucose-amended soil, *Soil Sci. Soc. Am. Proc.* **38:**591–594.

Ben-Yephet, Y., and Pinkas, Y., 1977, Germination of individual microsclerotia of *Verticillium dahliae, Phytoparasitica* **5:**159–166.

Berthe, M. C., Bonaly, R., and Reisinger, O., 1976, Dégradation chemique et enzymatique des parois de *Helminthosporium spiciferum.* Rôle protecteur des pigments, *Can. J. Microbiol.* **22:**929–936.

Bhattacharya, M., and Samaddar, K. R., 1977, Fungistasis, lysis and survival of *Pyricularia oryzae* and *Helminthosporium oryzae* in soil, *Phytopathol. Z.* **89:**60–67.

Bird, C. W., and Lynch, J. M., 1974, Formation of hydrocarbons by micro-organisms, *Chem. Soc. Rev.* **3:**309–382.

Black, R. L. B., and Dix, N. J., 1976, Spore germination and germ hyphal growth of microfungi from litter and soil in the presence of ferulic acid, *Trans. Br. Mycol. Soc.* **66:**305–311.

Blakeman, J. P., and Atkinson, P., 1976, Evidence for a spore germination inhibitor co-extracted with wax from leaves, in: *Microbiology of Aerial Plant Surfaces* (C. H. Dickinson and T. F. Preece, eds.), pp. 441–449, Academic Press, New York.

Blakeman, J. P., and Brodie, I. D. S., 1977, Competition for nutrients between epiphytic micro-organisms and germination of spores of plant pathogens on beetroot leaves, *Physiol. Plant Pathol.* **10:**29–42.

Blakeman, P. B., and Parbery, D. G., 1977, Stimulation of appressorium formation in *Colletotrichum acutatum* by phylloplane bacteria, *Physiol. Plant Pathol.* **11:**313–325.

Blakeman, J. P., and Sztejnberg, A., 1973, Effects of surface wax on inhibition of germination of *Botrytis cinerea* spores on beetroot leaves, *Physiol. Plant Pathol.* **3**:269–278.

Bloomfield, B. J., and Alexander, M., 1967, Melanins and resistance of fungi to lysis, *J. Bacteriol.* **93**:1276–1280.

Bonner, J. T., Kane, K. K., and Levey, R. H., 1956, Studies on the mechanisms of growth in the common mushroom, *Agaricus campestris, Mycologia* **48**:13–19.

Bowen, G. D., 1979, Integrated and experimental approaches to the study of growth of organisms around roots, in: *Soil-Borne Plant Pathogens* (B. Schippers and W. Gams, eds.), pp. 209–227, Academic Press, New York.

Bowen, G. D., and Rovira, A. D., 1973, Are modelling approaches useful in rhizosphere biology? in: *Modern Methods in the Study of Microbial Ecology* (T. Rosswall, ed.), pp. 443–450, Swedish Natural Science Research Council, Stockholm.

Bowen, G. D., and Rovira, A. D., 1976, Microbial colonization of plant roots, *Annu. Rev. Phytopathol.* **14**:121–144.

Bremner, J. M., and Banwart, W. L., 1976, Sorption of sulfur gases by soils, *Soil Biol. Biochem.* **8**:79–83.

Bremner, J. M., and Steele, C. G., 1978, Role of microorganisms in the atomospheric sulfur cycle, in: *Advances in Microbial Ecology*, Vol. 2 (M. Alexander, ed.), pp. 155–201, Plenum Press, New York.

Bristow, P. R., and Lockwood, J. L., 1975a, Soil fungistasis: Role of spore exudates in the inhibition of nutrient-independent propagules, *J. Gen. Microbiol.* **90**:140–146.

Bristow, P. R., and Lockwood, J. L., 1975b, Soil fungistasis: Role of the microbial nutrient sink and of fungistatic substances in two soils, *J. Gen. Microbiol.* **90**:147–156.

Brodie, I. D. S., and Blakeman, J. P., 1975, Competition for carbon compounds by a leaf surface bacterium and conidia of *Botrytis cinerea, Physiol. Plant Pathol.* **6**:125–135.

Brodie, I. D. S., and Blakeman, J. P., 1976, Competition for exogenous substrates in vitro by leaf surface micro-organisms and germination of conidia of *Botrytis cinerea, Physiol. Plant Pathol.* **9**:227–239.

Brodie, I. D. S., and Blakeman, J. P., 1977, Effect of nutrient leakage, respiration and germination of *Botrytis cinerea* conidia caused by leaching with water, *Trans. Br. Mycol. Soc.* **68**:445–448.

Brown, M. E., 1973, Soil bacteriostasis limitation in growth of soil and rhizosphere bacteria, *Can. J. Microbiol.* **19**:195–199.

Bruehl, G. W., Millar, R. L., and Cunfer, B., 1969, Significance of antibiotic production by *Cephalosporium gramineum* to its saprophytic survival, *Can. J. Plant Sci.* **49**:235–246.

Bull, A. T., 1970, Inhibition of polysaccharases by melanin: Enzyme inhibition in relation to mycolysis, *Arch. Biochem. Biophys.* **137**:345–356.

Bumbieris, M., and Lloyd, A. B., 1967a, Influence of soil fertility and moisture on lysis of fungal hyphae, *Aust. J. Biol. Sci.* **20**:103–112.

Bumbieris, M., and Lloyd, A. B., 1967b, Influence of nutrients on lysis of fungal hyphae in soil, *Aust. J. Biol. Sci.* **20**:1169–1172.

Burges, A., 1967, The decomposition of organic matter in the soil, in: *Soil Biology* (A. Burges and F. Raw, eds.), pp. 479–492, Academic Press, New York.

Carter, H. P., and Lockwood, J. L., 1957, Lysis of fungi by microorganisms and fungicides including antibiotics, *Phytopathology* **47**:154–158.

Chinn, S. H. F., and Ledingham, R. J., 1961, Mechanisms contributing to the eradication of spores of *Helminthosproium sativum* from amended soil, *Can. J. Bot.* **39**:739–748.

Chinn, S. H. F., and Ledingham, R. J., 1967, Influence of substances and soil treatments on the germination of spores of *Cochliobolus sativus, Phytopathology* **57**:580–583.

Chinn, S. H. F., and Tinline, R. D., 1964, Inherent germinability and survival of spores of *Cochliobolus sativus, Phytopathology* **54**:349–352.

Christias, C., and Lockwood, J. L., 1973, Conservation of mycelial constituents in four sclerotium-forming fungi in nutrient-deprived conditions, *Phytopathology* **63**:602–605.

Chu, S. B., and Alexander, M., 1972, Resistance and susceptibility of fungal spores to lysis, *Trans. Br. Mycol. Soc.* **58**:489–497.

Clarholm, M., and Rosswall, T., 1980, Biomass and turnover of bacteria in a forest soil and a peat, *Soil Biol. Biochem.* **12**:49–57.

Clark, F. E., 1965, The concept of competition in microbial ecology, in: *Ecology of Soil-Borne Plant Pathogens* (K. F. Baker and W. C. Snyder, eds.), pp. 339–345, University of California Press, Berkeley.

Clark, F. E., 1967, Bacteria in soil, in: *Soil Biology* (A. Burges and F. Raw, eds.), pp. 15–49, Academic Press, London.

Clark, F. E., and Paul, E. A., 1970, The microflora of grassland, *Adv. Agron.* **22**:375–435.

Considine, P. J., and Patching, J. W., 1975, Ethylene production by microorganisms grown on phenolic acids, *Ann. Appl. Biol.* **81**:115–118.

Cook, R. J., 1973, Influence of low plant and soil water potentials on diseases caused by soilborne fungi, *Phytopathology* **63**:451–458.

Cook, R. J., 1977, Management of the associated biota, in: *Plant Disease: An Advanced Treatise*, Vol. I (J. G. Horsfall and E. B. Cowling, eds.), pp. 145–166, Academic Press, New York.

Cook, R. J., and Schroth, M. N., 1965, Carbon and nitrogen compounds and germination of chlamydospores of *Fusarium solani* f. *phaseoli, Phytopathology* **55**:254–256.

Cook, R. J., and Snyder, W. C., 1965, Influence of host exudates on growth and survival of germlings of *Fusarium solani* f. *phaseoli* in soil, *Phytopathology* **55**:1021–1025.

Cooke, R. C., and Mitchell, D. T., 1967, Germination pattern and capacity for repeated stroma formation in *Claviceps purpurea, Trans. Br. Mycol. Soc.* **50**:275–283.

Cornforth, I. S., 1975, The persistence of ethylene in aerobic soils, *Plant Soil* **42**:85–96.

Cowley, G. T., and Whittingham, W. F., 1961, The effect of tannin on the growth of selected soil microfungi in culture, *Mycologia* **53**:539–542.

Davis, R. D. 1976, Soil bacteriostasis: Relation to bacterial nutrition and active soil inhibition, *Soil Biol. Biochem.* **8**:429–433.

deBont, J. A. M., 1975, Oxidation of ethylene by bacteria, *Ann. Appl. Biol.* **81**:119–120.

Dix, N.J., 1967, Mycostasis and root exudation: Factors influencing the colonization of bean roots by fungi, *Trans. Br. Mycol. Soc.* **50**:23–31.

Dix, N. J., 1972, Effect of soil fungistasis on spore germination and germ-tube growth in *Penicillium* species, *Trans. Br. Mycol. Soc.* **58**:59–66.

Dobbs, C. G., and Gash, M. J., 1965, Microbial and residual mycostasis in soils, *Nature* **207**:1354–1356.

Dobbs, C. G., and Hinson, W. H., 1953, A widespread fungistasis in soils, *Nature* **172**:197–199.

El-Abyad, M. S. H., and Webster, J., 1968a, Studies on pyrophilous discomycetes. I. Comparative physiological studies, *Trans. Br. Mycol. Soc.* **51**:353–367.

El-Abyad, M. S. H., and Webster, J., 1968b, Studies on pyrophilous discomycetes. II. Competition, *Trans. Br. Mycol. Soc.* **51**:369–375.

Emmett, R. W., and Parbery, D. G., 1975, Appressoria, *Annu. Rev. Phytopathol.* **13**:147–167.

Eno, C. F., Blue, W. G., and Good, J. M., Jr., 1955, The effect of anhydrous ammonia on nematodes, fungi, bacteria, and nitrification in some Florida soils, *Soil Sci. Soc. Amer. Proc.* **19**:55–58.

Eveleigh, D. E., Sietsma, J. H., and Haskins, R. H., 1968, The involvement of cellulase and laminarinase in the formation of *Pythium* protoplasts, *J. Gen. Microbiol.* **52**:89–97.

Farley, J. D., Wilhelm, S., and Snyder, W. C., 1971, Repeated germination and sporulation of *Verticillium albo-atrum* in soil, *Phytopathology* **61**:260–264.

Farwell, S. O., Sherrard, A. E., Pack, M. R., and Adams, D. F., 1979, Sulfur compounds volatilized from soils at different moisture contents, *Soil Biol. Biochem.* **11**:411–415.

Fencl, Z., 1978, Cell ageing and autolysis, in: *The Filamentous Fungi,* Vol. III, *Developmental Mycology* (J. E. Smith and D. R. Berry, eds.), pp. 389–405, John Wiley & Sons, New York.

Filonow, A. B., and Lockwood, J. L., 1979, Conidial exudation by *Cochliobolus victoriae* on soils in relation to soil mycostasis, in: *Soil-Borne Plant Pathogens* (B. Schippers and W. Gams, eds.), pp. 107–119, Academic Press, London.

Finstein, M. S., and Alexander, M., 1962, Competition for carbon and nitrogen between *Fusarium* and bacteria, *Soil Sci.* **94:**334–339.

Flanagan, P. W., and Bunnell, F. L., 1976, Decomposition models based on climatic variables, substrate variables, microbial respiration and production, in: *The Role of Terrestrial and Aquatic Organisms in Decomposition Processes* (J. M. Anderson and A. Macfadyen, eds.), pp. 437–457, Blackwell, Oxford.

Flanagan, P. W., and Van Cleve, K., 1977, Microbial biomass, respiration and nutrient cycling in a black spruce taiga ecosystem, in: *Soil Organisms as Components of Ecosystems* (U. Lohm and T. Persson, eds.), *Proc. 6th Coll. Soil Zool., Ecol. Bull. (Stockholm),* pp. 261–273, Swedish Natural Science Research Council, Stockholm.

Fokkema, N. J., 1971a, Influence of pollen on saprophytic and pathogenic fungi on rye leaves, in: *Ecology of Leaf Surface Micro-organisms* (T. F. Preece and C. H. Dickinson, eds.), pp. 277–282, Academic Press, London.

Fokkema, N. J., 1971b, The effect of pollen in the phyllosphere of rye on colonization by saprophytic fungi and on infection by *Helminthosporium sativum* and other leaf pathogens, *Neth. J. Plant Pathol.* **77,** Suppl. No. 1, 60 pp.

Fokkema, N. J., 1973, The role of saprophytic fungi in antagonism against *Drechslera sorokiniana (Helminthosporium sativum)* on agar plate and on rye leaves with pollen, *Physiol. Plant Pathol.* **3:**195–205.

Fokkema, N. J., 1978, Fungal antagonisms in the phyllosphere, *Ann. Appl. Biol.* **89:**115–119.

Fokkema, N. J., and Lorbeer, J. W., 1974, Interactions between *Alternaria porri* and the saprophytic mycoflora on onion leaves, *Phytopathology* **64:**1128–1133.

Fokkema, N. J., and Van der Meulen, F., 1976, Antagonism of yeastlike phyllosphere fungi against *Septoria nodorum* on wheat leaves, *Neth. J. Plant Path.* **82:**13–16.

Fokkema, N. J., Van de Laar, J. A. J., Nelis-Blomberg, A. L., and Schippers, B., 1975, The buffering capacity of the natural mycoflora of rye leaves to infections by *Cochliobolus sativus,* and its susceptibility to benomyl, *Neth. J. Plant Path.* **81:**176–186.

Ford, E. J., Gold, A. H., and Snyder, W. C., 1970a, Soil substances and chlamydospore formation by *Fusarium, Phytopathology* **60:**124–128.

Ford, E. J., Gold, A. H., and Snyder, W. C., 1970b, Induction of chlamydospore formation in *Fusarium solani* by soil bacteria, *Phytopathology* **60:**479–484.

Ford, E. J., Gold, A. H., and Snyder, W. C., 1970c, Interaction of carbon nutrition and soil substances in chlamydospore formation by *Fusarium, Phytopathology* **60:**1732–1737.

Francis, A. J., Adamson, J., Duxbury, J. M., and Alexander, M., 1973, Life detection by gas chromatography-mass spectrometry of microbial metabolites, in: *Modern Methods in the Study of Microbial Ecology* (T. Rosswall, ed.), pp. 485–488, Swedish Natural Science Research Council, Stockholm.

Frankland, J. C., 1974, Decomposition of lower plants, in: *Biology of Plant Litter Decomposition,* Vol. I (C. H. Dickinson and G. J. F. Pugh, eds.), pp. 3–36, Academic Press, London.

Fraser, A. K., 1971, Growth restrictions of pathogenic fungi, in: *Ecology of Leaf Surface Micro-organisms* (T. F. Preece and C. H. Dickinson, eds.), pp. 529–535, Academic Press, London.

Freebairn, H. T., and Buddenhagen, I. W., 1964, Ethylene production by *Pseudomonas solanacearum, Nature* **202:**313–314.

Freney, J. R., 1967, Sulfur-containing organics, in: *Soil Biochemistry,* Vol. I (A. D. McLaren and G. H. Peterson, eds.), pp. 229–259, Marcel Dekker, New York.

Fries, N., 1973, Effects of volatile organic compounds on the growth and development of fungi, *Trans. Br. Mycol. Soc.* **60:**1–21.

Garcia Acha, I., and Villanueva, J. R., 1963, The use of *Streptomyces* enzyme in preparation of protoplasts from mold spores, *Can. J. Microbiol.* **9**:139–140.

Garcia Acha, I., Rodriguez Aguirre, M. J., and Villanueva, J. R., 1964, "Protoplasts" from conidia of *Fusarium culmorum, Can. J. Microbiol.* **10**:99–101.

Garcia Mendoza, C., and Villanueva, J. R., 1962,Production of yeast protoplasts by an enzyme preparation of *Streptomyces* sp., *Nature* **195**:1326–1327.

Garcia Mendoza, C., and Villanueva, J. R., 1964, Stages in the formation of yeast protoplasts with streptozyme, *Nature* **202**:1241–1242.

Garrett, S. D., 1970, *Pathogenic Root-Infecting Fungi,* Cambridge University Press, London.

Garrett, S. D., 1976, Influence of nitrogen on cellulolysis rate and saprophytic survival in soil of some cereal foot-rot fungi, *Soil Biol. Biochem.* **8**:229–234.

Gilbert, R. G., and Griebel, G. E., 1969, The influence of volatile substances from alfalfa on *Verticillum dahliae* in soil, *Phytopathology* **59**:1400–1403.

Gilpatrick, J. D., 1969a, Role of ammonia in the control of avocado root rot with alfalfa meal soil amendment, *Phytopathology* **59**:973–978.

Gilpatrick, J. D., 1969b, Effect of soil amendments upon inoculum survival and function in *Phytophthora* root rot of avocado, *Phytopathology* **59**:979–985.

Goodlass, G., and Smith, K. A., 1978a, Effect of pH, organic matter content and nitrate on the evolution of ethylene from soils, *Soil Biol. Biochem.* **10**:193–199.

Goodlass, G., and Smith, K. A., 1978b, Effects of organic amendments on evolution of ethylene and other hydrocarbons from soil, *Soil Biol. Biochem.* **10**:201–205.

Goring, C. A. I., and Hamaker, J. W., 1972, *Organic Chemicals in The Soil Environment*, Vol. II, Marcel Dekker, New York.

Gottlieb, D., 1976, The production and role of antibiotics in soil, *J. Antibiot.* **24**:987–1000.

Gray, T. R. G., and Williams, S. T., 1971, Microbial productivity in soil, in: *Microbes and Biological Productivity* (D. E. Hughes and A. H. Rose, eds.), pp. 255–286, Cambridge University Press, London.

Gray, T. R. G., Hissett, R., and Duxbury, T., 1974, Bacterial populations of litter and soil in a deciduous woodland. II. Numbers, biomass and growth rate, *Rev. Ecol. Biol. Sol.* **11**:15–26.

Gregory, P. H., 1973, *The Microbiology of the Atmosphere,* Leonard Hill, Aylesbury.

Griebel, G. E., and Owens, L. D., 1972, Nature of the transient activation of soil microorganisms by ethanol or acetaldehyde, *Soil Biol.* Biochem. **4**:1–8.

Griffin, D. M., 1972, *Ecology of Soil Fungi,* Syracuse University Press, Syracuse.

Griffin, G. J., and Ford, R. H., 1974, Soil fungistasis: Fungus spore germination in soil at spore densities corresponding to natural population levels, *Can. J. Microbiol.* **20**:751–754.

Griffin, G. J., and Roth, D. A., 1979, Nutritional aspects of soil mycostasis, in: *Soil-Borne Plant Pathogens* (B. Schippers and W. Gams, eds.). pp. 79–96, Academic Press, London.

Griffin, G. J., Hale, M. G., and Shay, F. J., 1976, Nature and quantity of sloughed organic matter produced by roots of axenic peanut plants, *Soil Biol. Biochem.* **8**:29–32.

Griffiths, D. A., 1966, Sensitivity of Malayan isolates of *Fusarium* to soil fungistasis, *Plant Soil* **24**:269–278.

Griffiths, D. A., and Dobbs, C. G., 1963, Relationships between mycostasis and free monosaccharides in soils, *Nature* **199**:408.

Hale, M. G., Moore, L. D., and Griffin, G. J., 1978, Root exudates and exudation, in: *Interactions between Non-pathogenic Soil Microorganisms and Plants* (Y. R., Dommergues and S. V. Krupa, eds.), pp. 163–203, Elsevier, Amsterdam.

Hancock, J. G., 1979, Occurrence of soils suppressive to *Pythium ultimum,* in: *Soil-Borne Plant Pathogens* (B. Schippers and W. Gams, eds.), pp. 183–189, Academic Press, London.

Harrison, A. F., 1971, The inhibitory effect of oak leaf litter tannins on the growth of fungi, in relation to litter decomposition, *Soil Biol. Biochem.* **3**:167–172.

Hattori, T., and Hattori, R., 1976, The physical environment in soil microbiology: An attempt to extend principles of microbiology to soil microorganisms, *CRC Crit. Rev. Microbiol.* **4:**423–461.

Heagle, A. S., 1973, Interactions between air pollutants and plant parasites, *Annu. Rev. Phytopath.* **11:**356–388.

Henis, Y., and Chet, I., 1967, Mode of action of ammonia on *Sclerotium rolfsii, Phytopathology* **57:**425–427.

Holden, J., 1975, Use of nuclear staining to assess rates of cell death in cortices of cereal roots, *Soil Biol. Biochem.* **7:**333–334.

Hora, T. S., and Baker, R., 1974, Abiotic generation of a volatile fungistatic factor in soil by liming, *Phytopathology* **74:**624–629.

Hornby, D., 1979, Take-all decline: A theorist's paradise, in: *Soil-Borne Plant Pathogens* (B. Schippers and W. Gams, eds.), pp. 132–156, Academic Press, London.

Horikoshi, K., and Iida, S., 1958, Lysis of fungal mycelia by bacterial enzymes, *Nature* **181:**917–918.

Hsu, S. C., and Lockwood, J. L., 1971, Responses of fungal hyphae to soil fungistasis, *Phytopathology* **61:**1355–1362.

Hsu, S. C., and Lockwood, J. L., 1973, Chlamydospore formation in *Fusarium* in sterile salt solutions, *Phytopathology* **63:**597–602.

Hurst, H. M., and Wagner, G. H., 1969, Decomposition of ^{14}C-labeled cell wall and cytoplasmic fractions from hyaline and melanic fungi, *Soil Sci. Soc. Am. Proc.* **33:**707–711.

Hutchinson, S. A., 1973, Biological activities of volatile fungal metabolites, *Annu. Rev. Phytopath.* **11:**223–246.

Hwang, S. C., and Ko, W. H., 1974, Germination of *Calonectria crotalariae* conidia and ascospores on soil, *Mycologia* **66:**1053–1055.

Ilag, L., and Curtis, R. W., 1968, Production of ethylene by fungi, *Science* **159:**1357–1358.

Inman, R. E., Ingersoll, R. B., and Levy, E. A., 1971, Soil: A natural sink for carbon monoxide, *Science* **172:**1229–1231.

Ioannou, N., Schneider, R. W., and Grogan, R. G., 1977a, Effect of oxygen, carbon dioxide, and ethylene on growth, sporulation and production of microsclerotia by *Verticillium dahliae, Phytopathology* **67:**645–650.

Ioannou, N., Schneider, R. W., and Grogan, R. G., 1977b, Effect of flooding on the soil gas composition and production of microsclerotia by *Verticillium dahliae* in the field, *Phytopathology* **67:**651–656.

Jones, D., 1970, Ultrastructure and composition of the cell walls of *Sclerotinia sclerotiorum, Trans. Br. Mycol. Soc.* **54:**351–360.

Jones, D., and Webley, D. M., 1967, Lysis of the cell walls of yeast *(Saccharomyces cerevisiae)* by soil fungi, *Trans. Br. Mycol. Soc.* **50:**149–154.

Jones, D., Bacon, J. S. D., Farmer, V. C., and Webley, D. M., 1968, Lysis of cell walls of *Mucor ramannianus* Moller by *Streptomyces* sp., *Antonie van Leeuwenhoek J. Microbiol. Serol.* **34:**173–182.

Jones, D., Bacon, J. S. D., Farmer, V. C., and Webley, D. M., 1969, A study of the microbial lysis of the cell walls of soil yeast (*Cryptococcus* spp.), *Soil Biol. Biochem.* **1:**145–151.

Jones, D., Gordon, A. H., and Bacon, J. S. D., 1974, Co-operative action by endo- and exo-β-(1 → 3)-glucanases from parasitic fungi in the degradation of cell-wall glucans by *Sclerotinia sclerotiorum* (Lib.) de Bary, *Biochem. J.* **140:**47–55.

Jordan, V. W. L., Sneh B., and Eddy, B. P., 1972, Influence of organic amendments on *Verticillium dahliae* and on the microbial composition of the strawberry rhizosphere, *Ann. Appl. Biol.* **70:**139–148.

Kendrick, W. B., 1962, Soil fungi of a copper swamp, *Can. J. Microbiol.* **8:**639–647.

King, J. E., and Coley-Smith, J. R., 1969, Suppression of sclerotial germination in *Sclerotium cepivorum* Berk. by water expressed from four soils, *Soil Biol. Biochem.* **1:**83–87.

Kirk, P. W., Jr., 1980, The mycostatic effect of seawater on spores of terrestrial and marine higher fungi, *Bot. Mar.* **23**:233–238.

Ko, W. H., and Chow, F. K., 1977, Characteristics of bacteriostasis in natural soils, *J. Gen. Microbiol.* **102**:295–298.

Ko, W. H., and Chow, F. K., 1978, Soil fungistasis: Role of volatile inhibitors in two soils, *J. Gen. Microbiol.* **104**:75–78.

Ko, W. H., and Hora, F. K., 1971, Fungitoxicity in certain Hawaiian soils, *Soil Sci.* **112**:276–279.

Ko, W. H., and Hora, F. K., 1972a, Identification of an Al ion as a soil fungitoxin, *Soil Sci.* **113**:42–45.

Ko, W. H., and Hora, F. K., 1972b, The nature of a volatile inhibitor from alkaline soils, *Phytopathology* **62**:574–575.

Ko, W. H., and Lockwood, J. L., 1967, Soil fungistasis: Relation to fungal spore nutrition, *Phytopathology* **57**:894–901.

Ko, W. H., and Lockwood, J. L., 1970, Mechanism of lysis of fungal mycelia in soil, *Phytopathology* **60**:148–154.

Ko, W. H., Hora, F. K., and Herlicska, E., 1974, Isolation and identification of a volatile fungistatic substance from alkaline soil, *Phytopathology* **64**:1398–1400.

Kouyeas, V., and Balis, C., 1968, Influence of moisture on the restoration of mycostasis in air dried soils, *Ann. Inst. Phytopathol. Benaki* (N.S.) **8**:123–144.

Kuć, J. A., 1976, Phytoalexins, in: *Physiological Plant Pathology,* Vol. IV, *Encyclopedia of Plant Physiology* (A. Pirson and M. H. Zimmerman, eds.), pp. 632–652, Springer-Verlag, New York.

Kucera, C. L., and Kirkham, D. R., 1971, Soil respiration studies in tallgrass prairie in Missouri, *Ecology* **52**:912–915.

Kuo, M.-J., and Alexander, M., 1967, Inhibition of the lysis of fungi by melanins, *J. Bacteriol.* **94**:624–629.

Lahoz, R., and Ibeas, J. G., 1968, The autolysis of *Aspergillus flavus* in an alkaline medium, *J. Gen. Microbiol.* **53**:101–108.

Lahoz, R., Reyes, F., and Beltrá, R., 1966, Some chemical changes in the mycelium of *Aspergillus flavus* during autolysis, *J. Gen. Microbiol.* **45**:42–49.

Lahoz, R., Reyes, F., Beltrá, R., and García-Tapia, C., 1967, The autolysis of *Aspergillus terreus* in a physiologically acid medium, *J. Gen. Microbiol.* **49**:259–265.

Lahoz, R., Beltrá, R., and Ballesteros, A. M., 1970, Biochemical changes in cultures of *Nectria galligena* during the autolytic phase of growth, *Ann. Bot. (London)* **34**:205–210.

Leach, L. D., and Davey, A. E., 1935, Toxicity of low concentrations of ammonia to mycelium and sclerotia of *Sclerotium rolfsii, Phytopathology* **25**:957–959.

Lenne, J., and Parbery, D. G., 1976, Phyllosphere antagonists and appressorium formation in *Colletotrichum gloeosporioides, Trans. Br. Mycol. Soc.* **66**:334–336.

Levi, M. P., and Cowling, E. B., 1969, Role of nitrogen in wood deterioration. VII. Physiological adaptation of wood-destroying fungi to substrates deficient in nitrogen, *Phytopathology* **59**:460–468.

Levi, M. P., Merrill, W., and Cowling, E. B., 1968, Role of nitrogen in wood deterioration. VI. Mycelial fractions and model nitrogen compounds as substrates for growth of *Polyporus versicolor* and other wood-destroying and wood-inhabiting fungi, *Phytopathology* **58**:626–634.

Lewis, J. A., 1976, Production of volatiles from decomposing plant tissues and effect of these volatiles on *Rhizoctonia solani* in culture, *Can. J. Microbiol.* **22**:1300–1306.

Lewis, J. A., and Papavizas, G. C., 1970, Evolution of volatile sulfur-containing compounds from decomposition of crucifers in soil. *Soil Biol. Biochem.* **2**:239–246.

Lewis, J. A., and Papavizas, G. C., 1971, Effect of sulfur-containing volatile compounds and vapors from cabbage decomposition on *Aphanomyces euteiches, Phytopathology* **61**:208–214.

Lewis, J. A., and Papavizas, G. C., 1974, Effect of volatiles from decomposing plant tissues on pigmentation, growth, and survival of *Rhizoctonia solani, Soil Sci.* **118**:156–163.

Lindberg, T., Granhall, U., and Berg, B., 1979, Ethylene formation in some coniferous forest soils, *Soil Biol. Biochem.* **11**:637–643.

Linderman, R. G., and Gilbert, R. G., 1969, Stimulation of *Sclerotium rolfsii* in soil by volatile components of alfalfa hay, *Phytopathology* **59**:1366–1372.

Lingappa, B. T., and Lockwood, J. L., 1962, Fungitoxicity of lignin monomers, model substances, and decomposition products, *Phytopathology* **52**:295–299.

Lingappa, B. T., and Lockwood, J. L., 1964, Activation of soil microflora by fungus spores in relation to soil fungistasis, *J. Gen. Microbiol.* **35**:215–227.

Lloyd, A. B., 1968, Behavior of streptomycetes in soil, *J. Gen. Microbiol.* **56**:165–170.

Lloyd, A. B., and Lockwood, J. L., 1966, Lysis of fungal hyphae in soil and its possible relation to autolysis, *Phytopathology* **56**:595–602.

Lloyd, A. B., Noveroske, R. L., and Lockwood, J. L., 1965, Lysis of fungal mycelium by *Streptomyces* spp. and their chitinase systems, *Phytopathology* **55**:871–875.

Locke, J. C., and Green, Jr., R. J., 1977, The effect of various factors on germination of sclerotia of the charcoal rot fungus, *Macrophomina phaseolina, Proc. Am. Phytopathol. Soc.* **4**:97.

Lockwood, J. L., 1960, Lysis of mycelium of plant-pathogenic fungi by natural soil, *Phytopathology* **50**:787–789.

Lockwood, J. L., 1968, The fungal environment of soil bacteria, in: *The Ecology of Soil Bacteria* (T. R. G. Gray and D. Parkinson, eds.), pp. 44–65, Liverpool University Press, Liverpool.

Lockwood, J. L., 1975, Quantitative evaluation of a leaching model system for soil fungistasis, *Phytopathology* **65**:460–464.

Lockwood, J. L., 1977, Fungistasis in soils, *Biol. Rev.* **52**:1–43.

Lockwood, J. L., 1979, Soil mycostasis: Concluding remarks, in: *Soil-Borne Plant Pathogens* (B. Schippers and W. Gams, eds.), pp. 121–129, Academic Press, London.

Lynch, J. M., 1972, Identification of substrates and isolation of micro-organisms responsible for ethylene production in the soil, *Nature* **240**:45–46.

Lynch, J. M., 1974, Mode of ethylene formation by *Mucor hiemalis, J. Gen. Microbiol.* **83**:407–411.

Lynch, J. M., 1975, Ethylene in soil, *Nature* **256**:576–577.

Lynch, J. M., and Harper, S. M. T., 1974a, Formation of ethylene by a soil fungus, *J. Gen. Microbiol.* **80**:187–195.

Lynch, J. M., and Harper, S. M. T., 1974b, Fungal growth rate and the formation of ethylene in soil, *J. Gen. Microbiol.* **85**:91–96.

Lynch, J. M., and Panting, L. M., 1980, Cultivation and the soil biomass, *Soil Biol. Biochem.* **12**:29–33.

Macko, V., Staples, R. C., Yaniv, Z., and Granados, R. R., 1976, Self-inhibitors of fungal spore germination, in: *The Fungal Spore: Form and Function* (D. J. Webber and W. M. Hess, eds.), pp. 73–98, John Wiley and Sons, New York.

MacMillan, A., 1956, The entry of ammonia into fungal cells, *J. Exp. Bot.* **7**:113–126.

Madelin, M. F., 1956, Studies of the nutrition of *Coprinus lagopus* Fr., especially as affecting fruiting, *Ann. Bot. (London)* (N.S.) **20**:307–330.

Malo, B. A., and Purvis, E. R., 1964, Soil adsorption of atmospheric ammonia, *Soil Sci* **97**:242–247.

Marr, A. G., Nilson, E. H., and Clark, D. J., 1963, The maintenance requirement of *Escherichia coli, Ann. N.Y. Acad. Sci* **102**:536–548.

Marshall, K. C., and Alexander, M., 1960, Competition between soil bacteria and *Fusarium, Plant Soil* **12**:143–153.

Martin, J. K., 1975, ^{14}C-labelled material leached from the rhizosphere of plants supplied continuously with $^{14}CO_2$, *Soil Biol. Biochem.* **7**:395–399.

Martin, J. K., 1977a, The chemical nature of carbon-14-labelled organic matter released into soil from growing wheat roots: Effects of soil micro-organisms, in: *Soil Organic Matter Studies,* Vol. I pp. 197–203, International Atomic Energy Agency, Vienna.

Martin, J. K., 1977b, Factors influencing the loss of organic matter from wheat roots, *Soil Biol. Biochem.* **9**:1–7.

Martin, J. K., 1977c, Effect of soil moisture on the release of organic carbon from wheat roots, *Soil Biol. Biochem.* **9**:303–304.

Maurer, C. L., and Baker, R., 1965, Ecology of plant pathogens in soil. II. Influence of glucose, cellulose, and organic nitrogen amendments on development of bean root rot, *Phytopathology* **55**:69–72.

Mayfield, C. I., Williams, S. T., Ruddick, S. M., and Hatfield, H. L., 1972, Studies on the ecology of actinomycetes in soil. IV. Observations on the form and growth of streptomycetes in soil. *Soil Biol. Biochem.* **4**:79–91.

McCallan, S. E. A., and Weedon, F. R., 1940, Toxicity of ammonia, chlorine, hydrogen cyanide, hydrogen sulphide, and sulfur dioxide gases. II. Fungi and bacteria, *Contrib. Boyce Thompson Inst.* **11**:331–342.

McDougall, B. M., and Rovira, A. D., 1970, Sites of exudation of ^{14}C-labelled compounds from wheat roots, *New Phytol.* **69**:999–1007.

Merrill, W., and Cowling, E. B., 1966, Role of nitrogen in wood deterioration: Amount and distribution of nitrogen in fungi, *Phytopathology* **56**:1083–1090.

Meyers, J. A., and Cook, R. J., 1972, Induction of chlamydospore formation in *Fusarium solani* by abrupt removal of the organic carbon substrate, *Phytopathology* **62**:1148–1153.

Mircetich, S. M., and Zentmyer, G. A., 1969, Effect of carbon and nitrogen compounds on germination of chlamydospores of *Phytophthora cinnamomi* in soil, *Phytopathology* **59**:1732–1735.

Mitchell, C. P., and Dix, N. J., 1975, Growth and germination of *Trichoderma* spp. under the influence of soil fungistasis, *Trans. Br. Mycol. Soc.* **64**:235–241.

Mitchell, R., and Alexander, M., 1963, Lysis of soil fungi by bacteria, *Can. J. Microbiol.* **9**:169–177.

Mitchell, R., and Sabar, N., 1966, Autolytic enzymes in fungal cell walls, *J. Gen. Microbiol.* **42**:39–42.

Mortland, M. M., 1958, Reactions of ammonia in soils. *Adv. Agron.* **10**:325–348.

Musílková, M., and Fencl, Z., 1968, Some factors affecting the formation of protoplasts in *Aspergillus niger, Folia Microbiol.* (Prague) **12**:235–239.

Naiki, T., and Ui, T., 1972, Lytic phenomenon of the hyphae of *Rhizoctonia* in soil, *Trans. Mycol. Soc. Jpn.* **13**:140–148.

Nash, S. M., Christou, T., and Snyder, W. C., 1961, Existence of *Fusarium solani* f. *phaseoli* as chlamydospores in soil. *Phytopathology* **51**:308–312.

Neal, D. C., and Collins, E. R., 1936, Concentrations of ammonia necessary in a low-lime phase of Houston clay soil to kill the cotton root rot fungus, *Phytopathology* **26**:1030–1032.

Niederpruem, D. J., and Wessels, J. G. H., 1969, Cytodifferentiation and morphogenesis in *Schizophyllum commune, Bacteriol. Rev.* **33**:505–535.

Ochoa, A. G., Garcia Acha, I., Gascon, S., and Villanueva, J. R., 1963, The use of lytic enzymes of *Micromonospora* spp. to prepare protoplasts of yeasts, *Experientia* **19**:1–3.

Old, K. M., and Robertson, W. M., 1970, Effects of lytic enzymes and natural soil on the fine structure of conidia of *Cochliobolus sativus, Trans. Br. Mycol. Soc.* **54**:343–350.

Old, K. M., and Schippers, B., 1973, Electron microscopical studies of chlamydospores of *Fusarium solani* f. *cucurbitae* formed in natural soil, *Soil Biol. Biochem.* **5**:613–620.

Papavizas, G. C., 1966, Suppression of *Aphanomyces* root rot of pea by cruciferous soil amendments, *Phytopathology* **56**:1071–1075.

Papavizas, G. C., 1967, Comparison of treatments suggested for control of *Aphanomyces* root rot of peas, *Plant Dis. Rep.* **51:**125–129.

Papavizas, G. C., and Adams, P. B., 1969, Survival of root-infecting fungi in soil. XII. Germination and survival of endoconidia and chlamydospores of *Thielaviopsis basicola* in fallow soil and in soil adjacent to germinating bean seed, *Phytopathology* **59:**371–378.

Parbery, D. B., and Blakeman, J. P., 1978, Effect of substances associated with leaf surfaces on appressorium formation by *Colletotrichum acutatum, Trans. Br. Mycol. Soc.* **70:**7–19.

Park, D., 1963, Evidence for a common fungal growth regulator, *Trans. Br. Mycol. Soc.* **46:**541–548.

Park, D., 1976, Cellulose decomposition by a pythiaceous fungus, *Trans. Br. Mycol. Soc.* **66:**65–70.

Park, D., and Robinson, P. M., 1964, Isolation and bioassay of a fungal morphogen, *Nature* **203:**988–989.

Pavlica, D. A., Hora, T. S., Bradshaw, J. J., Skogerboe, R. K., and Baker, R., 1978, Volatiles from soil influencing activities of soil fungi, *Phytopathology* **68:**758–765.

Payne, W. J., 1970, Energy yields and growth of heterotrophs, *Annu. Rev. Microbiol.* **24:**17–52.

Pengra, R. M., Cole, M. A., and Alexander, M., 1969, Cell walls and lysis of *Mortierella parvispora* hyphae, *J. Bacteriol.* **97:**1056–1061.

Pirt, S. J., 1965, The maintenance energy of bacteria in growing cultures, *Proc. R. Soc. London, Ser.* B **163:**224–231.

Pitt, D., 1964, Studies on sharp eyespot of cereals. II. Viability of sclerotia: persistence of the causal fungus, *Rhizoctonia solani, Ann. Appl. Biol.* **54:**231–240.

Polacheck, R., and Rosenberger, R. F., 1975, Autolytic enzymes in hyphae of *Aspergillus nidulans:* Their action on old and newly formed walls, *J. Bacteriol.* **121:**332–337.

Potgieter, H. J., and Alexander, M., 1966, Susceptibility and resistance of several fungi to microbial lysis, *J. Bacteriol.* **91:**1526–1532.

Primrose, S. B., 1979, Ethylene and agriculture: The role of the microbe, *J. Appl. Bacteriol.* **46:**1–25.

Primrose, S. B., and Dilworth, M. J., 1976, Ethylene production by bacteria, *J. Gen. Microbiol.* **93:**177–181.

Qureshi, A. A., and Page, O. T., 1970, Observations on chlamydospore formation by *Fusarium* in a two-salt solution, *Can. J. Microbiol.* **16:**29–32.

Reyes, F., Lahoz, R., and Cornago, P., 1977, Autolysis of *Neurospora crassa* in different culture conditions and release of β-N-acetyl-glucosamidase and chitinase, *Trans. Br. Mycol. Soc.* **68:**357–361.

Robinson, P. M., Park, D., and Garrett, M. K., 1968, Sporostatic products of fungi, *Trans. Br. Mycol. Soc.* **51:**113–124.

Rotem, J., Clare, B. G., and Carter, M. V., 1976, Effects of temperature, leaf wetness, leaf bacteria, and leaf and bacterial diffusates on production and lysis of *Rhynchosporium secalis* spores, *Physiol. Plant Pathol.* **8:**297–305.

Rovira, A. D., and Davey, C. B., 1974, Biology of the rhizosphere, in: *The Plant Root and Its Environment* (E. W., Carson, ed.), pp. 153–204, University Press of Virginia, Charlottesville.

Rovira, A. D., and Vendrell, M., 1972, Ethylene in sterilized soil: its significance in studies of interactions between microorganisms and plants, *Soil Biol. Biochem.* **4:**63–69.

Schippers, B., 1972, Reduced chlamydospore formation and lysis of macroconidia of *Fusarium solani* f. *cucurbitae* in nitrogen-amended soil, *Neth. J. Plant Pathol.* **78:**189–197.

Schippers, B., and deWeyer, W. M. M. M., 1972, Chlamydospore formation and lysis of macroconidia of *Fusarium solani* f. *cucurbitae* in chitin-amended soil, *Neth. J. Plant Pathol.* **78:**45–54.

Schippers, B., and Old, K. M., 1974, Factors affecting chlamydospore formation by *Fusarium solani* f. *cucurbitae* in pure culture, *Soil Biol. Biochem.* **6:**153–160.

Schippers, B., and Palm, L. C., 1973, Ammonia, a fungistatic volatile in chitin-amended soil, *Neth. J. Plant Path.* **79**:279–281.

Schippers, B., Boerwinkel, D. J., and Konings, H., 1978, Ethylene not responsible for inhibition of conidium germination by soil volatiles, *Neth. J. Plant. Path.* **84**:101–107.

Schneider, E. F., and Seaman W. L., 1974, Development of conidial chlamydospores of *Fusarium sulphureum* in distilled water, *Can. J. Microbiol.* **20**:247–254.

Schroth, M. N., and Hendrix, F. F., 1962, Influence of nonsusceptible plants on the survival of *Fusarium solani* f. *phaseoli* in soil, *Phytopathology* **52**:906–909.

Schroth, M. N., and Hildebrand, D. C., 1964, Influence of plant exudates on root-infecting fungi, *Annu. Rev. Phytopathol.* **2**:101–132.

Schüepp, H., and Frei, E., 1969, Soil fungistasis with respect to pH and profile, *Can. J. Microbiol.* **15**:1273–1279.

Schüepp, H., and Green, R. J., Jr., 1968, Indirect assay methods to investigate soil fungistasis with special consideration of soil pH, *Phytopathol. Z.* **61**:1–28.

Sequeira, L., 1963, Effect of urea applications on survival of *Fusarium oxysporum* f. *cubense* in soil, *Phytopathology* **53**:332–336.

Shields, J. A., Paul, E. A., and Lowe, W. E., 1973, Turnover of microbial tissue in soil under field conditions, *Soil Biol. Biochem.* **5**:753–764.

Shields, J. A., Paul, E. A., and Lowe, W. E., 1974, Factors influencing the stability of labelled microbial materials in soils, *Soil Biol. Biochem.* **6**:31–37.

Sietsma, J. H., Eveleigh, D. E., Haskins, R. H., and Spencer, J. F. T., 1967, "Protoplasts" from *Pythium* sp. PRL 2142, *Can. J. Microbiol.* **13**:1701–1704.

Sivasithamparam, K., Stukely, M., and Parker, C. A., 1975, A volatile factor inducing transmissable lysis in *Gaeumannomyces graminis* (Sacc.) Arx and Olivier var. *tritici* Walker, *Can. J. Microbiol.* **21**:293–300.

Skujins, J. J., Potgieter, H. J., and Alexander, M., 1965, Dissolution of fungal cell walls by a streptomycete chitinase and β-(1 $\rightarrow$ 3)-glucanase, *Arch. Biochem. Biophys.* **111**:358–369.

Smiley, R. W., Cook, R. J., and Papendick, R. I., 1970, Anhydrous ammonia as a soil fungicide against *Fusarium* and fungicidal activity in the ammonia retention zone, *Phytopathology* **60**:1227–1232.

Smiley, R. W., Cook, R. J., and Papendick, R. I., 1972, *Fusarium* foot rot of wheat and peas as influenced by soil applications of anhydrous ammonia and ammonia-potassium azide solutions. *Phytopathology* **62**:86–91.

Smith, A. M., 1973, Ethylene as a cause of soil fungistasis, *Nature* **246**:311–313.

Smith, A. M., 1976a, Ethylene in soil biology, *Annu. Rev. Phytopathol.* **14**:53–73.

Smith, A. M., 1976b, Ethylene production by bacteria in reduced microsites in soil and some implications to agriculture, *Soil Biol. Biochem.* **8**:293–298.

Smith, A. M., and Cook, R. J., 1974, Implications of ethylene production by bacteria for biological balance of soil, *Nature* **252**:703–705.

Smith, K. A., 1978, Ineffectiveness of ethylene as a regulator of soil microbial activity, *Soil Biol. Biochem.* **10**:269–272.

Smith, K. A., and Restall, S. W. F., 1971, The occurrence of ethylene in anaerobic soil, *J. Soil Sci.* **22**:430–433.

Smith, K. A., and Russell, R. S., 1969, Occurrence of ethylene, and its significance in anaerobic soil, *Nature* **222**:769–771.

Smith, S. N., and Snyder, W. C., 1972, Germination of *Fusarium oxysporum* chlamydospores in soils favorable and unfavorable to wilt establishment, *Phytopathology* **62**:273–277.

Sneh, B., and Lockwood, J. L., 1975, Quantitative evaluation of the microbial nutrient sink in soil in relation to a model system for soil fungistasis, *Soil Biol. Biochem.* **8**:65–69.

Sneh, B., Katan, J., and Henis, Y., 1971, Mode of inhibition of *Rhizoctonia solani* in chitin-amended soil, *Phytopathology* **61**:1113–1117.

Sneh, B., Holdaway, B. F., Hooper G. R., and Lockwood, J. L., 1976, Germination-lysis as a mechanism for biological control of *Thielaviopsis basicola* pathogenic on soybean, *Can. J. Bot.* **54**:1499–1508.

Snyder, W. C., Schroth, M. N., and Christou, T., 1959, Effect of plant residues on root rot of bean, *Phytopathology* **49**:755–756.

Stanghellini, M. E., and Hancock, J. G., 1971, The sporangium of *Pythium ultimum* as a survival structure in soil, *Phytopathology* **61**:157–164.

Steiner, G. W., and Lockwood, J. L., 1969, Soil fungistasis: Sensitivity of spores in relation to germination time and size, *Phytopathology* **59**:1084–1092.

Steiner, G. W., and Lockwood, J. L., 1970, Soil fungistasis: Mechanism in sterilized, reinoculated soil, *Phytopathology* **60**:89–91.

Stevenson, I. L., and Becker, S. A. W. E., 1972, The fine structure and development of chlamydospores of *Fusarium oxysporum, Can. J. Microbiol.* **18**:997–1002.

Stevenson, L. H., 1978, A case for bacterial dormancy in aquatic systems, *Microb. Ecol.* **4**:127–133.

Stotzky, G., and Norman, A. G., 1961a, Factors limiting microbial activities in soil. I. The level of substrate, nitrogen, and phosphorus, *Arch. Mikrobiol.* **40**:341–369.

Stotzky, G., and Norman, A. G., 1961b, Factors limiting microbial activities in soil. II. The effect of sulfur, *Arch. Mikrobiol.* **40**:370–382.

Stotzky, G., and Schenck, S., 1976, Volatile organic compounds and microorganisms, *CRC Crit. Rev. Microbiol.* **4**:333–382.

Sutherland, J. B., and Cook, R. J., 1980, Effects of chemical and heat treatments on ethylene production in soil, *Soil Biol. Biochem.* **12**:357–362.

Swinburne, T. R., 1976, Stimulants of germination and appressoria formation by *Colletotrichum musae* (Berk. & Curt) Arx. in banana leachate, *Phytopathol. Z.* **87**:74–90.

Sztejnberg, A., and Blakeman, J. P., 1973, Studies on leaching of *Botrytis cinerea* conidia and dye absorption by bacteria in relation to competition for nutrients on leaves. *J. Gen. Microbiol.* **78**:15–22.

Tabak, H. H., and Cooke, W. B., 1968, The effects of gaseous environments on the growth and metabolism of fungi, *Bot. Rev.* **34**:126–252.

Takeuchi, S., 1976, Scanning electron microscopical observations of *Fusarium oxysporum* f. sp. *raphani* in soil, *Ann. Phytopathol. Soc. Jpn.* **42**:49–52.

Terman, G. L., 1979, Volatilization losses of nitrogen as ammonia from surface-applied fertilizers, organic amendments, and crop residues, *Adv. Agron.* **31**:189–223.

Trinci, A. P. J., 1971, Exponential growth of germ tubes of fungal spores, *J. Gen. Microbiol.* **67**:345–348.

Trinci, A. P. J., and Righelato, R. C., 1970, Changes in constituents and ultrastructure of hyphal compartments during autolysis of glucose-starved *Penicillium chrysogenum, J. Gen. Microbiol.* **60**:239–249.

Tsao, P. H., and Zentmyer, G. A., 1979, Suppression of *Phytophthora cinnamomi* and *Ph. parasitica* in urea-amended soils, in: *Soil-Borne Plant Pathogens* (B. Schippers and W. Gams, eds.), pp. 191–199, Academic Press, London.

Tschudi, S., and Kern, H., 1979, Specific lysis of the mycelium of *Gaeumannomyces graminis* by enzymes of *Streptomyces lavendulae,* in: *Soil-Borne Plant Pathogens* (B. Schippers and W. Gams, eds.), pp. 611–615, Academic Press, London.

Vaartaja, O., 1974, Inhibition of *Pythium ultimum* in molecular fractions from gel filtration of soil extracts, *Can. J. Microbiol.* **20**:1273–1280.

Vaartaja, O., 1977, Responses of *Pythium ultimum* and other fungi to a soil extract containing an inhibitor with low molecular weight, *Phytopathology* **67**:67–71.

Vaartaja, O., 1978, Production of inhibitors of *Pythium* in soil, *Bimonthly Research Notes* **34**(5)(Sept.–Oct):32–33, Forestry Service, Fisheries, and Environment of Canada.

Van Eck, W. H., 1976, Ultrastructure of forming and dormant chlamydospores of *Fusarium solani* in soil, *Can. J. Microbiol.* **22**:1634–1642.

Van Eck, W. H., 1978, Chemistry of cell walls of *Fusarium solani* and the resistance of spores to microbial lysis, *Soil Biol. Biochem.* **10**:155–157.

Van Eck, W. H., and Schippers, B., 1976, Ultrastructure of developing chlamydospores of *Fusarium solani* f. *cucurbitae* in vitro, *Soil Biol. Biochem.* **8**:1–6.

Walker, J. C., Morell, S., and Foster, H. H., 1937, Toxicity of mustard oils and related sulfur compounds to certain fungi, *Am. J. Bot.* **24**:536–541.

Warren, R. C., 1972, Interference by common leaf saprophytic fungi with the development of *Phoma betae* lesions on sugarbeet leaves, *Ann. Appl. Biol.* **72**:137–144.

Wicklow, D. T., 1975, Fire as an environmental cue triggering ascomycete development in prairie soils, *Mycologia* **67**:852–862.

Wicklow, D. T., and Hirschfield, B. J., 1979, Competitive hierarchy in post-fire ascomycetes, *Mycologia* **71**:47–54.

Wicklow, D. T., and Zak, J. C., 1979, Ascospore germination of carbonicolous ascomycetes in fungistatic soils: An ecological interpretation, *Mycologia* **71**:238–242.

Widden, P., and Parkinson, D., 1975, The effects of a forest fire on soil microfungi, *Soil Biol. Biochem.* **7**:125–138.

Williams, G. H., and Western, J. H., 1965, The biology of *Sclerotinia trifoliorum* Erikss. and other species of sclerotium-forming fungi, *Ann. Appl. Biol.* **56**:261–268.

Witt, W. W., and Weber, J. B., 1975, Ethylene adsorption and movement in soils and adsorption by soil constitutents, *Weed Sci.* **23**:302–307.

Wright, J. M., 1956, The production of antibiotics in soil. IV. Production of antibiotics in coats of seed sown in soil, *Ann. Appl. Biol.* **44**:561–566.

Yoder, D. L., and Lockwood, J. L., 1973, Fungal spore germination on natural and sterile soil, *J. Gen. Microbiol.* **74**:107–117.

Zak, J. C., and Wicklow, D. T., 1978a, Factors influencing patterns of ascomycete sporulation following simulated "burning" of prairie soils, *Soil Biol. Biochem.* **10**:533–535.

Zak, J. C., and Wicklow, D. T., 1978b, Response of carbonicolous ascomycetes to aerated steam temperatures and treatment intervals, *Can. J. Bot.* **56**:2313–2318.

Zakaria, M. A., Lockwood, J. L., and Filonow, A. B., 1980, Reduction in *Fusarium* population density in soil by volatile degradation products of oilseed meal amendments, *Phytopathology* **70**:495–499.

Zentmyer, G. A., and Bingham, F. T., 1956, The influence of nitrite on the development of *Phytophthora* root rot of avocado, *Phytopathology* **46**:121–124.

Zevenhuizen, L. P. T. M., and Bartnicki-Garcia, S., 1970, Structure and role of a soluble cytoplasmic glucan from *Phytophthora cinnamomi*, *J. Gen. Microbiol.* **61**:183–188.

2

Oligotrophy

Fast and Famine Existence

JEANNE S. POINDEXTER

1. Introduction

According to A. Koch's interpretation (Koch, 1971), *Escherichia coli* has evolved a strategy for surviving a "feast and famine" existence. Oligotrophic bacteria, in contrast, are conceived of as those never invited to a feast; their properties should include microbial adaptations to uninterrupted nutrient limitation.

The concept of oligotrophy and the oligotrophic habitat was recently reviewed (Kuznetsov *et al.,* 1979), and oligotrophs were tentatively identified as those bacteria that can be isolated on media containing 1–15 mg organic-C/liter and can be subcultured on such media. They may also be able to grow on richer media on subsequent cultivation, either by "spontaneous" (mutational?) acquisition of such ability or as a consequence of physiologic adaptation. In nature, the concentration of nutrients in a given habitat varies with a large number of factors—e.g., season (depending on latitude), time of day, availability of water, erratic deposition of animal excreta and of carcasses of all types of organisms, and distance from bottom sediment and from shore, and it is more appropriate to define a habitat in terms of the average flux of nutrients. For example, the flux of organic-C in a eutrophic freshwater lake appears to be at least 5 mg C/liter per day, whereas in an oligotrophic lake the flux does not exceed 0.1 mg C/liter per day (Hood, 1970). Similar fluxes seem appropriate for distinguishing oligotrophic regions of seawaters (Bulion, 1977).

JEANNE S. POINDEXTER ● The Public Health Research Institute of New York, Inc., New York, New York 10016.

Oligotrophic bacteria can then be conceived of as those whose survival in nature depends on their ability to multiply in habitats of low nutrient flux (from near zero to—arbitrarily, for the present—a fraction of a mg C/liter per day), in contrast to organisms whose survival depends on their ability to exploit an environment in which the nutrient flux is at least 50-fold higher and does not drop to zero for prolonged periods.

The open ocean seems to be the clearest example of an oligotrophic habitat. The *in situ* generation times of bacteria adapted to that habitat have been estimated to range from 20 to 200 hr (Jannasch, 1969). This range overlaps, but also exceeds that estimated for soil oligotrophs (less than 24 hr; Gray and Williams, 1971); it does not overlap the 12-hr *in situ* generation time estimated for *E. coli* in the intestine, a eutrophic habitat (Koch, 1971).

There is growing evidence of the existence of bacteria especially suited for the exploitation of low nutrient flux habitats, and that they can be distinguished from organisms especially suited for high nutrient flux environments. These "others"—the principal participants in the past century of bacteriological research—are variously and not often distinctively referred to as "eutrophs" (Kuznetsov *et al.,* 1979), "heterotrophs" (Akagi *et al.,* 1977; Mallory *et al.,* 1977), "saprophytes" (Kuznetsov *et al.,* 1979), or "organisms able to grow on rich media." I have proposed the term *copiotrophic* to refer to such types (Poindexter, 1981), since they grow and multiply in the presence of an abundance of nutrients, and their natural distribution implies that nutrient abundance favors their survival and is an important factor in their competitiveness.

The present discussion is an extension of an earlier presentation (Hirsch, 1979). Following a brief description of approaches to the isolation of oligotrophic bacteria, the concept of oligotrophy as distinct from copiotrophy will be examined in two ways. First, hypothetical characteristics that could reasonably be expected to contribute to the ability of bacteria to exploit low nutrient flux environments will be set forth; probable consequences of such features for oligotrophic organisms will be considered. Second, certain of those characteristics will be examined as they are exhibited by bacterial groups whose natural distribution is consistent with their being nominated as example oligotrophs. The latter exercise is entered upon with the qualification that there may not exist any obligate oligotrophs; an oligotrophic habitat may be transiently enriched with nutrients, and successful oligotrophs would need to survive such enrichment even if inadequately prepared to exploit it. The exceptional habitat in this regard is (or once was) most of the open ocean, wherein any strict oligotrophs that may exist are most likely to be found.

1.1. Isolation of Oligotrophic Bacteria

Bacteria capable of growing and multiplying in very dilute media can be isolated in two ways: by plating directly on dilute media and by chemostat selection at very low nutrient flux. The first method challenges the cells to achieve a net synthesis of cell material by accumulating, in excess of their maintenance requirements, nutrients that are widely dispersed in space. The second challenges them to accumulate nutrients that are widely dispersed in time. The latter method, reviewed by Jannasch and Mateles (1974) and by Harder *et al.* (1977), appears to approximate the oligotrophic environment more closely. Both methods have been applied more frequently to marine than to soil or freshwater samples, which is consistent with the recognition that the ocean environment is the most constant oligotrophic habitat in the biosphere.

1.1.1. Direct Plating

The most stringent application of direct plating is exemplified by the study reported by Akagi *et al.* (1977), in which the oligotrophic medium contained glass fibers, rather than agar, as the solidifying agent. The total organic content of the oligotrophic medium provided 16.8 mg C/liter; for comparative purposes, counts were also performed on a richer medium qualitatively similar in composition but providing 2 g organic-C/liter, plus agar. Each sample was plated on both media, and the numbers of oligotrophs and copiotrophs (called "heterotrophs") were compared and related to the dissolved carbohydrate content of the sample. In general, the counts of oligotrophic colonies were at least ten times greater than the counts on rich medium for samples containing less than 0.5 mg soluble carbohydrate-C/liter; the proportion was reversed for sediment samples and for water samples containing more than 0.5 mg soluble carbohydrate-C/liter.

Three sets of samples in which copiotrophs exceeded oligotrophs were also plated on an agar medium prepared in charcoal-treated seawater without added nutrients. The viable counts on this medium closely matched the counts on the oligotrophic glass-fiber medium, implying that the colonies that appeared on the latter medium represented bacteria that were active in low-nutrient seawater. In the majority of platings, the counts of copiotrophs exceeded the counts on the seawater agar, implying that such bacteria did not multiply in seawater of low organic content. The ability of the oligotrophic populations to grow on the rich medium was not tested, and further description of isolates was not reported.

Carlucci and Shimp (1974) described an oligotrophic marine bacterium able to multiply in unsupplemented seawater. The maximum population density was determined by the amount of organic-C in the natural seawater sample. Surface water containing slightly less than 1 mg glucose-C/liter supported the accumulation of 2×10^6 colony-forming units (cfu)/ml, while water from a depth of 3000 m, containing no more than 0.5 mg glucose-C/liter, supported only 1×10^5 cfu/ml. In rich media (0.5–5 g peptone/liter), the population reached 10^9 cfu/ml. Generation times were 3–6 hr, depending mainly on incubation temperature. This report illustrates a major problem in the study of oligotrophy: media containing nutrients at concentrations approximating the standing concentrations in oligotrophic habitats do not support the development of visible turbidity, and while molar growth yields are probably not greatly different from those achieved with copiotrophs, the cell yield/ml of oligotrophic populations grown oligotrophically is not suited to present methods of biochemical and physiological research with bacteria (Jannasch, 1967b).

1.1.2. Chemostat Enrichment Cultivation

Enrichment of marine oligotrophs in chemostat vessels is achieved by providing no more than 0.1 mg C/liter at high dilution rates (D) (0.2–0.3/hr; Hamilton et al., 1966) or higher nutrient levels (up to 50 mg C/liter) at lower D (down to 0.01/hr; Jannasch, 1965, 1967a). Chemostat enrichment has also been successful in isolation of freshwater oligotrophs, as enrichment employing 0.2 g lactate-C/liter and a dilution rate of 0.05/hr (Matin and Veldkamp, 1978).

Chemostat cultivation requires that the population not only accumulate nutrients and grow, but also reproduce at the rate demanded by dilution. Wash-out resulting from continuous dilution of the population may approximate natural population attrition due to factors such as predation, settling, and parasitism, and so may mimic events of the natural habitat. In oligotrophic habitats, however, very low bacterial population densities may be insufficient to support predation or to communicate parasites, so that only infrequent reproductive events may be sufficient to maintain a nutrient-accumulating population during periods of extremely low nutrient flux. Accordingly, even when very low dilution rates are employed, the chemostat may exert an attritional pressure on the population that is not a factor in the natural environment. This consideration limits the chemostat approach to the study of reproducing populations, and chemostat enrichment cultures to selection of oligotrophs that maintain their reproductive as well as nutrient-accumulation rate in the face of both nutrient limitation and population dilution (Jannasch and Mateles, 1974). Oligotrophs whose strategy is accumulation rather than maintenance of reproduction might have a relatively high rate of survival in nature but would be indiscriminately removed with depleted, moribund cells in the chemostat. The lack of discrimination between dead and viable cells in chemostat attrition may be responsible for such enrichment procedures yielding a smaller variety of oligotrophs than those obtained by direct plating methods (cf. Mallory et al., 1977, with Hamilton et al., 1966, and Jannasch, 1967a).

Results of enrichment would probably differ in a "chemostat" so designed that it would allow a standing bacterial population while the nutrient liquid alone flowed through the system. In such an apparatus, also, the viable population might not be reduced below that apparently critical density (Postgate, 1973) needed for mutual support of the cells through their combined influence [e.g., reducing action (Jannasch, 1963, 1967b)] on physical–chemical qualities of their environment.

1.2. Oligotrophs as Aerobes

It will be assumed in this discussion that the model oligotroph is a respiring, oxybiontic organism. This would allow efficient and complete dissimilation of organic substrates in the production of metabolic energy; an oligotroph that excreted oxidizable organic substances as metabolic by-products would not be competitive in an environment of low organic flux. The low organic content of oligotrophic habitats also implies that reducing action due to spontaneous chemical reactions would not be great, and since bacterial population densities also tend to be far lower than the densities found in regions of high organic content, O_2 consumption by the population would not be sufficient to maintain a low redox potential (E_h).

On the other hand, since O_2-sensitivity is related principally to reduced intermediates of O_2 metabolism rather than to diatomic oxygen itself (Morris, 1976; Fridovich, 1978), it is not clear that oligotrophs would require protective agents such as superoxide dismutases and catalase. Formation of the reduced intermediates is proportional to the concentration of rapidly utilizable organic nutrients and to metabolic rates, as well as to O_2 tension. In habitats of low organic content and low bacterial population densities, the deleterious oxygen species may not often reach significant, selective levels. The possession of catalase is not universal among aerobic oligotrophs (Mallory *et al.*, 1977; Kuznetsov *et al.*, 1979), but superoxide dismutases have not been sought in any group of oligotrophic isolates. If these organisms also lacked superoxide dismutases, they might prove disappointing exceptions to the generalization that O_2-utilizing bacteria possess one or more of such enzymes (Morris, 1975). Kuznetsov *et al.* (1979) have suggested that lack of such protection may explain the inability of some oligotrophic isolates to grow on rich organic media. Accordingly, model oligotrophic chemoheterotrophs should be respiring aerobes but need not be fully armed for the detoxification of oxygen radicals.

2. Proposed Characteristics of Oligotrophs

Since oligotrophy is conceived of as a way of life particularly suited to uninterrupted nutrient limitation, the physiologic properties central to this existence are those directly related to nutrient uptake and to a broad category of physiologic and regulatory processes that can be designated nutrient management. Uptake characteristics must be suitable for acquisition of growth substrates against steep gradients between the cell and its milieu. Acquired nutrients must be so managed that the cell can maintain its integrity and a composition suitable for the accomplishment of growth and—despite long intervals between a full complement of the resources necessary—reproduction.

2.1. Nutrient Uptake

To achieve a higher uptake capacity relative to nutrient-consuming processes than is achieved by copiotrophs, the oligotroph should exhibit several characteristics. Two physical means of increasing uptake capacity would be (a) high surface:volume (S/V) ratio; and (b) high density of transport sites per unit surface area. For both rod-shaped and spherical cells, the S/V ratio increases with decreasing diameter; shortening or elongating a cylinder does not change its S/V ratio so dramatically as does decreasing its diameter. Accordingly, oligotrophs, particularly during periods of very low nutrient flux, should have the form of either relatively slender rods or small spheres. Alternatively, the surface could be extended by envelope outgrowth in the form of slender appendages containing little or no cytoplasm.

Increasing the density of uptake sites would involve the composition of the cytoplasmic membrane and, in gram-negative bacteria, also the outer membrane, which is known to exert an influence on the differential permeability of the gram-negative cell to solutes (reviewed by Nikaido and Nakae, 1979). The cytologic consequences of such modifications are not yet clear, since the chemical nature of uptake sites and their structural relationships to surrounding membrane components are still largely hypothetical.

It would further be to the advantage of the oligotroph to possess uptake systems of relatively high affinity for substrate molecules. In order for a substrate molecule to enter the cell via an uptake site, it must collide effectively with that site. Since the K_m values of many sugar and amino acid uptake systems known in copiotrophs are in the range of 10 μM or greater (summarized by Tempest and Neijssel, 1978), the implication is that the great majority of probable collisions are ineffective, i.e., do not lead to transport. (A 10-μM solution contains 6×10^{15} molecules/ml, allowing multiple collisions per cell per minute at a cell density of 10^8–10^9/ml.) Accordingly, "high affinity" may reflect a higher proportion of substrate-site collisions being effective. This would imply that the uptake site (a) had a lower energy of activation for binding a substrate molecule; (b) possessed a higher number of binding sites per molecule; or (c) efficiently maintained the site(s) in a position allowing effective collision and/or rapidly and with minimal energy expenditure returned the site(s) to effective position after each transport event. Since the occurrence of both high- and low-affinity uptake systems for a given substrate are known to occur in single strains of bacteria and to be synthesized preferentially under different cultural conditions, it is at least clear that the primary structure of a transport molecule influences its affinity for the substrate. However, precisely by which mechanism (not limited to the three mentioned above) is not yet clear in any case.

While an obligate oligotroph might conceivably synthesize only high-affinity uptake systems, and those largely constitutively, a more advantageous regulatory pattern for uptake systems would be inducible low-affinity systems and repressible high-affinity systems. Further, the threshold level for the inducible systems should be relatively low so that uptake capacity could be expanded promptly in response to the appearance of a utilizable substrate. In contrast, the threshold for repression of uptake systems should be fairly high, so that only in the presence of an abundant or sustained supply of the repressing substance would the high-affinity systems not be synthesized.

In the oligotroph, efficient transport should not require modification of the substrate, e.g., by phosphorylation, that would cause the process to be dependent on the availability of a specific cofactor or metabolite. This implies that the cell must maintain a capacity for active transport, presently interpreted as dependent on the maintenance of a proton gradient across the cell boundary (reviewed by Dills *et al.,* 1980). In aerobic organisms, this maintenance depends on respiration to produce the proton gradient directly or via oxidative phosphorylation and subsequent hydrolysis of the ATP formed. Thus, the oligotroph would need to maintain, even increase, its respiratory capacity when under conditions of severe nutrient limitation. In addition, the energy supply to transport activities must be allowed a high priority relative to biosynthetic demands. A greater dependence of transport directly on respiration as compared to ATP hydrolysis (which it would share with biosynthesis) would be advantageous to the oligotroph.

A final feature of substrate uptake to be considered in this context is specificity. Low specificity of the transport systems would be of considerable advantage for the oligotroph because it would allow a given site to serve in the uptake of several different species of substrates. Collision between a nutrient molecule and any of a relatively large number of sites on the cell surface could be effective. Low specificity would, accordingly, have the same net effect as increasing the number of transport sites per unit of surface area.

Similarly important would be a lack of exclusion of slowly metabolized (less desirable?) substrates and of substrates in excess relative to other nutrients. It would be to the oligotroph's advantage to take in practically all utilizable substrates whenever they were available, even though this would occasionally result in "unbalanced growth." If a nutrient present in excess could be suitably stored so that its presence within the cell would not interfere with osmotic or electromotive equilibria or balances among metabolites, it would remain exclusively available for the cell's use and not be lost through diffusion into the environment or to uptake by another cell. Thus, not only specificity but substrate selectivity should be low in the oligotroph.

In practical terms of possible experimental determinations, the combined effects of high S/V ratio, high density of effective (and/or low specificity) uptake sites, high-affinity uptake systems, and a dependable supply of respiratory energy to the uptake systems would be twofold: the oligotroph should exhibit a low K_m for uptake of (although not necessarily growth on) a variety of substrates, as well as a low threshold concentration of substrate adequate to support metabolic activity, some growth, and occasionally a reproductive cycle.

2.2. Nutrient Management

2.2.1. Reserve Formation and Utilization

Periods of starvation are probably experienced by all free-living bacteria. Such periods alternate frequently with periods of nutrient abundance in copiotrophic habitats. In contrast, starvation is infrequently interrupted in oligotrophic habitats, and then by only small and short-lived quantities of nutrients. An additional problem, particularly in aquatic oligotrophic habitats, is that much of the time, nutrients will be available only as solutes. The carcass of an organism provides a reasonably well-balanced supply of nutrients; organic material provided by the environment predominantly in soluble form will be less diverse, and organic substances may not be available simultaneously with required mineral nutrients. Thus, even when nutrients reappear in the oligotrophic habitat, they may not be in sufficient quantity or diversity to support a complete doubling in mass and subsequently in number of the bacterial population. The cells must acquire and store the still-inadequate supply until a reserve sufficient for a reproductive cycle has been accumulated.

Three major types of storage polymers are known in bacteria: polyglucose (PS); lipids, principally poly-β-hydroxybutyrate (PHB); and polyphosphate. (Intracellular nitrogen-storage polymers are not known among chemoheterotrophic bacteria; cellular RNA and proteins appear to serve as nitrogen reserves.) While accumulation of such materials is widespread among free-living bacteria, the capacity for storage of polymeric reserves and for tightly regulated mobilization of such materials should be especially important to and characteristic of oligotrophs.

The major problem for the starving bacterium generally seems to be the provision of sufficient energy for maintenance of structural integrity, internal and membrane-associated macromolecules, proton gradients (Hueting et al., 1979), and energy charge (Nazly et al., 1980). Cells containing PS or PHB reserves are, in almost all instances examined, better able to survive periods of starvation than polymer-poor cells (Dawes and Senior, 1973; Dawes, 1976), principally through the mechanism of utilization of the reserves as a source of energy, which spares the degradation of more critical cell components such as RNA and protein. Some (copiotrophic) bacteria use their reserves solely as sources of energy (Dawes and Senior, 1973), and light spares the reserve energy of polymers in phototrophic bacteria (Breznak et al., 1978).

Accordingly, the following features of reserve formation and mobilization would be advantageous to oligotrophs. First, reserve formation should serve as a principal means of nutrient assimilation, particularly when nutrient exhaustion is imminent. Second, both synthesis and mobilization should be regulated by uptake systems, rather than via intermediary metabolites, allowing synthesis when uptake rates were high, but mobilization only when uptake rates were low. [This is not the type of regulation elucidated in bacilli and azotobacters for PHB storage, and effectors of PS synthesis appear to be different in *Arthrobacter* from those in copiotrophs such as enterics and corynebacteria (Dawes and Senior, 1973).] Utilization should occur only when nutrient uptake rates approach zero, i.e., during starvation. Third, utilization should lead almost exclusively to complete oxidation and the production of energy via respiration, allowing biosynthesis only for essential repair. However, when a limiting nutrient that is not a potential energy source (e.g., nitrogen, sulfur, or a cation) becomes available to a well-stocked cell, it should be possible to use the storage polymer as a source both of energy and of carbon for biosynthesis.

Of the two major carbon polymers known in bacteria, PHB has several possible advantages over PS for the oligotroph. In contrast to PS formation, PHB synthesis does not require direct participation of ATP. Similarly, its mobilization does not require inorganic phosphate, a nutrient commonly limiting in aquatic oligotrophic habitats. Its oxidation state is lower, making it a more concentrated "electron sink," and the immediate product of mobilization (β-hydroxybutyrate, the free acid) can be dehydrogenated and its reducing power transferred to NAD. Further metabolism of the acetoacetate thus produced can lead in only two steps to the tricarboxylic acid cycle and terminal oxidation. Thus, a fourth feature of reserve metabolism in oligotrophs should be synthesis of PHB as the principal carbonaceous energy reserve.

The third reserve polymer common in bacteria is polyphosphate. The physiologic function of this polymer, if any, beyond its service as a phosphorus reserve, is unclear, although it could potentially serve as an energy reserve (Harold, 1966). Solely as a phosphorus reserve, however, it should be expected in aquatic oligotrophs, since phosphate may be limiting when carbon is available—due not only to consumption of the phosphate by the photosynthetic organisms providing the organic-C but also to the low solubility of calcium phosphate.

2.2.2. Regulation of Net Biosynthesis

An oligotroph must not just survive but it must also grow and reproduce in the low nutrient flux environment, and its management of nutrients with respect to biosynthesis of materials other than reserves must also be appropriate to the oligotrophic habitat. As in the case of polymer synthesis, net biosynthesis of functional macromolecules such as structural components (lipoprotein and peptidoglycan of the envelope, for example), nucleic acids, and enzymes should be regulated according to nutrient uptake rates. In particular, net biosynthesis should be halted when uptake rates begin to decline, signaling imminent nutrient exhaustion, and before metabolite pools decline significantly. In this way, remaining metabolites would be available for the completion of syntheses in progress (e.g., DNA replication) whose interruption would be wasteful of the materials already committed to the synthetic process, if not perilous to the cell line.

The regulation of catabolic systems should include three sets of dissimilatory systems: one for nutrients frequently available, whose synthesis should be constitutive; one for nutrients occasionally available in high flux (large quantities and/or prolonged availability), whose synthesis might be inducible; and a third set for the catabolism of nutrients not often available, and then in small amounts for short periods, whose synthesis should be repressible.

These three sets are well known among copiotrophs, and the terms for referring to them have been provided by elucidation of these regulatory mechanisms in such bacteria. However, regulation in oligotrophs should contrast quantitatively with that characteristic of copiotrophs in the following ways. The first set—constitutive catabolic systems—should be small. The energy and material costs of synthesizing and maintaining a large number of infrequently used enzymes would be inappropriate for the oligotroph. In contrast to regulation of uptake systems, both inducible and repressible catabolic systems should be influenced only by relatively high levels of inducers or repressor-inactivators. In the case of inducible systems, this would result in their (enhanced) synthesis only when the supply of inducer/substrate was sufficiently high to justify both synthesis of the catabolic system and its consequent expanded access to substrate that otherwise would be stored. In the case of repressible systems, synthesis of a battery of catabolic enzymes would begin while a utilizable substrate and its metabolites (one or more of which might effect repression) were still available as a resource for the derepressed syntheses.

In all instances, the oligotroph should commit only a modest amount of biosynthesis to any catabolic system, producing moderate amounts of constitutive enzymes and responding to induction or derepression by limited syntheses. Increases of several-hundred-fold in one or a few enzymes, not uncommon in copiotrophs, would be inappropriate, since a single extensive system is not likely to be substrate-saturated in the oligotrophic habitat, nor to serve as the principal means of catabolism for a period many times longer than the life span of a single cell generation. As already proposed, the metabolic priority in the oligotroph should be reserve formation, and vigorous catabolism should occur only when uptake systems are effectively saturated and reserve mobilization is not required to support the material and energy demands of net biosynthesis.

2.3. The Ideal Oligotroph and Its Problems

Having so far proposed properties that may contribute to the ability of bacteria to exploit a low nutrient flux environment, we can examine the implications of these properties with respect to the total physiotype of an ideal oligotroph. It is also important to attempt to discern the ecophysiologic limitations on such an organism resulting from these properties, particularly to assess the probability that such a physiotype would have evolved.

As stated earlier in this discussion, it is presumed that the ideal oligotroph would be a respiring, oxybiontic organism dependent on the availability of O_2. In low nutrient flux environments such as the open ocean and clean bodies of freshwater, O_2 is typically near saturation and would not be a limiting factor for oligotrophs. The upper horizons of soils of low organic content are also well aerated (Dommergues *et al.*, 1978). The typical oligotrophic habitats known are aerobic, and the presumption seems a safe one at present.

According to the above proposals, the ideal oligotroph would have the following traits, most of which would distinguish it from copiotrophs by quantitative, rather than qualitative parameters. It would be a small sphere, a slender rod or an envelope-appendaged (*prosthecate;* Staley, 1968) bacterium. It would exhibit a high affinity (low K_m) for uptake of a variety of utilizable substrates, which it would transport usually without chemical modification by means of active transport closely supported by respiration. Its transport systems of this type would exhibit low substrate specificity and high substrate affinity, and they would be repressible, but only in the presence of high levels of substrates or metabolites. Transport would lead primarily to reserve formation, and the enzymes of polymer synthesis would be constitutive. Its dissimilatory capacity would be versatile; however, it would not operate at high rates due to limited biosynthesis even of inducible catabolic systems. Consequently, its rate of "balanced growth" would not be high; it would not be expected to grow and multiply at doubling times comparable to those achieved by copiotrophs such as the enteric bacteria. Nevertheless, generation times approaching 24 hr would be seriously inappropriate, since nutrient fluxes in natural habitats occur diurnally, and failure to reproduce with each period of nutrient availability could seriously reduce competitiveness. Regulation of net biosynthesis would be effected by uptake rates rather than by intermediates of dissimilatory metabolism; for economy, a central regulatory substance could be a cofactor or intermediate of the pathway leading to reserve polymer formation.

The principal problems of this physiotype arise from its active, yet not very discriminating uptake systems. Low substrate specificity, in particular, would have two adverse consequences: (a) competitive exclusion of a highly suitable substrate by a less readily utilizable, relatively abundant substance of similar composition; and (b) lowered ability to restrict the entry of potentially toxic substances. The oligotroph would consequently find it difficult to manage an abundant supply of unbalanced nutrients. Repressibility of its high-affinity systems would, however, offer some protection against internal imbalance in the presence of a single excessive nutrient. An oligotroph without such protection would probably occur in nature *only* in oligotrophic habitats and could be considered an obligate oligotroph. The obvious consequence of inability to restrict the entry of potential toxic materials would be sensitivity to bioactive substances, including metabolic products of other organisms. This property, if not alterable by regulatory mechanisms, would further restrict the distribution of the oligotroph, and would probably exclude it from any dense microbial community.

Secondary to these problems are those arising from the proposed metabolic predominance of nutrient storage over dissimilation and biosynthesis of functional cell components. When an adequate supply of nutrients appeared in the oligotroph's environment, it would rapidly acquire—but store—nutrients and, as a delayed response, resume net biosynthesis and subsequent reproduction. Thus, its reproductive response to nutrient availability would be slower than that of bacteria whose metabolism was not so regulated (viz., copiotrophs). If the nutrient supply was sustained, expansion of the copiotrophic population could exhaust the nutrients and transiently allow the numerical predominance of the rapidly multiplying copiotrophs over the oligotrophs. If, however, the period of availability was brief relative to the reproductive cycle time of the copiotrophs, the oligotrophs could succeed in acquiring sufficient nutrient reserves so that their capacity for survival following nutrient exhaustion would be significantly greater than that of the copiotrophs.

3. Characteristics of Probable Oligotrophs

To evaluate the proposed properties of an ideal oligotroph, they should be compared with the properties of bacterial groups whose natural distribution, behavior in enrichment procedures, and apparent preferences (as pure cultures) for low nutrient concentrations identify them as oligotrophs. At present, this can be done only in an erratic manner, since oligotrophic physiology, as such, has not yet been explored in any one group to the extent of the proposals. It is therefore necessary to select a few examples of probable oligotrophic bacteria and compare available information regarding some of their properties with the proposed characteristics.

For this purpose, three groups have been selected: *Arthrobacter* as an example of a soil oligotroph; *Caulobacter* as an oligotroph widely distributed in both fresh and seawaters but also found in soils; and an oligotrophic *Spirillum* sp. obtained by chemostat enrichment at low nutrient flux. This minute selection does not represent a lack of diversity among oligotrophs; rather, it reflects the present early stage of studies on oligotrophic physiology. It is hoped that the demonstration of one or more of the proposed properties in known organisms will encourage further comparative studies of this physiotype with copiotrophy.

3.1. *Arthrobacter*

Several aspects of the physiology of *Arthrobacter* spp. relevant to this discussion have been studied with the aim of understanding their morphogenesis and/or their ability to survive in the vegetative state during prolonged periods of starvation or desiccation in the soil. The studies of particular relevance in this context are those describing the effect of growth rate on morphology (Luscombe and Gray, 1974), regulation of substrate utilization (Krulwich and Ensign, 1969), survival and endogenous metabolism (Boylen and Ensign, 1970; Ensign, 1970), and resistance to desiccation (Robinson *et al.*, 1965; Boylen, 1973).

Morphogenesis in *Arthrobacter* consists of an alternation between rod shape and coccoid form. Generally, this is observed during the course of incubation of batch cultures as an alternation between rods during exponential growth that divide to produce coccoid cells during transition to stationary phase (Keddie, 1974). Although this phenomenon is useful taxonomically and to recognize *Arthrobacter* isolates, it is not considered likely that the rod form occurs in nature (Mulder, 1963). Growth as rods depends on the composition of the medium, particularly on the nature of the carbon source. Compounds that are readily utilized and support higher exponential growth rates tend to promote rod morphology, whereas in media containing a compound that supports only slower growth, the cells are coccoid throughout the growth phases. Initially, certain compounds were regarded as "inducers" of rod morphology (Ensign and Wolfe, 1964). However, the prevailing influence of growth rate alone as the determinant of morphology was demonstrated by cultivating each of four species of *Arthrobacter* in carbon-limited chemostats with either glucose (a "noninducer") or succinate (an "inducer") (Luscombe and Gray, 1974). While specific quantities varied with species, in all cases coccoid cells predominated at lower dilution rates (D) and rods at higher D. At lower rates, the volume per cell was also significantly reduced; for example, for *A. globiformis*, the average volume decreased from 0.45–0.60 μm^3 at $D \geq 0.25$/hr to 0.16–0.24 μm^3 at $D \leq 0.20$/hr. The reduction in volume was accompanied by conversion from rod to coccus shape; assuming the length of the rods was approximately twice their diameter, the increase in S/V ratio would have been at least twofold. Thus, the S/V ratio of each *Arthrobacter* sp. increased as nutrient flux decreased. On the basis of the growth rates at which the morphologic transition occurred and the estimated generation times of arthrobacters in soil, it was again concluded that the morphology of these bacteria in nature would be coccoid.

Another finding in these experiments relevant to oligotrophic physiology was the determination of maximum and minimum specific growth rates (μ). The maximum rate was 0.37/hr, representing a doubling time of 1.87 hr and a cell generation time of 2.7 hr. This is considerably in excess of the generation time [somewhat less than 24 hr (Gray and Williams, 1971)] estimated for *Arthrobacter* in soil and reveals that the capacity for relatively rapid growth exists. More significantly, the lowest growth rate allowed in the chemostat was 0.01/hr, which was 0.027 μ_{max}. At this rate, viability was maintained at about 90%. In contrast, viability of copiotrophic bacteria decreases significantly at $\mu = 0.06 \, \mu_{max}$. On the basis of studies with such bacteria, it has been proposed

(Tempest *et al.*, 1967; Pirt, 1972) that bacteria have a "minimum growth rate" consistent with maintenance of viability, and that μ_{min} approximates 0.06 μ_{max}. In the cultivation of oligotrophic bacteria in chemostats, the problems in maintaining steady-state populations at very low growth rates ($D \leq 0.01/hr$) seem due to tendencies of the cells to attach to vessel walls or to clump (Jannasch, 1967b; Jannasch and Mateles, 1974) rather than to loss of viability (Postgate, 1973). Thus, while proposing (above) on a hypothetical basis that the highest μ_{max} among oligotrophs will be less than the μ_{max} of the fastest growing copiotrophs, it will probably also prove true that minimum growth rate in oligotrophs will be a far smaller fraction of μ_{max} than in copiotrophs.

In the earlier study of "morphogenesis-inducing compounds," Krulwich and Ensign (1969) examined in detail the effects of succinate on glucose utilization. Succinate is rapidly utilized, and its uptake is constitutive (but enhanced in succinate-grown cells). Glucose supports much slower growth, and its uptake by active transport is inducible; the affinity ($K_m = 0.8$ mM) of the uptake system is significantly lower than in bacteria, such as enterics, that grow relatively rapidly on sugars (K_m values 0.050–0.100 mM). In mixed-substrate media, the system for glucose uptake by *Arthrobacter* was repressed by succinate, and a distinct diauxie was observed; glucose was not utilized in the presence of succinate. The authors noted that this type of diauxie was similar to that observed in pseudomonads and contrasted with the influence of sugars on substrate utilization in enteric bacteria. Their discussion implies that the regulation of catabolic systems in free-living soil and water bacteria may differ in the identity of the effectors involved. It should also stand as a cautionary note that while the exploration of regulatory patterns in oligotrophs may depend in principle on mechanisms elucidated by studies with copiotrophs, some biochemical imagination may be required in order to identify catabolites that serve in repression in oligotrophs.

Arthrobacter spp. synthesize a polyglucose material as their energy-storage reserve, and they do not seem to store PHB (Boylen and Ensign, 1970). During starvation in liquid suspension, the glycogen-like material was initially utilized more rapidly than other cell components (protein, amino acid pools, and RNA) by cocci, but RNA was utilized rapidly by rods. In both forms, however, endogenous metabolism decreased markedly by the second day of starvation, and remained at about 0.03% of cell carbon respired to CO_2 per hour. Endogenous metabolism was reduced to a level less than one-tenth that commonly observed in starving copiotrophs in similar experiments, and this low metabolic rate seemed to account for the ability of *Arthrobacter* to survive 10–100 times longer (50% viability after 80 days of starvation) than has been reported for enteric bacteria, gram-positive cocci, and pseudomonads (see Table 1 in Ensign, 1970). These studies implied that the availability of reserve materials was less important in the survival of *Arthrobacter* than the ability of starving cells to restrict the rate of reserve utilization and of degradation of other cellular components. According to Dawes (1976), it is particularly this type of reserve conservation that contributes to the survival of starving bacteria in general. Ensign (1970) concluded that while *Arthrobacter* spp. may not compete with other soil organisms, such as pseudomonads, with respect to maximum growth rate, their ability to persist as viable vegetative cells in the absence of nutrients could account for the ubiquity of arthrobacters in nutrient-poor soils.

In further contrast to survival of copiotrophs, it was found that exponentially-growing *Arthrobacter* cells exhibited far greater survival during starvation than stationary phase cells. Whether this was due to accumulation of toxic products in stationary phase at high cell densities and subsequent adverse effects on the cells prior to starvation, or whether this will prove to be characteristic of oligotrophs awaits similar studies with other oligotrophs.

As soil organisms, arthrobacters experience not only nutrient-poor conditions but also periods of desiccation. As demonstrated by the studies of Robinson *et al.* (1965) and of Boylen (1973), the survival of *Arthrobacter* cells (both rods and coccoids) was enhanced by desiccation relative to the proportions observed in the studies (above) carried out in liquid suspension. Particularly in desiccated soil, arthrobacters exhibited an ability to survive that greatly exceeded that of pseudomonads and was comparable to that of *Bacillus* endospores (100% viability after 6 months of desiccation). Again, as in liquid suspension, survival of growing cells was significantly greater than that of cells in stationary phase at the time of desiccation. The physiological basis of *Arthrobacter* resistance to desiccation has not been elucidated. While not necessarily related physiologically to oligotrophy, such resistance confers considerable survival capacity on the individual cell, a capacity which is appropriate to organisms whose nutrient supply does not support the development of dense populations.

In summary, arthrobacters exhibit several traits that should promote their survival in oligotrophic soils. (a) The cells are small spheroids, particularly under conditions of nutrient limitation. (b) Their uptake and catabolic systems preferentially utilize organic acids and amino acids rather than sugars; this would reduce their competition with fast-growing aerobic copiotrophs (with the probable exception of pseudomonads) and of fermentative bacteria, both of which generally prefer sugars. (c) They accumulate large amounts (up to 40% of cell dry weight) of reserve material that is used, by the natural morphologic type, preferentially to RNA and protein during starvation. (d) They respond to starvation conditions by a rapid and drastic decrease in endogenous metabolism and survive for periods measurable in weeks rather than hours. (e) Their maximum growth rate is moderate (approximately 3 hr per doubling), and their minimum growth rate is a smaller fraction of the maximum than is found in copiotrophs. (f) As vegetative cells, they may be as resistant to desiccation in soil as *Bacillus* endospores.

3.2. *Caulobacter*

Caulobacter and the related genus *Asticcacaulis* are represented by prosthecate bacteria whose occurrence in waters and soils of low organic content is widely recognized (reviewed by Poindexter, 1981). They are distinguished from other prosthecate genera fundamentally by the unique morphology of the dividing cell, which is asymmetric and possesses a flagellum at one pole and a banded outgrowth of the envelope at (or near) the opposite pole. The latter appendage bears adhesive material that can mediate attachment of the *Caulobacter* cell to solid surfaces; the appendage of *Asticcacaulis* is not adhesive.

In natural materials, enrichment cultures, and pure cultures in dilute media (not more than 0.05% organic material), the length of the prostheca, or stalk, exceeds the cell length by 5–40 times. In richer media (at least 0.2% organic material), the length of the stalk typically is not more than that of the cell. The concentration of at least one inorganic nutrient, viz., phosphate, is likewise inversely proportional to the length of the appendage, a relationship seen in other prosthecate bacteria (Whittenbury and Dow, 1977). Accordingly, stalk elongation is regarded as a morphological response to nutrient limitation and can be interpreted as a means of increasing the S/V ratio of the cell in dilute environments (Pate and Ordal, 1965; Jordan *et al.*, 1974; Porter and Pate, 1975; Larson and Pate, 1976). A stalked cell whose appendage is ten times the cell length has a S/V ratio that is twice that of the cell alone. This is comparable to the increase resulting from rod-to-sphere alternation in *Arthrobacter* (see above) but is achieved without reproduction and without reduction in cell volume. Even more important, with respect to increasing the ratio of potential uptake sites to metabolically active cytoplasm, the caulobacter appendages are composed entirely of membranes (Poindexter and Cohen-Bazire, 1964; Pate and Ordal, 1965); they are generally inactive as sites of energy-consuming biosynthesis (Schmidt and Stanier, 1966) and lack complete catabolic systems (Jordan *et al.*, 1974). The bands peculiar to caulobacter prosthecae may serve to restrict the entry of cytoplasm into the stalk so that its contribution as an uptake organelle is not reduced by substrate-consuming reactions (Poindexter, 1981).

The appendage should exhibit at least two further characteristics if it is to serve as an uptake organelle. First, it should be capable of active transport of nutrients, preferably supported by its own respiratory system. Second, the nutrients it accumulates should be translocated to the cytoplasm, predominantly for storage but also possibly for consumption. Only the first of these two properties has been tested, and only with respect to organic nutrients.

Prosthecae of *A. biprosthecum* were separated from their cells by treatment in a Sorvall Omnimixer, purified by differential centrifugation, and used in studies of the rates, requirements, and inhibition of nutrient uptake into the isolated organelles (Jordan *et al.*, 1974; Porter and Pate, 1975; Larson and Pate, 1976). The isolated appendages were found capable of energy-yielding respiration and of active transport of sugars and amino acids, in some cases achieving an intraorganelle accumulation that was 200 times the ambient concentration. Glucose uptake involved two systems. The low-affinity system exhibited a K_m (34 μM) of the same order of magnitude as that found in many copiotrophs; the high-affinity system exhibited a K_m 20-fold lower (1.8 μM), an affinity which is rarely but nevertheless sometimes observed in copiotrophs. The affinity for proline was similarly high ($K_m = 11$ μM).

Glucose analogues (methyl α-D-glucoside and 2-deoxyglucose) were not transported. However, the D-aldoses D-xylose and D-galactose inhibited D-glucose uptake by both the high- and the low-affinity systems, leading Larson and Pate (1976) to conclude that a variety of substrates could be transported by a relatively small number of highly efficient systems located in the prostheca. Transport was inhibited by p-chloromercuribenzoic acid (reversed by dithiothreitol), azide, cyanide, and O_2 deprivation, but not by arsenate or N,N'-dicyclohexylcarbodiimide. Uncoupling agents (5-chloro-3-$tert$-butyl-2'-nitrosalicylanilide, 5-chloro-3-(p-chlorophenyl)-4'-chlorosalicylanilide, 2,4-dinitrophenol, and carbonyl cyanide m-chlorophenyl hydrazone) inhibited uptake whether or not they stimulated respiration. These observations led to the inference that uptake by the appendage was dependent on an energy supply from respiration and some coupled process, but not on the formation and hydrolysis of phosphorylated intermediates such as ATP.

In any given caulobacter isolate (of the very few so far studied), enzymes for the catabolism of a utilizable sugar are typically present in detectable amounts whether or not the corresponding sugar was present during growth. However, the specific activity of key degradative enzymes typically increases about two-fold when cells are grown on the sugar (Poindexter, 1964; Kurn *et al.*, 1978); similar (1.4-fold) increases in sugar-uptake activities have been reported (Kurn *et al.*, 1978). The response of the most widely studied species, *C. crescentus*, appears exceptionally high and may be 10- to 40-fold, depending on the sugar (Poindexter, 1964; Shedlarski, 1974; Kurn *et al.*, 1977). In this species, glucose-6-phosphate dehydrogenase levels were higher in cells grown in a medium containing glucose as the sole carbon source than in cells grown in a complex medium containing glucose, implying that the presence of other utilizable carbon sources exerted some repressive influence on the enhancement of glucose catabolism. The other species were tested in a sugar-supplemented complex medium, and their lower apparent responses to sugars may have been influenced by such repression. Glucose repression of β-galactosidase synthesis was not observed in *A. biprosthecum* (Larson and Pate, 1975) but did occur in *C. crescentus* (Kurn *et al.*, 1977). In contrast to regulation in copiotrophic bacteria, however, glucose repression was not relieved by cAMP even though lactose induction of the enzyme was accelerated by this nucleotide in the absence of glucose (Kurn *et al.*, 1977).

These studies, like those with *Arthrobacter,* imply that regulation of the synthesis of catabolic enzymes in these oligotrophs involves mechanisms basically similar to those that serve in copiotrophs, but that such mechanisms may be effected by different regulatory molecules. In the present context of oligotrophic physiology, it is even more significant that the extent of change in specific enzyme activities seen in caulobacters is smaller, by 10- to 100-fold, than the responses to inducers of catabolic enzymes exhibited by copiotrophs. This, also, is a proposed property of oligotrophs.

Caulobacters have not been systematically studied under conditions of starvation, as has *Arthrobacter*. However, their occurrence in such habitats as the open ocean and distilled water, where they are a minor nuisance to electron microscopists, clearly implies that they can multiply in environments of such low nutrient concentrations that visible bacterial growth does not accumulate.

Caulobacter cells are known to accumulate PHB under conditions of nitrogen (Poindexter, 1964) or phosphate (Schmidt and Stanier, 1966) limitation, to the extent of 26% of the cell dry weight. Cells provided with glucose but without a nitrogen source increased in dry weight by 21% in 12 hr, with 90% of the increase being accounted for by the synthesis of PHB (77%) and of polyglucose (13%) (Poindexter, 1964). Earlier cytologic studies had revealed that under such conditions (nitrogen starvation in sugar-phosphate medium), the cells also accumulated polyphosphate reserve granules (Grula and Hartsell, 1954). Accordingly, *Caulobacter* has the capacity to form all three principal types of reserve polymers simultaneously and so may be able to survive during periods of exhaustion of any of a variety of nutrients required for growth and reproduction at maximum rate.

The only continuous cultivation of caulobacters so far reported employed *C. crescentus* and standard PYE medium diluted $1:10$ (PYE/10:0.02% peptone, 0.01% yeast extract, and 0.001% $MgSO_4 \cdot 7H_2O$; phosphate is present at 0.03 mM when not added separately) and a single dilution rate (0.06–0.07/ hr) (Haars and Schmidt, 1974). The rate of stalk elongation was determined in steady-state populations growing with and without supplementation of the medium with phosphate. This rate was nearly twice as high in unsupplemented PYE/10 (0.14 μm/hr) as in PYE/10 containing 1 mM phosphate (0.065 μm/ hr), indicating a significant reduction in stalk elongation per generation in the higher phosphate flux.

In the author's laboratory (Poindexter, unpublished), a defined medium (HIGg; Poindexter, 1978) has been employed in chemostat cultivation of *C. crescentus*. At a constant dilution rate (0.07/hr), population density (as cell mass) was determined by the reservoir concentrations of the combined carbon sources (glucose plus glutamate). Morphology, however, was determined not by the concentration of the organic components of the medium but by the ratio of phosphate-P to organic-C (P:C). At an input ratio of 0.008 mg phosphate-P/mg organic-C, the population was carbon-limited; the mean cell length was 1.7 μm, the mean stalk length was 6.3 μm, and 65–70% of the population was accounted for by stalked cells. When the ratio was decreased to 0.002 by increasing the glucose and glutamate concentrations fourfold, the population was phosphate-limited. As in phosphate-limited batch cultures (Schmidt and Stanier, 1966), the cells were longer (mean cell length 2.7 μm) and densely packed with PHB and bore stalks often more than 20 μm in length. The mean stalk length of those measurable in electron micrographs was 8.6 μm; however, many stalks extended beyond the micrographic field (17.5 × 15 μm) and could not be measured. Stalked cells accounted for 83–85% of the population. The dry weight of cells per milliliter was four times as great in the second steady-state population, but the number of cells per milliliter remained constant at 2 × 10⁹, reflecting the increase in cell size. Viability was constant and approximately 100% as estimated from the agreement between microscopic and viable counts.

Restoration of the P:C ratio to 0.008 by further addition of phosphate to the reservoir medium reversed the morphology of the population; however, as the large, previously phosphate-limited cells returned to the smaller size exhibited during carbon limitation, their reproductive rate exceeded the dilution rate for the four volume changes available prior to exhaustion of the reservoir; a third steady state was achieved only in terms of cell mass (approximately 10% higher than in the second steady state), while cell number had increased to 6 × 10⁹/ml and was still rising. Cells from the three stages of this chemostat are illustrated in Fig. 1. It is evident in the micrographs that PHB accumulation began immediately after the shift from carbon to phosphate limitation, and that utilization of the reserve was so gradual that PHB had diminished but had not yet been exhausted at the end of cultivation.

This initial experiment was based on preliminary studies in batch culture that had implied that morphology was determined principally by the P:C ratio, rather than by phosphate concentration alone (Poindexter, 1979). These studies are continuing in order to test the hypothesis that stalk development assists caulobacter cells in the accumulation of nutrients from an unbalanced supply, which is unsuitable for balanced growth, and that the resulting reserves are subsequently available for biosynthesis during periods of availability of the transiently insufficient nutrient(s).

In these presumptive oligotrophs, it is especially phosphate limitation that stimulates the peculiar morphogenesis that markedly increases the S/V ratio; during such limitation, the cells can accumulate and store considerable quantities of organic carbon. The central role of phosphate in enhancing their oligotrophic properties is consistent with the wide distribution of caulobacters in aquatic habitats, where phosphate is commonly the nutrient that limits total biomass of the community. It would be of considerable interest to determine the role of phosphate in the regulation of induced enzyme systems in these bacteria.

In summary, caulobacters are known to exhibit several of the proposed oligotrophic properties. (a) Their appendage is capable of nutrient accumulation, and its development is enhanced by nutrient limitation. (b) The quantitative response to inducers of catabolic enzyme synthesis is limited. Although not yet tested with specific compounds such as amino acids, components of peptone and/or yeast extract appear to interfere with induction of enzymes of

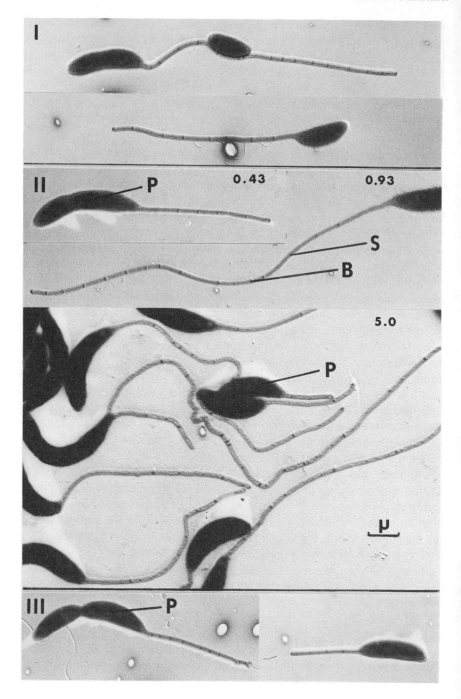

sugar catabolism. (c) They respond to specific nutrient limitation by accumulating reserves of still available nutrients. (d) Their maximum growth rate is never greater than 2 hr per doubling; for many isolates, the minimum doubling time is 4–5 hr (Poindexter, 1964). (e) Ammonia assimilation depends on the high-affinity glutamine synthetase system; glutamate dehydrogenase has not been detected in caulobacters (Ely *et al.*, 1978).

3.3. Chemostat-Selected Oligotrophs

Chemostats have been especially useful for selective enrichment of oligotrophic bacteria in studies of marine samples. Low nutrient flux chemostats generally favor *Spirillum, Vibrio,* and *Achromobacter* types (Jannasch and Mateles, 1974; Harder *et al.*, 1977). However, the most detailed comparative study of isolates selected by chemostat enrichment at different dilution rates has been the work of Matin and Veldkamp (1978) with two freshwater organisms. Using samples of a single source of pond water (whose composition was not described), a low dilution rate (*D*) (0.05/hr) selectively enriched a *Spirillum* species, whereas a higher *D* (0.30/hr) selectively enriched a *Pseudomonas* species. The bacteria were isolated by streaking on a moderately rich medium (0.05% lactate, equivalent to 0.2 g organic-C/liter) and subsequently characterized with respect to growth constants and competitiveness in lactate-limited chemostats at various dilution rates. Of this pair, the *Spirillum* can be regarded as a relative oligotroph, the *Pseudomonas* as a copiotroph.

The respective characteristics of these two isolates are presented in Table I. Although the maximum growth rate of *Spirillum* was only 55% of the maximum growth rate of the *Pseudomonas* on lactate, in every other feature listed the *Spirillum* exhibited traits that would promote its competitiveness with the *Pseudomonas* at low lactate flux. It removed lactate from the medium more efficiently, whereas the residual lactate in the *Pseudomonas* culture was still greater than the K_s for growth of the *Spirillum*. The ability of the *Spirillum* to predominate over the *Pseudomonas* at $D = 0.24$/hr could be attributed largely to the ability of the *Spirillum* to grow at the expense of lactate unavailable for *Pseudomonas* assimilation. A somewhat greater increase in S/V ratio also occurred in *Spirillum* at lower dilution rates (48%, compared with a 23%

←──

Figure 1. *Caulobacter crescentus* cells from a three-stage chemostat at $D = 0.07$/hr. The reservoir medium (HIGg; Poindexter, 1978) contained: Stage I, 0.1% glucose plus glutamate and 0.1 mM phosphate; Stage II, 0.4% glucose plus glutamate and 0.1 mM phosphate; Stage III, 0.4% glucose plus glutamate and 0.4 mM phosphate. I. Cells from stage I steady state. II. Cells from stage II at 0.43 volume change, 0.93 volume change, and steady state (5.0 volume changes). III. Cells from stage III at 3.4 volume changes. (S) stalk; (B) stalk band; (P) PHB granule. Bar equals 1 μm.

Table I. Characteristics of Isolates of *Spirillum* and *Pseudomonas*[a] Selectively Enriched from Pond Water by Lactate Limitation[b]

Characteristic	*Spirillum*	*Pseudomonas*
μ_{max}	0.35/hr	0.64/hr
μ_{min}	0.01/hr	0.05–0.08/hr
	(<3% of μ_{max})	(6–12% of μ_{max})
K_s (lactate)	23 μM	91 μM
Lactate uptake		
$\quad K_m$	5.8 μM	20 μM
$\quad V_{max}$	29 nmol/mg protein per min	20 nmol/mg protein per min
Residual lactate at $D = 0.15$/hr	17 μM	28 μM
Y (yield constant)	0.38 g cell dry wt/g lactate consumed	0.41 g cell dry wt/g lactate consumed
Energy of maintenance	0.016 g lactate/g cell dry wt/hr	0.066 g lactate/g cell dry wt/hr

[a]The selective dilution rates were 0.05/hr and 0.30/hr for *Spirillum* and *Pseudomonas*, respectively.
[b]Matin and Veldkamp (1978).

increase in the *Pseudomonas* ratio), achieved mainly through a reduction in cell diameter.

In contrast to the uptake of lactate, the capacity of the *Pseudomonas* to catabolize lactate (measured as enzyme activities) exceeded that of the *Spirillum* at all dilution rates. Since the yield coefficients in the two organisms were similar, this implies that synthesis of catabolic enzymes by the *Spirillum* was more economical and that speed is not a measure of efficiency. Catabolic capacity increased in both organisms at lower dilution rates [a general phenomenon in heterotrophs (Matin, 1979)], but the two other activities were enhanced at low dilution rates only in the *Spirillum:* PHB accumulation and respiratory capacity. Increase in the latter capacity may have maximized energy generation for active uptake against an increasingly steep gradient; it may also have been required for maintenance of ionic gradients across the cell membrane, a problem that increases with decreasing rate of cation supply (Hueting *et al.,* 1979).

Accumulation of PHB by the *Spirillum* (Matin *et al.,* 1979) suggested that either the cells were not carbon-limited at low dilution rates (the flux of another nutrient may have become limiting) or that reserve mobilization was reduced relative to synthesis. If, as proposed above for oligotrophs, reserve accumulation depends on rate of nutrient uptake—being so regulated that reserve synthesis takes precedence over catabolism and mobilization is activated by minimum concentrations of catabolites—then the low level of lactate available at low dilution rates would result in an increased proportion of the lactate being shunted into reserves, whereas a decrease in the level of catabolites would reduce reserve mobilization. The result would be the accumulation of reserves derived from the limiting nutrient. Survival of the *Spirillum* during starvation was proportional to the amount of PHB present at the onset of starvation (Matin *et al.*, 1979), so that at low dilution rates, survival capacity increased with decreasing D. This was consistent with the earlier observation (Matin and Veldkamp, 1978) that at very low dilution rates ($D \leq 0.05/\text{hr}$), the viability of the *Spirillum* population was not reduced, whereas that of the *Pseudomonas* began to decrease at $D = 0.05/\text{hr}$.

In this study, the freshwater *Spirillum* enriched by chemostat cultivation at low dilution rates was capable of growth on a moderately rich (0.2 g organic-C/liter) medium but nevertheless exhibited physiologic properties entirely consistent with oligotrophy. Similarly thorough comparative studies of these properties with marine isolates are not yet available. It may be found that freshwater oligotrophs will be relatively flexible since their habitat is less constant, and marine oligotrophs will prove more strictly oligotrophic.

In the studies of Matin and Veldkamp (1978), nutrient flux was determined by varying the dilution rate while the reservoir concentration of substrate (S_R) remained constant. In this way, a nearly constant population density (*ca.* $10^9/\text{ml}$) was maintained, regardless of the nutrient flux, and the amount of metabolic activity required to maintain each chemostat population was therefore constant per unit volume of culture. Alternatively, nutrient flux can be varied for populations in different chemostats by employing a constant D and varying S_R. When populations are subjected to different nutrient fluxes in this manner, the rate of growth (dx/dt) is constant from one population to another, but the population density ($\bar{x}$) varies among them in proportion to S_R. Although it is theoretically possible to adjust D and S_R so that the average nutrient flux is the same under both types of limitation, there is some evidence that the $\mu - K_s$ relationship is not constant over a wide range of D at constant S_R (Matin and Veldkamp, 1978), whereas variation in S_R at constant D results in effects that may be due to population density rather than to nutrient flux *per se* (Jannasch, 1963, 1967b; Postgate, 1973).

Employing the second manner of varying nutrient flux to chemostat populations, Jannasch (1967b) detected a further difference between oligotrophic and copiotrophic populations, viz., that isolates exhibiting the combination of growth constants low K_s–low μ_{max} (presumptive oligotrophs) also exhibit lower threshold S_R values at which growth and multiplication of the population can occur. Although discussed by Jannasch primarily as correspondence between S_R and initiation of growth, his comparative data also demonstrate that oligotrophic strains can initiate substrate utilization, growth, and reproduction at significantly (> 10-fold) lower population densities than copiotrophic (high K_s–high μ_{max}) strains. This implies that the growth of oligotrophs is less dependent on conditioning of the environment by their own metabolic activities. In a natural oligotrophic habitat such as the open ocean, where metabolic products are lost principally by diffusion rather than by continuous, unremitting dilution, population densities lower than those demanded by chemostat conditions may be sufficient to allow assimilatory activities by oligotrophs, and threshold concentrations may be considerably lower than can be recognized by chemostat steady-state studies. By determining the relationship between washout and dilution rates in transition state studies, Jannasch (1969) was able to estimate generation times of natural populations that were far longer than retention times in the chemostat. Even with this technique, however, minimum population density allowing multiplication was nearly 10^5 cells/ml, still higher than viable counts of oligotrophs in many ocean samples (Jannasch and Jones, 1959).

4. Concluding Remarks

An attempt has been made in this essay to project the properties of oligotrophic bacteria and to assess the concept of oligotrophy by examining certain properties of presumptive oligotrophic bacteria. The primary purpose has been to imply that oligotrophy does exist, as a specific set of properties that are especially suitable for exploitation of habitats extremely poor in organic nutrients.

The proposed properties comprise principally physiologic and morphologic characteristics that maximize the ability of the cell to gather nutrients (a) across steep spatial gradients and (b) over extended periods of time through conservative utilization of nutrients once they are inside the cell. The latter property, in particular, would contribute to survival of vegetative cells during starvation and may be even more important to the oligotroph than high-affinity uptake systems. The collective properties proposed would not, however, be advantageous under conditions of nutrient abundance, particularly sustained abundance, and could account for the inability of oligotrophs to compete with copiotrophs in such circumstances. High-affinity uptake systems do not necessarily exhibit greater velocities of uptake and so do not improve uptake at substrate levels well above saturation. Further, as long as nutrients are abundant, any reserves accumulated by the oligotroph will be superfluous, and if their synthesis competes with biosynthesis of functional cell components, the rates of balanced growth and reproduction of the oligotroph will predictably lag behind those of copiotrophs that are not storing nutrients. Finally, the possibility that some oligotrophs are not enzymatically protected against the ravages of oxygen radicals would predict their decline, not simply their relative retardation, when under conditions that supported a dense population of copiotrophs.

It has been argued earlier in this series (Tempest and Neijssel, 1978) that studies with nutrient-sufficient (batch) cultures do not adequately reveal the physiologic and regulatory mechanisms operative in nutrient-limited (chemostat or natural) environments. In consonant spirit, it is suggested here that to fully appreciate the physiology and regulation of nutrient-limited bacteria, the studies must include those bacteria for which, throughout their existence and probably through much of their history, low nutrient conditions are not and have not been interrupted by nutrient sufficiency. Accordingly, the secondary purpose of this discussion has been to imply that the proper study of oligotrophy is oligotrophs. Experimental study of such bacteria should clarify whether and to what extent they affect the concentrations and the composition of the organic matter that is diluted into the vast oligotrophic regions of the biosphere.

References

Akagi, Y., Taga, N., and Simidu, U., 1977, Isolation and distribution of oligotrophic marine bacteria, *Can. J. Microbiol.* **23**:981–987.

Boylen, C. W., 1973, Survival of *Arthrobacter crystallopoietes* during prolonged periods of extreme desiccation, *J. Bacteriol.* **113**:33–37.

Boylen, C. W., and Ensign, J. C., 1970, Intracellular substrates for endogenous metabolism during long-term survival of rod and spherical cells of *Arthrobacter crystallopoietes, J. Bacteriol.* **103**:578–587.

Breznak, J. A. Potrikus, C. J., Pfennig, N., and Ensign, J. C., 1978, Viability and endogenous substrates used during starvation survival of *Rhodospirillum rubrum, J. Bacteriol.* **134**:381–388.

Bulion, W. W., 1977, Extracellular production of phytoplankton, *Usp. Sovrem. Biol.* **84**:294–304 (in Russian).

Carlucci, A. F., and Shimp, S. L., 1974, Isolation and growth of a marine bacterium in low concentration of substrate, in: *Effect of the Ocean Environment on Microbial Activities* (R. R. Colwell and R. Y. Morita, eds.), pp. 363–367, University Park Press, Baltimore.

Dawes, E. A., 1976, Endogenous metabolism and the survival of starved prokaryotes, *Symp. Soc. Gen. Microbiol.* **26**:19–53.

Dawes, E. A., and Senior, P. J., 1973, The role and regulation of energy reserve polymers in micro-organisms, *Adv. Microbial Physiol.* **10**:135–266.

Dills, S. S., Apperson, II., Schmidt, M. R., and Saier, M. H., Jr., 1980, Carbohydrate transport in bacteria, *Microbiol. Rev.* **44**:385–418.

Dommergues, Y. R., Belser, L. W., and Schmidt, E. L., 1978, Limiting factors for microbial growth and activity in soil, in: *Advances in Microbial Ecology,* Vol. 2 (M. Alexander, ed.), pp. 49–104, Plenum Press, New York.

Ely, B., Amarasinghe, A. B. C., and Bender, R. A., 1978, Ammonia assimilation and glutamate formation in *Caulobacter crescentus, J. Bacteriol.* **133**:225–230.

Ensign, J. C., 1970, Long-term starvation survival of rod and spherical cells of *Arthrobacter crystallopoietes, J. Bacteriol.* **103**:569–577.

Ensign, J. C., and Wolfe, R. S., 1964, Nutritional control of morphogenesis in *Arthrobacter crystallopoietes, J. Bacteriol.* **87**:924–932.

Fridovich, I., 1978, The biology of oxygen radicals, *Science* **201**:875–880.

Gray, T. R. G., and Williams, S. T., 1971, Microbial productivity in soil, *Symp. Soc. Gen. Microbiol.* **21**:256–286.

Grula, E. A., and Hartsell, S. E., 1954, Intracellular structures in *Caulobacter vibrioides, J. Bacteriol.* **68**:498–504.

Haars, E. G., and Schmidt, J. M., 1974, Stalk formation and its inhibition in *Caulobacter crescentus, J. Bacteriol.* **120**:1409–1416.

Hamilton, R. D., Morgan, K. M., and Strickland, J. D. H., 1966, The glucose uptake kinetics of some marine bacteria, *Can. J. Microbiol.* **12**:995–1003.

Harder, W., Kuenen, J. G., and Matin, A., 1977, Microbial selection in continuous culture, *J. Appl. Bacteriol.* **43**:1–24.

Harold, F. M., 1966, Inorganic polyphosphates in biology: Structure, metabolism, and function, *Bacteriol. Rev.* **30**:772–794.

Hirsch, P., 1979, Life under conditions of low nutrient concentrations, in: *Strategies of Microbial Life in Extreme Environments* (M. Shilo, ed.), pp. 357–372, Dahlem Conference Life Sciences Research Report 13, Berlin.

Hood, D. W. (ed.), 1970, *Organic Matter in Natural Waters,* Institute of Marine Science, Alaska.

Hueting, S., deLange, T., and Tempest, D. W., 1979, Energy requirement for maintenance of the transmembrane potassium gradient in *Klebsiella aerogenes* NCTC418: A continuous culture study, *Arch. Microbiol.* **123**:183–188.

Jannasch, H. W., 1963, Bacterial growth at low population densities (II), *Nature (London)* **197**:1322.

Jannasch, H. W., 1965, Eine Notiz über die Anreicherung von Mikroorganismen in Chemostaten, in: *Anreicherungskultur und Mutantenauslesen* (H. G. Schlegel, ed.), Suppl. 1 to *Zentralbl. Bakteriol.,* I. Abt., pp. 498–502.

Jannasch, H. W., 1967a, Enrichment of aquatic bacteria in continuous culture, *Arch. Mikrobiol.* **59**:165–173.

Jannasch, H. W., 1967b, Growth of marine bacteria at limiting concentrations of organic carbon in seawater, *Limnol. Oceanogr.* **12**:264–271.

Jannasch, H. W., 1969, Estimation of bacterial growth rates in natural waters, *J. Bacteriol.* **99**:156–160.

Jannasch, H. W., and Jones, G. E., 1959, Bacterial populations in sea water as determined by different methods of enumeration, *Limnol. Oceanogr.* **4**:128–139.

Jannasch, H. W., and Mateles, R. I., 1974, Experimental bacterial ecology studied in continuous culture, *Adv. Microbial Physiol.* **11**:165–212.

Jordan, T. L., Porter, J. S., and Pate, J. L., 1974, Isolation and characterization of prosthecae of *Asticcacaulis biprosthecum, Arch. Mikrobiol.* **96**:1–16.

Keddie, R. M., 1974, *Arthrobacter,* in: *Bergey's Manual of Determinative Bacteriology,* 8th ed. (R. E. Buchanan and N. E. Gibbons, eds.), pp. 618–625, Williams & Wilkins Company, Baltimore.

Koch, A. L., 1971, The adaptive responses of *Escherichia coli* to a feast and famine existence, *Adv. Microbial Physiol.* **6**:147–217.

Krulwich, T. A., and Ensign, J. C., 1969, Alteration of glucose metabolism of *Arthrobacter crystallopoietes* by compounds which induce sphere to rod morphogenesis, *J. Bacteriol.* **97**:526–534.

Kurn, N., Shapiro, L., and Agabian, N., 1977, Effect of carbon source and the role of cyclic adenosine 3′,5′-monophosphate on the *Caulobacter* cell cycle, *J. Bacteriol.* **131**:951–959.

Kurn, N., Contreras, I., and Shapiro, L., 1978, Galactose catabolism in *Caulobacter crescentus, J. Bacteriol.* **135**:517–520.

Kuznetsov, S. I., Dubinina, G. A., and Lapteva, N. A., 1979, Biology of oligotrophic bacteria, *Annu. Rev. Microbiol.* **33**:377–387.

Larson, R. J., and Pate, J. L., 1975, Growth and morphology of *Asticcacaulis biprosthecum* in defined media, *Arch. Microbiol.* **106**:147–157.

Larson, R. J., and Pate, J. L., 1976, Glucose transport in isolated prosthecae of *Asticcacaulis biprosthecum, J. Bacteriol.* **126**:282–293.

Luscombe, B. M., and Gray, T. R. G., 1974, Characteristics of arthrobacter grown in continuous culture, *J. Gen. Microbiol.* **82**:213–222.

Mallory, L. M., Austin, B., and Colwell, R. R., 1977, Numerical taxonomy and ecology of oligotrophic bacteria isolated from the estuarine environment, *Can. J. Microbiol.* **23**:733–750.

Matin, A., 1979, Microbial regulatory mechanisms at low nutrient concentrations as studied in chemostat, in: *Strategies of Microbial Life in Extreme Environments* (M. Shilo, ed.), pp. 323–339, Dahlem Conference Life Sciences Research Report 13, Berlin.

Matin, A., and Veldkamp, H., 1978, Physiological basis of the selective advantage of a *Spirillum* sp. in a carbon-limited environment, *J. Gen. Microbiol.* **105:**187–197.

Matin, A., Veldhuis, C., Stegeman, V., and Veenhuis, M., 1979, Selective advantage of a *Spirillum* sp. in a carbon-limited environment. Accumulation of poly-β-hydroxybutyric acid and its role in starvation, *J. Gen. Microbiol.* **112:**349–355.

Morris, J. G., 1975, The physiology of obligate anaerobiosis, *Adv. Microbial Physiol.* **12:**169–246.

Morris, J. G., 1976, Oxygen and the obligate anaerobe, *J. Appl. Bacteriol.* **40:**229–244.

Mulder, E. G., 1963, *Arthrobacter,* in: *Principles and Applications in Aquatic Microbiology* (H Henkelekian and N. C. Dondero, eds.), pp. 254–279, John Wiley & Sons, New York.

Nazly, N., Carter, I. S., and Knowles, C. J., 1980, Adenine nucleotide pools during starvation of *Beneckea natriegens, J. Gen. Microbiol.* **116:**295–303.

Nikaido, H., and Nakae, T., 1979, The outer membrane of gram-negative bacteria, *Adv. Microbial Physiol.* **20:**163–250.

Pate, J. L., and Ordal, E. J., 1965, The fine structure of two unusual stalked bacteria, *J. Cell Biol.* **27:**133–150.

Pirt, S. J., 1972, Prospects and problems in continuous flow culture of microorganisms, *J. Appl. Chem. Biotechnol.* **22:**55–64.

Poindexter, J. S., 1964, Biological properties and classification of the *Caulobacter* group, *Bacteriol. Rev.* **28:**231–295.

Poindexter, J. S., 1978, Selection for nonbuoyant morphological mutants of *Caulobacter crescentus, J. Bacteriol.* **135:**1141–1145.

Poindexter, J. S., 1979, Morphological adaptation to low nutrient concentrations, in: *Strategies of Microbial Life in Extreme Environments* (M. Shilo, ed.), pp. 341–356, Dahlem Conference Life Sciences Research Report 13, Berlin.

Poindexter, J. S., 1981, The caulobacters: Ubiquitous unusual bacteria, *Microbiol. Rev.* **45:**123–179.

Poindexter, J. S., and Cohen-Bazire, G., 1964, The fine structure of stalked bacteria belonging to the family Caulobacteraceae, *J. Cell Biol.* **23:**587–597.

Porter, J. S., and Pate, J. L., 1975, Prosthecae of *Asticcacaulis biprosthecum:* system for the study of membrane transport, *J. Bacteriol.* **122:**976–986.

Postgate, J. R., 1973, The viability of very slow-growing populations: A model for the natural ecosystem, *Bull. Ecol. Res. Comm. (Stockholm)* **17:**287–292.

Robinson, J. B., Salonius, P. O., and Chase, F. E., 1965, A note on the differential response of *Arthrobacter* spp. and *Pseudomonas* spp. to drying in soil, *Can. J. Microbiol.* **11:**746–748.

Schmidt, J. M., and Stanier, R. Y., 1966, The development of cellular stalks in bacteria, *J. Cell Biol.* **28:**423–436.

Shedlarski, J. G., Jr., 1974, Glucose-6-phosphate dehydrogenase from *Caulobacter crescentus, Biochim. Biophys. Acta* **358:**33–43.

Staley, J. T., 1968, *Prosthecomicrobium* and *Ancalomicrobium*: New prosthecate freshwater bacteria, *J. Bacteriol.* **95:**1921–1942.

Tempest, D. W., and Neijssel, O. M., 1978, Eco-physiological aspects of microbial growth in aerobic nutrient-limited environments, in: *Advances in Microbial Ecology,* Vol. 2 (M. Alexander, ed.), pp. 105–153, Plenum Press, New York.

Tempest, D. W., Herbert D., and Phipps, P. J., 1967, Studies on the growth of *Aerobacter aerogenes* at low dilution rates in a chemostat, in: *Microbial Physiology and Continuous Culture* (E. O. Powell, C. G. T. Evans, R. E. Strange, and D. W. Tempest, eds.), pp. 240–261, H. M. Stationery Office, London.

Whittenbury, R., and Dow, C. S., 1977, Morphogenesis and differentiation in *Rhodomicrobium vannielii* and other budding and prosthecate bacteria, *Bacteriol. Rev.* **41:**754–808.

3

Water and Microbial Stress

D. M. GRIFFIN

1. General Introduction

Reference to a dictionary reveals that an object under stress is subjected to "demand upon energy" or is "constrained." It is precisely these two aspects of microbial ecology that I have considered in this review, the stress being imposed by some factor directly associated with the water regime of the immediate environment. Such a topic is relevant to many branches of microbiology. Thus, those working in microbial physiology, soil microbiology, plant pathology, food preservation, biodegradation, and marine and lacustrine microbiology all find water and its biological availability to be an important factor influencing microbial activity and hence microbial ecology.

For many reasons, often of a historical or personal nature, the conceptual framework concerning water that exists in the various subdisciplines differs considerably. Even more obvious is the widely differing units of measurement, few of which are easily interconvertible by mental arithmetic. The combined effect is to reduce considerably the interaction between the various subdisciplines and to make the results obtained by one group of workers not readily intelligible to others.

It might be expected that review articles, of which there are many, would have broken down these barriers, but this is scarcely so. There still remains a considerable gulf between those who work with organisms (mainly bacteria, algae, and some fungi, of which the yeasts are the obvious example) in systems where stress arises through the presence of solutes and those who work largely

---

D. M. GRIFFIN ● Department of Forestry, Australian National University, Canberra, A.C.T., Australia.

with filamentous fungi and bacteria in systems where stress arises mainly through water–solid interactions. The differences of approach were explored at the Dahlem Workshop on Strategy of Life in Extreme Environments held in Berlin in 1978 and are reflected in the publication resulting from it (Shilo, 1979).

My aim in this article is to provide a synthesis of the information available from the various subdisciplines, and I have therefore often cited reviews in those subdisciplines rather than the original accounts of experiments. In this way, the reader can be guided both to a more detailed evaluation of a particular issue than I have been able to give here and to the relevant experimental methods and results. Because of my own background, emphasis on fungi as organisms and on soil as an environment is inevitable, but data from other groups of organisms and environments will not be ignored. I should also mention in passing that my own position on many matters pertinent to this article have evolved significantly over the last twenty years so that my treatment here often differs from that given in my earlier publications.

Fundamental to any consideration of the ecological effects of water is the concept of water moving along gradients of potential energy. In the absence of a pumping mechanism, water will be available to an organism only if it can so adjust the potential energy of its existing water that the appropriate gradient is produced. Such adjustment will involve the organism in the expenditure of chemical energy. If the fundamental concept, then, is potential energy, a consideration of thermodynamics is unavoidable. Without it, any analysis of the role of water will lack a firm foundation. The approach of this review is therefore thermodynamic, even though thermodynamics is not the favorite subject of most microbiologists.

In the section on thermodynamics, there is nothing new except in the particular assemblage and ordering of matters eventually pertinent to microbiology. My aim there has been to present simply but in appropriate detail the essential concepts and to derive the necessary equations with a degree of rigor satisfactory for the purposes of this review. Some complex areas have been skated over, but reference has been given to more profound treatments. The monograph by Klotz (1950) is both clear and authoritative, while useful treatments of thermodynamics for biologists are those of Spanner (1964) and Slatyer (1967). This theoretical introduction, concerning the thermodynamics of water is, I believe, more detailed than in any other microbiological review. It forms, I hope, an appropriate basis for the subsequent review of experimental work and for a better understanding between workers in the future.

If a strictly thermodynamic approach is to be adopted, it is appropriate to use basic units of measurement, in that the same units can be applied readily and unambiguously to the widest range of systems. I therefore have used the Système International d'Unités (SI). This may appear to be a stumbling block because little of the existing literature uses SI units. I believe them, however, to be the units of the future and so appropriate to a review of this nature.

2. Thermodynamic Considerations

2.1. Chemical Potential

The First Law of Thermodynamics is given by the equation

$$\Delta U = W + Q^*$$ (1)

where ΔU is the increase in internal energy of the system; W is the work done *on* the system; and Q is the heat received *by* the system.

The Second Law of Thermodynamics has many alternative statements. For our purposes it may best be represented by the equation

$$\Delta S = Q_{rev}/T$$ (2)

where ΔS is the increase in entropy of the system; Q_{rev} is the heat received by the system, under reversible conditions; and T is the temperature (°K). From equation (1),

$$W = \Delta U - Q$$ (3)

and so, using equation (2),

$$-W_{max} = -\Delta U + T\Delta S$$ (4)

where $-W_{max}$ is the maximum work that can be done *by* the system, i.e., when the conditions are reversible. Further, if the work done by the system is pressure–volume work,

$$-W_{max} = PdV$$ (5)

where P and V are the pressure and volume of the system, respectively. Hence, in a reversible system,

$$dU = TdS - PdV$$ (6)

---

*All symbols used are listed in Table I.

Table I. Symbols and SI Units

Main symbols

a	activity
c	concentration (mol/m^3)
d	differential of a variable
G	Gibbs free energy (J)
g	gravitational constant (m/sec^2)
H	enthalpy (ΔH — enthalpy of fusion) (J)
h	vertical distance (m)
i	chemical species whose concentrations vary
j	chemical species whose concentrations are constant
ln	natural logarithm
M	molecular weight (kg/mol)
m	molality (mol/kg solvent)
N	mole fraction
n	number of moles
P	pressure (Pa)
p	partial gas or vapor pressure (Pa)
Q	heat received (J)
R	molar gas constant (J/mol/K)
r	radius (m)
S	entropy (J/K)
T	temperature (K)
U	internal energy (J)
V	volume (m^3)
W	work done (J)
w	water
x	independent variable
y	dependent variable
γ	activity coefficient
Δ	difference between values of a variable
∂	partial differential
θ	water content $(m^3/m^3$ or kg/kg)
μ	chemical potential (J/mol)
ν	ions/molecule
π	osmotic pressure (Pa)
ρ	density $(kg/m^3$ or kg/liter in some traditional equations)
σ	surface tension (N/m)
τ	matric suction (Pa)
ϕ	osmotic coefficient
Ψ	total water potential (Pa)
ψ	water potential (Pa); specific water potential (J/kg)

Superscripts and overbar

A	of absorbed water
C	of a cell
E	of the environment of a cell

Table I. *(Cont.)*

Superscripts and overbar (Continued)

P	of the protoplasm of a cell
$P1$	of one portion of the protoplasm
$P2$	of a second portion of the protoplasm
W	of the wall of a cell
0 (zero)	of a standard state
overbar (e.g., $\overline{V}$)	partial molal quantity [e.g., volume (m^3/mol)]

Subscripts

a	of a given solute
g	as affected by a difference in elevation, as in gravitational potential
i	of chemical species whose concentrations vary
j	of chemical species whose concentrations are constant
m	as affected by the presence of a matrix, as in matric potential
p	as affected by a difference in pressure, as in pressure potential
s	as affected by the presence of solutes, as in solute potential
w	of water
max	maximal, in a thermodynamic sense
rev	reversible, in a thermodynamic sense
0 (zero)	of the freezing point

The Gibbs free energy (G) of a system is defined by

$$G = U + PV - TS \tag{7}$$

Differentiating,

$$dG = dU + PdV + VdP - TdS - SdT \tag{8}$$

Substituting equation (6) into equation (8),

$$dG = VdP - SdT \tag{9}$$

The above treatment assumes that the number of moles of substances in the system remains constant, i.e., that the system is "closed." Assume now an "open" system in which the moles of substance i (n_i) are variable but in which the moles of all other substances, denoted by n_j, are constant. The total differential of the internal energy (U) as a function of S, V, and n_i is given by

$$dU = (\partial U/\partial S)_{V,n_i}dS + (\partial U/\partial V)_{S,n_i}dV + (\partial U/\partial n_i)_{V,S,n_j}dn_i \tag{10}$$

From equations (6) and (10), it can be seen that

$$(\partial U/\partial S)_{V,n_i} = T \tag{11}$$

and

$$(\partial U/\partial V)_{S,n_i} = -P \tag{12}$$

Further, the function $(\partial U/\partial n_i)_{S,V,n_j}$ is defined as being the chemical potential μ_i of the substance i. Thus, from equation (10),

$$dU = TdS - PdV + \mu_i dn_i \tag{13}$$

and, from equation (8),

$$dG = VdP - SdT + \mu_i dn_i \tag{14}$$

Hence

$$(\partial G/\partial n_i)_{T,P,n_j} = \mu_i = \overline{G}_i \tag{15}$$

where $\overline{G}_i$ is the partial molal Gibbs free energy of i.

If change is to be spontaneous, the system must be able to do work, i.e., $-W$ in equations (4) and (5) must have a positive value. Therefore $(-\Delta U + T\Delta S)$ and $P\Delta V$ must be positive, and ΔG (from equation (8), when P and T are constants) must be negative; for spontaneous change affecting the component i, $\Delta \mu_i$ must be negative.

2.1.1. Chemical Potential and Osmotic Pressure

Assume now an open system in which the number of moles (n_i) of substance i can vary. The total differential of μ_i is then

$$d\mu_i = (\partial \mu_i/\partial P)_{T,n_i} dP + (\partial \mu_i/\partial T)_{P,n_i} dT + (\partial \mu_i/\partial n_i)_{T,P,n_j} dn_i \tag{16}$$

By use of equations (9) and (15), this may be simplified to

$$d\mu_i = \overline{V}_i dP - \overline{S}_i dT + (\partial \mu_i/\partial n_i)_{T,P,n_j} dn_i \tag{17}$$

where $\overline{V}_i$ and $\overline{S}_i$ are the partial molal volume and entropy of i, respectively. It remains to evaluate

$$(\partial \mu_i/\partial n_i)_{T,P,n_j} dn_i$$

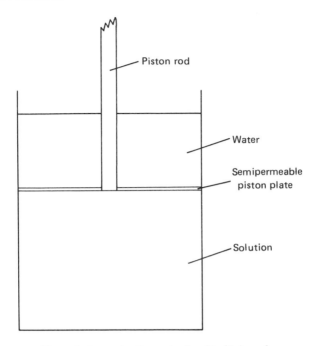

Figure 1. Apparatus for evaluating $(\partial \mu_i / \partial n_i)_{T,P,n_j} dn_i$.

Let the substance i be water, denoted by the subscript w. Further, assume that P and T are constant in the system depicted in Fig. 1. In such a system, the chemical potential of water in the solution will be less than that in the pure phase, and there will therefore be a tendency for water to move into the solution. If the piston plate consists of a semipermeable membrane, a downward pressure, π, on the piston will be required to maintain equilibrium. According to the standard definition, π is the osmotic pressure of the solution.

Now allow the piston to rise infinitesimally and reversibly. The pressure remains π, but the volume of the solution increases by $\overline{V}_w \partial n_w$, where $\overline{V}_w$ is the partial molal volume of water appropriate to the given solution and ∂n_w is the number of moles of water added to the solution. Thus, the work done by the system is

$$\pi \overline{V}_w \, \partial n_w = -\partial G \tag{18}$$

and

$$(\partial G / \partial n_w)_{T,P} = -\pi \overline{V}_w \tag{19}$$

[Should the system consist of two solutions of different concentrations, rather than a solution and pure water, then π in equation (18) signifies the difference in osmotic pressures between the two phases.]

From equations (15), (17), and (19),

$$d\mu_w = (\partial\mu_w/\partial n_w)_{T,P}dn_w = -\overline{V}_w d\pi \qquad (20)$$

which is the required relationship. Hence, in an open system where P and T may change,

$$d\mu_w = \overline{V}_w dP - \overline{S}_w dT - \overline{V}_w d\pi \qquad (21)$$

Further, in an open system where T is constant,

$$\mu_w - \mu_w^0 = \overline{V}_w(P - \pi) \qquad (22)$$

where P and π are the external and osmotic pressures, respectively, of the phase represented by w relative to the standard conditions represented by w^0.

It is instructive to consider in an alternative fashion the pressure differences existing between solutions separated by a semipermeable membrane. Consider the system depicted in Fig. 1 to be altered by the replacement of the movable piston by a fixed membrane. In this system, there will be a tendency for water to pass from the pure phase through the membrane into the solution. But matter tends to move down a pressure gradient. Hence, if the pressure of the pure water is taken as a reference such that $\pi_{water} = 0$, then the pressure of the solution must be $-\pi$. If osmotic pressure is conceived of in this way, the sign of π in the previous equations is changed and equation (22) becomes

$$\mu_w - \mu_w^0 = \overline{V}_w(P + \pi) \qquad (23)$$

2.1.2. Chemical Potential and Matric Suction

When pure water is held in a matrix, that is, in a porous solid, μ_w depends not only on T and P but also on the volumetric water content of the system (θ). Hence,

$$d\mu_w = (\partial\mu_w/\partial P)_{T,\theta}dP + (\partial\mu_w/\partial T)_{P,\theta}dT + (\partial\mu_w/\partial\theta)_{T,P}d\theta \qquad (24)$$

It remains to evaluate the new component $(\partial\mu_w/\partial\theta)_{T,P}\,d\theta$, which may be done in a way comparable to that used in connection with osmotic pressure. The historical development of the treatment of the matric component is, however, subtly different and leads to differences in terminology.

Assume the matrix is sand and that it is held in a sintered-glass Buchner funnel in the system depicted in Fig. 2. Further assume the temperature to be constant and the atmospheric pressure at the surface of the sand to be constant with a value of P^0. At equilibrium, the pressure in the water at the air–water interface in the sand (P) is given by

$$P = P^0 - h\rho g \tag{25}$$

where ρ is the density of water; g is the gravitational constant; and h is the vertical distance between the air–water interface in the sand and the meniscus of the water in the tube. The pressure of the water at the surface of the system relative to that of a reference pool of pure water at the same elevation is equal to $P - P^0$. If P^0, as the reference pressure, is taken to be zero, the pressure of the water will be negative and of magnitude $-P$. But a negative pressure is a positive suction. Hence, the matric suction (τ) is given by

$$\tau = -P = h\rho g \tag{26}$$

The introduction of the term "suction" may be confusing, and perhaps a brief comparison with osmotic pressure will be helpful. The osmotic component

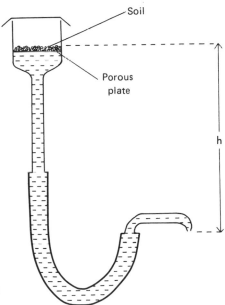

Figure 2. A tensiometer used to illustrate the significance of matric potential (h = height).

is termed a *pressure* because historically it was viewed as the *external pressure that must be applied to a system* to maintain equilibrium. If the external pressure required is $+\pi$, then the initial pressure of the water in the solution must be $-\pi$; that is, the water in the solution exerts a suction (negative pressure) equal to $+\pi$. Similarly, the matric component is termed a *suction* because historically it was viewed as that *property of the water in a given system* which expressed the tendency of that system to attract water from a reference pool at zero pressure (Marshall, 1959).

Now move the lower end of the tube downward, in the system of Fig. 2, reversibly and infinitesimally. The matric suction will remain unaltered (τ), but a volume of water ($\partial\theta$) will be withdrawn from the matrix. Thus, the work done by the system is

$$-\partial G = \tau \partial\theta \tag{27}$$

But

$$\partial\theta = \overline{V}^0_w \partial n_w \tag{28}$$

where $\overline{V}^0_w$ is the molal volume of pure water, solutes being absent, so that, from equation (15),

$$(\partial G / \partial n_w)_{T,P} = \mu_w = -\overline{V}^0_w \tau \tag{29}$$

and

$$d\mu_w = -\overline{V}^0_w d\tau \tag{30}$$

Hence in an open system where P and T may change,

$$d\mu_w = \overline{V}^0_w dP - \overline{S} dt - \overline{V}^0_w d\tau \tag{31}$$

and if T is constant,

$$\mu_w - \mu^0_w = \overline{V}^0_w (P - \tau) \tag{32}$$

where P and τ are the external pressure relative to the standard conditions and matric suction, respectively.

2.2. Potential Energy of Water and Water Potential

For a system in which a *solution* is held within a *matrix*, the combination of equations (22) and (33) yields

$$\mu_w - \mu_w^0 = \overline{V}_w(P - \pi - \tau) \tag{33}$$

This equation involves both osmotic pressure and matric suction. A comparable equation can be developed by concentrating on the potential energy of water in the system. This has many theoretical advantages and simplifies the convention concerning signs and avoids any confusion about the terms "pressure" and "suction."

When the movement, or equilibrium, of water between different phases in a system is the prime consideration, it is the energy per unit quantity of water that must be considered. Energy so considered is thus an extensive variable in thermodynamic terms. Movement of water results from spatial differences in energy per unit quantity of water, and this energy is thus a scalar point function. Further, the force acting in a given direction on a unit quantity of water can be found by differentiating the energy per unit quantity with respect to distance in that direction. Any scalar point function which upon differentiation yields a force is termed, by definition, a potential. Hence, the energy with which we are concerned is potential energy, or more simply, potential.

In the above discussion, the term *energy per unit quantity of water* has been used frequently. What is the unit quantity? The units of chemical potential (μ) are joule/mole, and these are the units of energy in all the fundamental equations. Energy per mole is not easily handled experimentally, however. Energy per unit mass (J/kg, specific energy) or per unit volume (J/m³) are generally far more convenient.

In the important equations (22), (32), and (33), by slight rearrangement, the term $(\mu_w - \mu_w^0)/\overline{V}_w$ occurs. This is an expression of energy per unit volume; more precisely, the difference in chemical potential (J/mol) per partial molal volume of water (m³/mol), between water in the given system (μ_w) and that at the reference conditions (μ_w^0), and has the units J/m³. Let

$$(\mu_w - \mu_w^0)/\overline{V}_w = \psi \tag{34}$$

where ψ is the water potential (per unit volume). From the arguments presented above, it can be seen that ψ will be determined by the sum of its components, i.e., pressure potential (ψ_p, the potential of water per unit volume as affected by external pressure), solute (or osmotic) potential (ψ_s, the potential of water per unit volume as affected by the presence of solutes), and matric potential (ψ_m, the potential of water per unit volume as affected by the presence of a solid matrix). Thus,

$$\psi = \psi_p + \psi_s + \psi_m \qquad (35)$$

and

$$\psi_p = P \qquad (36)$$
$$\psi_s = -\pi \qquad (37)$$
$$\psi_m = -\tau \qquad (38)$$

It should be noted in passing that some authors replace $\overline{V}_w$ (partial molal volume of water) by $\overline{V}_w^0$ (molal volume of pure water) in equation (34). The practical effect is negligible, but reference may be made to Spanner (1964, 1972) for the issues involved.

Water potential, as given by equations (34) and (35), can be defined as the amount of work that must be done per unit quantity of pure water in order to transport reversibly and isothermally an infinitesimal quantity of water from a pool of pure water at atmospheric pressure to a point in the system under consideration at the same elevation. *Matric potential* can be defined as the amount of work that must be done per unit quantity of pure water in order to transport reversibly and isothermally an infinitesimal quantity of water from a pool, containing a solution identical in composition to that of the solution in the system under consideration at the elevation and the external gas pressure of the point under consideration, to the system. Definitions of the same general form for solute, pressure, and other potentials have been given by Aslyng (1963), Rose (1966), and Slatyer (1967).

It has been shown above that the units of energy per unit volume are joules per cubic meter (J/m^3). But one joule is equivalent to one newton meter (Nm). Thus, the units of energy per unit volume may be simplified to N/m^2, which is by definition the unit of pressure, the *pascal*. It is thus usual to express measurements of water potential with the unit of the pascal (Pa) or its multiples (kPa, MPa). It has occasionally been argued that it is improper to use a pressure unit as a measure of energy, but it is clearly permissible and indeed inevitable when energy per unit volume is being considered.

2.2.1. Total Water Potential

Equation (35) assumes that only three factors affect the energy status of water. This is false if temperature is also a variable. If a temperature differential exists, a term must be introduced involving the entropy of the system [see equation (21)], but this is highly inconvenient. It has therefore become conventional to assume that no temperature differential exists. In microbiology this causes no problems, as the temperature of a microorganism will be that of its immediate environment.

Equation (35) also assumes the absence of a gravitational component, i.e., it assumes that there is no difference in elevation between the experimental system and the reference pool of water. Again, this is probably always true in microbial systems although for other large organisms, e.g., trees, it may not be so. If the gravitational component is included, equation (35) becomes

$$\Psi = \psi_g + \psi_p + \psi_s + \psi_m \tag{39}$$

where Ψ is the *total* water potential and ψ_g is the gravitational potential.

In many natural systems, e.g., soils or cells, both a solid matrix and an aqueous solution are present. Conventionally, the effects of matrix and solute potentials have been taken to be simply additive, as in equation (35). Noy-Meir and Ginzburg (1967) have critically analyzed the water potential isotherm of living plant tissue and the validity of equation (35). No attempt will be made here to discuss their rigorous and complex article, but it should be noted that they conclude that equation (35) is inadequate to account theoretically for the complexity of water potential components in a multiphasic and heterogeneous system. It remains to be seen, however, whether the error introduced by simplified theory exceeds those of normal experimental error. In this review, no term such as an interaction potential will therefore be introduced as a component of total water potential.

2.2.2. Water Potential and Water Activity

For an ideal gas

$$\overline{V}_i = RT/p_i \tag{40}$$

where p_i is the partial gas pressure of component i. Further, from equation (17), which is valid for both gases and liquids (and considering T and n_i to be constant),

$$(\partial \mu_i / \partial P)_{T,n_i} = \overline{V}_i \tag{41}$$

Hence, from equations (40) and (41)

$$(\partial \mu_w / \partial P)_{T,n_w} = RT/p_w \tag{42}$$

where the component i is water and p_w is the partial vapor pressure of water. Integrating,

$$\int_{\mu_w^0}^{\mu_w} d\mu_w = \int_{p_w^0}^{p_w} RT\, dp_w \tag{43}$$

and

$$\mu_w - \mu_w^0 = RT \ln p_w/p_w^0 = RT \ln a_w \tag{44}$$

where a_w is the water activity. But at equilibrium, the chemical potentials of the vapor and liquid phases of water are identical, that is,

$$\mu_w \text{ [liquid]} = \mu_w \text{ [vapor]} = \mu_w \tag{45}$$

Hence, equation (44) is valid for aqueous solutions as well as the vapor phase. Further,

$$(\mu_w - \mu_w^0)/\overline{V}_w = \psi = (RT \ln a_w)/\overline{V}_w \tag{46}$$

2.2.3. Water Potential and Solute Concentration

Now equation (46) is valid regardless of the way in which a change in ψ is brought about. The equation can, however, be developed further if an alteration in ψ is produced entirely by change in the concentration of solutes, i.e., if ψ_s is considered. From Raoult's Law,

$$p_w = N_w p_w^0 \tag{47}$$

where N_w is the mole fraction of water in the aqueous solution. Hence,

$$\mu_w - \mu_w^0 = RT \ln N_w \tag{48}$$

But

$$N_w = 1 - N_a \tag{49}$$

where N_a is the mole fraction of a given solute in the solution, and

$$-\ln N_w = -\ln(1 - N_a) \tag{50}$$
$$= N_a + \tfrac{1}{2}N_a^2 + \tfrac{1}{3}N_a^3 + \cdots \tag{51}$$

Now if N_a is small, all but N_a are negligible. Therefore,

$$-\ln N_w \doteq N_a \tag{52}$$

and, from equations (46) and (48),

$$\psi_s = -RTN_a/\overline{V}_w \tag{53}$$

Further,

$$N_u/\overline{V}_w \doteq c_a \tag{54}$$

and

$$\psi_s = -RTc_a \; [\text{Pa}] \tag{55}$$

where R is the molar gas constant (8.314 J/mol/°K), T is temperature (°K), and c_a is the solute concentration (mol/m^3). Equation (55) is valid only for ideal solutions of a nonelectrolyte.

In the case of nonideal solutions of a nonelectrolyte, equation (55) may be used after incorporation of either of two correction factors. Thus,

$$\mu_w - \mu_w^0 = RT \ln \gamma N_w = RT\phi \ln N_w \tag{56}$$

where γ and ϕ are the activity and osmotic coefficients of the solute, respectively, at the given molality and temperature. From equation (55)

$$\psi_s = -RT\phi c_a \; [\text{Pa}] \tag{57}$$

By further extension, for electrolytes,

$$\psi_s = -RT\nu\phi c_a \; [\text{Pa}] = -RTm\nu\phi \; (\text{J/kg}) \tag{58}$$

where m is molality (mol/kg solvent) and ν is the number of ions per molecule (taken as 1 for nonelectrolytes). [The osmotic coefficients of many solutes are given by Robinson and Stokes (1968).] Further, from equations (46) and (58),

$$RT \ln a_w/\overline{V}_w = -RTm\nu\phi\rho \; (\text{Pa}) \tag{59}$$

$$\ln a_w = -\nu m\phi\rho\overline{V}_w \tag{60}$$

$$= -(\nu m\phi)/55.51 \tag{61}$$

if it is assumed that $\overline{V}_w$ does not differ significantly in value from $\overline{V}_w^0$ and that the density of water (ρ) at the temperature of the experiment is 1000 kg/m^3. Equation (61) is that given by Scott (1957) in his very influential review article on water relations of food spoilage microorganisms.

The partial molal volume ($\overline{V}$) has appeared as a component of a number of equations, and it is featured in others to be developed subsequently. Its significance and evaluation, however, are usually neglected in the biological literature. The nature of this parameter may be appreciated by considering a solution of such high concentration that the addition of one mole of solute or solvent will have no appreciable effect on the concentration. If one mole of the solvent, say water, is added to this solution, then the increase in volume of the solution is the partial molal volume of water, that is,

$$\overline{V}_w = (\partial V / \partial n_w)_{T,P,n_a} \tag{62}$$

where n_w and n_a are the number of moles of solvent (water) and solute, respectively. $\overline{V}_w$ will be constant for a solution of a given substance and water only if the addition of one mole of water to a sea of solution *of any concentration* yields the same volume increase. This is unlikely, so that the value of $\overline{V}_w$ will therefore vary with concentration of solute and generally will differ from the molal volume of the pure solvent ($\overline{V}_w^0$).

As can be seen from equation (62), the partial molal volume is in fact a partial differential; it can be evaluated exactly only if the precise equation relating solution volume to solvent molality, at given constant values for temperature, external pressure, and solute molality, is known. In practice, this equation is never known, and $\overline{V}_w$ must be calculated using approximations. A number of methods exist giving, as might be expected, slightly different results.

As an example, the partial molal volume of water in a 21% by weight solution of KCl at 20°C will be considered. The gram molecular weight of water and KCl will be taken as 18.016 and 74.6, respectively. Using the method of Spanner (1964) and the chord-area plot method of Klotz (1950), values for $\overline{V}_w$ of 17.881 and 17.927, respectively, are obtained, compared with molal volume of pure water of 18.048 ml/mole. Such a discrepancy is worrisome, if the trouble has been taken in the first place to discriminate between $\overline{V}_w$ and $\overline{V}_w^0$. Lewis and Randall (1961) have discussed in detail various methods of calculation of $\overline{V}_w$; fortunately, one method described by them is both simple and acceptably accurate. The only data required are the densities (D_4^{20}) of a range of solutions of known composition. These can be obtained from standard tables, e.g., *Handbook of Physics and Chemistry:* from 1% to 24% solutions in the case of KCl. Percentage concentration (x axis) is then plotted against the reciprocal of the density, i.e., the partial specific volume (y axis). If $\overline{V}_w$ for a certain percentage concentration is required, a tangent to the curve is drawn from that concentration to the intercept with the y axis (zero concentration of solute). The value of the intercept multiplied by the gram molecular weight of water gives the value of $\overline{V}_w$.

Alternatively, with little if any loss of accuracy, a chord-intercept calculation may be made as follows, again for a 21% KCl solution. For a 20% solution, the density and the partial specific volume are 1.1328 kg/liter and 0.88277 liter/kg, respectively; for a 22% solution 1.1474 kg/liter and 0.87154 liter/kg, respectively. Hence, the chord through these two points on the concentration-partial specific volume curve would intercept the y axis at 0.88277 + (0.88277 − 0.87154) × 20/2, i.e., at 0.995088 liter/kg. $\overline{V}_w$ for a 21% solution of KCl is thus 0.9958 × 18.016, i.e., 17.9275 ml/mol. Because of the experimental error inherent in most biological experiments, no significant additional error will be introduced if the value of $\overline{V}_w$ for most solutions (and of $\overline{V}_w^0$) is taken as being 18 ml/mol, or in SI units, 1.8×10^{-5} m³/mol.

2.2.4. Freezing Point of Solutions

The properties of solutions that depend upon the concentrations of the components are known as colligative properties. Osmotic pressure is one such property. Another is the depression of the freezing point of the solvent which occurs when a nonvolatile solute is added to it. The relevant equation for an undissociated solute [the derivation of which is given by Slatyer (1967)] is

$$\Delta T = RT_0^2 N_a/\Delta H \tag{63}$$

where ΔT is the freezing point depression; T_0 is the freezing point of pure water; N_a is the mole fraction of solute; and ΔH is the molal enthalpy of fusion (6.012 kJ/mol). Hence,

$$\Delta T = 1.86m \quad [°C] \tag{64}$$

where m is molality.

From equations (53) and (63),

$$\psi_s = -(T\Delta H\Delta T)/\overline{V}_w T_0^2 \tag{65}$$
$$= -4.46 T\Delta T \; [\text{kPa}] \tag{66}$$
$$= -1330\Delta T \; [\text{kPa at } 25°C] \tag{67}$$

2.2.5. Choice and Equivalence of Units

Two major systems of expressing the energy status of water exist in the microbiological literature. The first is based on water activity (a_w) and is thus immediately compatible with other work specifying relative humidity (a_w =

p/p_0 = relative humidity/100). The relationship of water activity to the chemical potential is given by equation (44), which shows the relationship to be logarithmic and temperature-dependent. Such a system is conceptually simple and is particularly convenient in two circumstances. One is when the effects of solutions are being considered, as in studies of saline lakes or sugar solutions. The other is when only the totality of the water status of a microbial substrate, often expressed as its equilibrium relative humidity, is of significance, as with many stored food products.

The second system is based on water potential, which is related to chemical potential by equation (34). The relationship is linear and is independent of temperature so long as the reference pool of water is at the same temperature as the experimental system (which is normally the case in microbiology). At first sight, the concept of water potential appears complex. Even one of its great advantages, expression in SI units, may be seen as a disadvantage by those unfamiliar with units such as the pascal. Another advantage is that water potential can be partitioned into its components (solute, matric, etc.) in such a way that simple arithmetic addition of the values of the component potentials yields the value of the final potential. This cannot be done so simply with water activity. Water potential is thus of particular use in the analysis of complex systems containing important solid components, such as soil or cells. In this review, I have adopted the water potential system because of its significant theoretical and practical advantages.

Of the ways of expressing water potential, three use SI units. If equation (34) is taken to define ψ and hence to assume the measurement of energy per unit volume, then the molal water potential is equal to $\psi \overline{V}_w$ J/mol. Further, as the partial specific energy is defined as being equal to μ_i/M_i, where M_i is the (kg) molecular weight of the component i, then the specific water potential is given by $\psi \overline{V}_w/M_w$ J/kg. For nearly all practical purposes, therefore, a (volumetric) water potential of 1 kPa is equal to a specific water potential of 1 J/kg. The following linear relationships are also applicable:

$$1 \text{ MPa} = 1 \text{ kJ/kg} = 10 \text{ bar} = 10^7 \text{ dyne/cm}^2 = 9.87 \text{ atm} \qquad (68)$$
$$= 10{,}220 \text{ cm water pressure} = 750 \text{ cm mercury pressure}$$

The calculation is more complex for logarithmic relationships such as the old pF scale for matric suction, where $\text{pF}x = 10^x$ cm water. Equation (46) relates ψ to a_w; thus,

$$\psi = 8.314 \; T \ln a_w/\overline{V}_w \; [\text{Pa}] \qquad (69)$$

The equivalences in Table II are based on equation (69), with the assumptions that $\overline{V}$ is 1.8×10^{-5} m³/mol and T is 298°K.

Table II. Relationship between Water Activity (a_w) and Water Potential (ψ) at 25°C

a_w	$-\psi$	a_w	$-\psi$
0.999999	138 Pa	0.80	30.7 MPa
0.99999	1.38 kPa	0.75	39.6 MPa
0.9999	13.8 kPa	0.70	40.1 MPa
0.999	138 kPa	0.65	59.3 MPa
0.995	690 kPa	0.60	70.3 MPa
0.99	1.38 MPa	0.50	95.4 MPa
0.98	2.78 MPa	0.40	126 MPa
0.97	4.19 MPa	0.30	165 MPa
0.96	5.62 MPa	0.20	221 MPa
0.95	7.06 MPa	0.10	317 MPa
0.90	14.5 MPa	0.01	634 MPa
0.85	22.4 MPa	0.001	951 MPa

Occasionally the concept of osmolality is used in water relations research. Its utility is apparent particularly when solutions of unknown composition, such as cell sap or cytoplasm, are of importance. If the depression of freezing point is determined for such a solution, then equation (64) can be employed to calculate the molality of the solution, on the assumptions that all the solutes are undissociated, that the solution is ideal, and that matric potential is zero. None of these assumptions is of course likely to be valid, and it is therefore impossible to present the data realistically in terms of actual molality. Instead the hypothetical molality is referred to as *osmolality* with units of osmoles. Thus, the osmolality of a solution is that molal concentration of nonelectrolyte ($\nu = 1$) which would give the observed freezing point depression if it were present in an ideal solution. Also 1 osmol/kg water will develop a water potential of -2.48 MPa at 25°C (or an osmotic pressure of 24.4 atm in traditional non-SI units). The concept of osmolality may, however, be avoided if equation (66) is used to convert data on depression of freezing point directly to water potential.

A final related concept is that of the *osmosity* of a solution, which may be defined as the molar concentration of NaCl with the same freezing point or osmotic pressure as the solution. There seems to be no virtue in using osmosity in biology.

2.2.6. Convention Concerning Signs

Although the convention concerning signs is implicit in the above arguments, and has been referred to explicitly at a number of points, it may be useful to state it comprehensively here.

For spontaneous movement of water, $\Delta \overline{G}$ (and $\Delta\psi$) must be negative. It follows that water will move spontaneously from sites of higher to lower water potential. Now the water potential of pure water under reference conditions is assumed to be zero, yet water moves spontaneously into a solution or into an unsaturated matrix. Hence, the value of ψ_s for a solution is always negative; that for ψ_m for the water in an unsaturated matrix is also always negative. Further, water moves spontaneously from systems of less negative to more negative water potentials. Water at a less negative potential (say -40 kPa) has a higher potential than water at a more negative potential (say -80 kPa).

2.2.7. Measurement of Water Potential

The concentration of protoplasmic solutes in physiological studies has generally been found by measuring the depression of the freezing point. Variations of the isopiestic technique of Robinson and Sinclair (1934), e.g., Scott (1957), Chen and Griffin (1966), Ayerst (1969), Harris et al. (1970), have generally been used to equilibrate substrates with control solutions of known potential. Later, a variety of techniques were introduced, often to measure or equilibrate matric potentials (Marshall, 1959; Griffin, 1972). All such techniques have now been largely superseded, most notably by a number based upon thermocouple psychrometry which can be applied over the whole range of water potential from ca 50 kPa down (R. W. Brown and van Haveren, 1972; Papendick and Campbell, 1975, 1981). Another relatively new technique uses electric hygrometers to measure the conductance of a hygroscopic salt, itself in equilibrium with the experimental sample (Troller, 1977).

3. Matric Potential and Associated Factors

3.1. Matric Potential, Capillarity, and Absorption

For most purposes in microbial ecology, equation (35) ($\psi = \psi_p + \psi_s + \psi_m$) adequately represents the system in regard to the energy of water, and it is hoped that the meaning and relevance of water potential, pressure potential, and solute potential are sufficiently clear from the above discussions. The concept of matric potential is less widely understood, and a further consideration of it seems appropriate.

Consider a capillary tube with one end dipping into a pool of water. Water will rise in the tube until the forces associated with the interface between the water and the air (surface tension) balance the downward force generated by the raised column of water. The governing equation is

$$\pi r^2 h \rho g = 2\pi r \sigma \qquad (70)$$

where h is the height of water column (m), ρ is the density of water (kg/m^3), g is the gravitational constant (m/sec^2), r is the radius of tube (m), and σ is the surface tension (N/m). Thus, from equation (26),

$$\tau = h\rho g = 2\sigma/r \qquad (71)$$

Alternatively, with ψ_m expressed in pascals,

$$r(\text{meters}) = -2\sigma/\psi_m \doteq -0.147/\psi_m \qquad (72)$$

Not only is the water in the meniscus in equilibrium with the column of water subtended from it but it is also in equilibrium with water in the adjacent vapor phase. Expressing the equilibrium in terms of potential energy per unit mass, from equations (38), (46), and (71),

$$-2\sigma/\rho r = RT \ln a_w/M_w \qquad (73)$$

Equation (73) is a form of the famous Kelvin equation, first formulated over one hundred years ago. Some authors have doubted that it remains valid at extremely small values of r because of the possible failure of classical thermodynamics to express the behavior of the small volumes of water involved. Recently, however, experimental verification of the validity of the Kelvin equation has been obtained with menisci of radius at least as small as 4 nm (Fisher and Israelachvili, 1979). It can therefore be seen from equation (72) that the relationships hold to matric potentials at least as low as -37 MPa and probably through the whole range of potentials of biological interest (see Sections 4 and 5).

From equation (71), the height of capillary rise is inversely related to the radius of curvature of the meniscus and, more importantly, from equation (72), the radius of curvature of the meniscus governs the matric potential of the water. This still remains true if the capillary tube is removed from the pool; the matric potential of the water remaining in the tube is still controlled by the radius of curvature of the meniscus and hence by the radius of the tube in which the meniscus lies.

Few environments closely resemble a rigid-walled capillary tube, but the same principles apply wherever water, or an aqueous solution, is held within a rigid porous matrix. A naturally occurring example is a sand in which the particles are closely packed. If the radius of curvature of all water menisci is 1.5×10^{-5} m, then the consequent matric potential is $ca. -10$ kPa. If the sand loses water so that the radius of all menisci becomes 7.5×10^{-6} m, then the matric potential falls to -20 kPa.

Further, the relationship between the water content of a porous system and matric potential will depend upon the pore size distribution within the system. The relationship is best depicted by the water sorption isotherm of the system, and Fig. 3 provides examples from soils. The water sorption isotherms (moisture characteristic curves) for a sand, a loam, and a clay are conspicuously different.

Careful consideration shows that when a soil is drying, the controlling radii are those of the pore necks, whereas when it is filling with water, the controlling radii are those of the widest portions between necks. This difference leads to hysteresis in the water sorption isotherm for most matrices (Fig. 3) (Griffin, 1972).

The concept of the matric potential of water held within a rigid matrix is thus relatively simple. Unfortunately, few natural matrices are rigid. Instead, they tend to swell as water is added and to shrink as they dry. Their pore-size distribution thus changes with matric potential. Common examples are clay soils, timber, and dried fruit. Such changes complicate the concept of matric potential and emphasize the forces acting between surfaces and liquids, but only two examples will be considered further to illustrate the problems involved. As a soil with a moderate clay content commences to drain from saturation, its initial behavior is governed by the larger particles and thus by the simple concept of matric potential given above. As drying continues, however, it is the behavior of the colloidal clay particles that predominates. As is the case with many colloids, the clay particles carry a variable electrical charge, in this case negative. They thus tend to repel one another, and work is involved

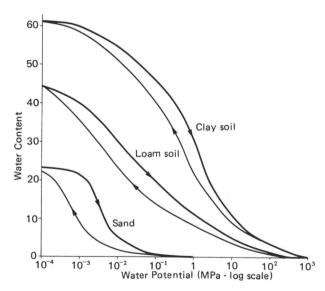

Figure 3. Water sorption isotherms for three soils, showing the hysteresis between the drying and wetting boundary curves (after Griffin, 1972).

in bringing them closer together. As a clay soil dries, it shrinks because water is lost from between the colloids. The matric potential of the remaining water is thus determined by the force of repulsion between the particles. In other systems, the behavior of water is affected not so much by charged particles as by the special association of water molecules with macromolecules, and this association has significant effects on matric potential (see Section 6.1.1).

Matric potential, therefore, sums all those variations in potential derived from liquid–gas and liquid–solid interfaces involving capillarity, repulsion between charged colloidal particles, and absorption by walls, membranes, and macromolecules.

3.2. Matric Potential and Water-Filled Pathways

As the matric potential of water in a rigid matrix decreases, the water content also decreases, the exact form of the association being given by the water sorption isotherm (see previous section). Thus, a declining matric potential causes, in general, reduction in (a) the volumetric water content, (b) the radius of the largest pore to remain water filled [equation (72)], and (c) the continuity of water-filled pathways. These associated changes have important consequences.

3.2.1. Solute Diffusion

The rate of diffusion of nonionized, nonpolar solutes in water held in a matrix is directly proportional to the volumetric water content. Diffusion rates for ions or charged molecules are determined in a more complex fashion but are still sensitive to volumetric water content. Thus, the diffusion of both stimulatory and inhibitory solutes within soil is markedly affected by water content and, in turn, by matric potential (Griffin, 1972, 1978). The low mobility of enzyme substrates in dried foods is a similar phenomenon (Acker, 1962, 1969).

3.2.2. Movement of Unicellular Structures

Unicellular organisms, such as bacteria, some algae, and zoospores of fungi, require water in which to swim or to be moved passively by Brownian movement. Clearly, they cannot move through pores that have drained nor through water-filled pores whose radii are less than that of the cell. Further, movement of appreciable distances through a matrix by such cells requires a continuity of water-filled pores of the requisite size. Matric potential can therefore act indirectly as an important constraint on the movement of certain groups of organisms or structures. The great ecological significance of this constraint is noted in Section 5.

3.2.3. Gaseous Diffusion

The rate of diffusion of O_2 through a gas-filled space is 2.1×10^{-1} cm^2/sec, whereas the comparable rate through water is 2.6×10^{-5} cm^2/sec. As air replaces water in the pores of a soil or some other matrix, as an accompaniment to decreased matric potential, there is therefore a great change in the rates of diffusion of O_2 and other gases. The analysis of the system is made more difficult because soils are not homogeneous; rather they consist of aggregates (or crumbs) of small, often colloidal particles, separated by intercrumb pores. Under many field situations, the matric potential will be such as to cause the intercrumb pores to drain, whilst the smaller pores within the crumbs remain saturated with water. Microorganisms situated on the surface of a crumb are, therefore, in a very different environment than those within a crumb. More detailed discussions of these and related topics are given by Griffin (1968, 1972). For the present purpose, it is enough to emphasize that the response of organisms to a change in matric potential in many systems may be complicated by simultaneous changes in gaseous diffusion.

4. Water Potential and Microbial Growth

Water moves spontaneously from regions of higher to regions of lower water potential. In the absence of a water pump, which would drive water against a potential gradient, cells can therefore absorb water only if their internal water potential is less than that of the environment. It is therefore instructive to grow microorganisms at various external water potentials and to note the effect on growth. In theory, external potential can be controlled by solutes or matrically, or by a combination of the two. All methods have their problems, however.

If the influence of external solute potential is to be investigated, it would be ideal if the solute used would affect the cell entirely through the colligative properties of the solution. In practice, however, specific solute effects are always likely to be present, their magnitude depending on the chemical properties of the solute. Historically, the choice of solutes has no doubt been on the basis of trial and error, and this has revealed a variety of preferred solutes which are now used in most experiments. Whenever possible, it is best to use a variety of solutes so that idiosyncratic behavior can be detected, thus increasing confidence that the cell's response to solute potential *per se* is really being measured when other solutes are used. An excellent example of this approach is provided by a study of various strains of *Fusarium roseum* (Wearing and Burgess, 1979). This showed that the radial growth of the fungus on agar, in the solute potential range from 0 to -12 MPa, was altered in the same way in the presence of potassium chloride, sodium chloride, sucrose, and a mixture of sodium sulfate, potassium chloride, and sodium chloride. Few microorganisms, however, react quite so uniformly, although it is still possible to discern with considerable accuracy the general form of the response to solute potential.

If the system is controlled by the alteration of matric potential, the problems of specific solute effects are avoided but only at the cost of introducing the complications noted in Section 3.2. The consequences are illustrated later in this section and in Section 5.

Solid substrates or gels may be equilibrated with atmospheres of known relative humidity so that the water potential of the substrates, as the sum of the solute and matric components, is controlled (Section 2.2.7). Although the rate of equilibration can be inconveniently long, and residual spatial heterogenicity of potential in the substrate may persist, this method has been extensively adopted.

The influence of water potential on the germination of fungal spores and on the growth of many microorganisms has been tested in a variety of experimental systems. Data on germination have been collected by Griffin (1963, 1972), Ayerst (1968, 1969), Byrne and Jones (1975), and Hocking and Pitt (1979). A striking feature is the increase in latent period of germination as potential declines. Thus, conidia of *Aspergillus chevalieri* germinate at 25°C within one day at −10 MPa but after 32 days at −50 MPa (Ayerst, 1969). This leads to problems in determining the minimum potential permitting germination because the longer the experiment is continued, the lower is likely to be the observed minimum. The latent period also increases with the age of spores. Curiously, germination under the more extreme conditions is not always followed by continued mycelial growth (Cother and Griffin, 1974; Pitt, 1975). This may be because the nutrient reserves of the spore are sufficient to permit production of the compatible solutes necessary for the required reduction in solute potential (see Section 6.3), whereas in hyphae there is a failure somewhere in the chain of events leading from nutrient absorption to production of compatible solutes.

The range of potentials over which growth occurs has been determined for many microorganisms, and representative data have been given by Ayerst (1969), Tresner and Hayes (1971), Pitt (1975), A. D. Brown (1976), Gould and Measures (1977), and Harris (1981). Such data do not, however, reveal the effects of water potential on growth rate in the range of potentials that permits growth. Many data on growth rates of fungi, bacteria, and some actinomycetes have been published, and these may be traced through articles by Scott (1957), Onishi (1963), Anand and Brown (1968), Ayerst (1968, 1969), Griffin (1972, 1977, 1978), Wong and Griffin (1974), Harrower and Nagy (1979), Hocking and Pitt (1979), Duniway (1976), Griffin and Luard (1979), and Wilson and Griffin (1979). The growth response to different solute potentials usually has the form of curve A in Fig. 4. For any given species, there is no continued growth at zero Pa solute potential because solutes, and therefore

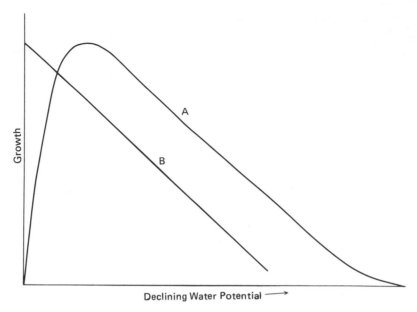

Figure 4. Idealized relationships between microbial growth and solute (A) and matric (B) potentials.

nutrients, are entirely lacking. With the addition of nutrients, and the consequent very small lowering of solute potential, the growth rate of most organisms increases rapidly. The potential at which growth is maximal and the precise shape of the response curve at potentials between the optimum and minimum are, however, dependent upon both the individual organism and other factors of the environment. The exact value of the minimum potential permitting growth is difficult to determine because of the extremely slow growth rates of most organisms under these conditions.

    With filamentous fungi, radial growth rates and increases in dry weight respond similarly to solute potential, but the dry weight measurements are always more sensitive, the slope of the response curve at potentials lower than the optimum being steeper. There is a strong interaction between potential and nutrient availability in determining both latent period of germination and growth response (Ayerst, 1968; Griffin, 1972, 1978; Wearing and Burgess, 1979). With increasing nutrient concentrations, the minimal potential for germination and growth decreases, in extreme cases by a factor of two. A strong interaction also exists between water potential and temperature in affecting growth. For many fungi and bacteria, the optimal temperature increases by about 5°C as potential decreases over the range permitting growth (Ayerst,

1968, 1969; Griffin, 1972, 1978; Hellebust, 1976). The influence of pH is, however, less marked (Pitt and Hocking, 1977; Griffin, 1978).

The rate of linear extension of some filamentous fungi at various matric potentials has been determined. In general, the response is of the form of line B in Fig. 4. The optimal matric potential is close to zero, and the minimal is as much as twice that for solute potential. Matric systems used in these experiments have, however, had very low nutrient concentrations, and it may well be that the reduced tolerance to low potentials is associated with nutrients rather than with matric potential *per se.*

With some degree of speculation, many microorganisms can now perhaps be grouped on the basis of their reported growth response to solute potential. Such groups are, however, in no way discrete; each merges into the next, and their only justification is to facilitate comprehension of the rapidly increasing information.

Group 1. Extremely sensitive species with an optimum growth response at *ca.* −0.1 MPa solute potential and with little growth below −2 MPa. Wood decay fungi and some soil basidiomycetes and gram-negative bacteria are characteristic of this group.

Group 2. Sensitive species with an optimum growth response at *ca.* −1 MPa and little growth at *ca.* −5 MPa. Most phycomycetous and coprophilous fungi probably fall into this group, along with many other filamentous fungi, most gram-negative rods, and many soil actinomycetes.

Group 3. Species of moderate sensitivity with an optimum at *ca.* −1 MPa and with little growth at *ca.* −10 to −15 MPa. Species of *Fusarium,* a number of other soil fungi (ascomycetes, basidiomycetes, and fungi imperfecti), *Saccharomyces cerevisiae,* and many soil actinomycetes and gram-positive bacteria are included in this group.

Group 4. Xerotolerant species with an optimum at or below −5 MPa and a minimum betwen *ca.* −20 and −40, or even −50 MPa. Many yeasts, e.g., *Saccharomyces rouxii, Debaryomyces hansenii,* and a number of filamentous fungi, of which those within the genera *Aspergillus, Eurotium, Penicillium,* and *Eremascus* and the species *Wallemia sebi* are characteristic. It may be that many specialized bacteria of saline lakes belong to this group (Brock, 1979), but their taxonomy and physiology are very poorly known.

Group 5. Xerophilic species that fail to grow at high potentials, say above −4 MPa, and have minima at −40 MPa or less. All isolates of the fungi *Monascus bisporus* and *Chrysosporium fastidium* and some isolates of *Aspergillus restrictus* and *Saccharomyces rouxii* fall into this group. The actinomycete *Actinospora halophila* is also a member. *M. bisporus* has the most extreme tolerance, its aleuriospores germinating at −69 MPa. The halophilic bacteria have been included here although their need is not for low potential *per se* but for the high concentrations of sodium chloride necessary for membrane stability and of potassium chloride for the functioning of their enzymes.

5. Water Potential and Microbial Ecology

The complex roles of water potential in plant pathology (Cook and Papendick, 1972; Cook, 1973; Baker and Cook, 1974; Papendick and Campbell, 1975; Ayres, 1978; Schoeneweiss, 1978; Duniway, 1979; Cook and Duniway, 1981) and as a selective factor in soil microbiology (Griffin, 1981) have been reviewed recently, and these topics will not be considered here in any detail. Rather, an attempt will be made to summarize more generally the data relating water potential to microbial ecology. Naturally occurring environments with low water potentials are very diverse and may be categorized as in Table III. By far the most important division is between group A and the others. In group A, unicellular organisms predominate, whereas in groups B and C, it is the filamentous fungi that are of greatest significance. As I have discussed in detail elsewhere (Griffin, 1972, 1981; Wong and Griffin, 1976a,b; Griffin and Luard, 1979), this difference is based upon an interaction between matric potential and microbial morphology. Unicellular structures depend for their movement on the existence of water-filled pathways of the requisite dimensions (Section 3.2.2). In most systems, their continued activity will be similarly dependent. But the existence of water-filled pathways in a porous solid is largely determined by the matric potential.

As shown by equation (72), no pore neck of radius large enough to permit the passage of a bacterium of radius 0.5 μm, without deformation, will remain water-filled at -0.3 MPa; no pore neck large enough to permit the passage of a fungal zoospore will remain filled at -15 kPa. By the time allowance is made for the necessity for continuity of pathways and for the helical (Allen and Newhook, 1973) or other nonlinear movement of the cells, it becomes apparent that far higher matric potentials are required if these cells are not to be significantly constrained (Duniway, 1976, 1979; Wong and Griffin, 1976a,b; Griffin, 1978, 1981; Young *et al.*, 1979). Thus in soil, infection of roots by zoospores of *Phytophthora* is maximal at matric potentials exceeding -100 Pa, and most bacterial activities are reduced at -100 kPa and are negligible at -500 kPa matric potential because the bacteria are then physically constrained. Similar limits probably exist in other systems within groups B and C. In group A, however, water is abundant, matric effects are nonexistent, and soluble nutrients diffuse readily; consequently, unicellular organisms are active at lower water potentials. In the contrasting environments of groups B and C, the filamentous structure of most fungi conveys the ability to bridge air-filled pores and to penetrate solid nutrient substrates. The very different microfloras

Table III. A Classification of Xeric Environments and their Characteristic Microfloras

Group	A		B	C	
	a	b		a	b
Major constituent	Liquid		Porous solid	Porous solid	
Solute potential	Low		High	Low	Low
Matric potential	High		Low	Low	Variable
Nutrient concentration	Low	High	Low	Low	High
Examples	Salt lake	Syrups: preservative brines	Soil of low water content	Saline soil of low water content	Stored food
Characteristic microorganisms	Bacteria and algae	Yeasts	Filamentous fungi	Filamentous fungi	Filamentous fungi
	(*Halobacterium, Halococcus, Dunaliela*)	(*Debaryomyces hansenii, Saccharomyces rouxii*)	(*Aspergillus, Penicillium*)	(*Aspergillus, Penicillium*)	(*Aspergillus, Penicillium, Chrysosporium, Eremascus, Monascus, Wallemia*)

of group A and of groups B and C environments are therefore easily understandable.

Growth rates in pure culture over a range of water potentials are not usually good indicators of performance at similar potentials in naturally existing systems. Thus, the characteristic microflora of seas could not be predicted from the response of organisms to solute potential. Were this to be so, it would be anticipated that the mycoflora of seas, saline soils, and salt marshes would be similar. In fact, the fungi of soils of high solute content are those found frequently in more normal soils but not at all in marine environments (Chen and Griffin, 1966; Moustafa, 1975). Indeed, marine fungi are not characterized by unusual tolerance to solute potentials of the order of -2 MPa, and many fungi apparently well-suited by their physiological response to reduced potential are not normal members of the marine flora. Clearly, many other poorly understood factors are involved, probably associated with spore dispersal and initial colonization of the substrate (Jones et al., 1971; Jennings, 1973).

In soil, most fungi are not active over the full range of water potential permitting growth in pure culture. Faced with competition, the activity of a given organism is restricted to water potentials departing not too far from its optimum. Thus, in a natural system at -1 MPa, a fungus in group 3 (Section 4) is likely to dominate one in group 4, but the reverse would be true at -10 MPa. Such a pattern of activity is revealed by a study of the colonization by fungi of sterile hair lying on various soils (Griffin, 1972). The reduced competition as potential decreases is shown by the numbers of genera/species recorded as colonists: 22 genera/51 species (water potential about -0.1 MPa), 11/45 (-7.1 MPa), 7/38 -14.5 MPa), 4/19 (-22.4 MPa), 2/11 (-30.7 MPa), and 2/7 (-39.6 MPa) (Chen and Griffin, 1966). At the lower potentials, only species of *Aspergillus* and *Penicillium* were recorded.

At water potentials less than -20 MPa, ability to grow in pure culture is a good indicator of activity in natural environments. Thus, in brines and sugar solutions of substrate group A (Table III), *Debaryomyces hansenii* and *Saccharomyces rouxii* are characteristic spoilage agents (Onishi, 1963), whereas the fungi in groups 4 and 5 (Section 4) are characteristic of solid substrates (groups B and C). Within substrate groups B and C, species of *Aspergillus* and *Penicillium* are ubiquitous, but *Chrysosporium fastidium, C. xerophilum, Eremascus albus, E. fertilis, Monascus bisporus,* and *Wallemia sebi* are apparently restricted to the substrates of high nutrient content that form subgroup Cb. These more restricted organisms, with the exception of *W. sebi,* grow in pure culture at reduced solute potentials only when that solute is an easily metabolized organic molecule. Salts are inhibitory.

Reduction in matric potential may cause significant physical changes in the substrate itself. A well-studied example concerns the degradation of wood (Griffin, 1977). As matric potential declines from zero, water is withdrawn from progressively smaller voids, first from the cell lumina and then from intercellular and membrane pores. At still lower potentials, water is withdrawn from the intramural transient voids, the voids disappear, and the wood shrinks. These transient voids are generally so small that enzyme molecules cannot enter them. Degradation of cellulose and lignin is therefore negligible at matric potentials below about −4 MPa, and for rapid degradation a value in excess of −1 MPa is required. Then the substrate holds sufficient water to permit effective diffusion first of degrading enzymes and second of the breakdown products. Terrestrial wood-decay fungi appear to be adapted to such a situation in that their minimal water potential for growth corresponds with that matric potential at which their substrate becomes effectively inaccessible to degrading enzymes. It might be expected that similar limits would apply to biodegradation of grain and milled food products. This is not so, and fungi are active in them at water potentials as low as −40 MPa. The difference between wood and grains in this regard possibly lies in the presence of sugar and other nutrients of low molecular weight in the latter but not in the former. The local diffusion of small molecules would be subject to far less geometrical constraint than that for large enzymes (Acker, 1962, 1969). Regardless of this, however, the mechanisms for absorption by hyphae at matric potentials of −40 MPa merit investigation.

It should perhaps be noted that the often-quoted value of −1.4 MPa as the permanent wilting point of higher plants is due more to the failure, at this matric potential, of hydraulic conductivity within soil to supply water at an adequate rate to replace transpirational losses than to any direct effects on higher plant cells at such a potential (see also Section 6.2).

6. Water Potential and Microbial Physiology

6.1. Components of Water Potential of Cells

Under equilibrium conditions, the following relationships hold:

$$\psi_p^E + \psi_s^E + \psi_m^E = \psi^E \tag{74}$$
$$\psi^E = \psi^C \tag{75}$$

where the superscripts E and C indicate environment and cell, respectively. If growth is to occur, uptake of water is necessary, requiring that ψ^C should be less than ψ^E. If the difference is large, however, the cell is likely to be ruptured. Where steady microbial growth occurs, the difference between ψ^E and ψ^C is likely to be very small, and for most purposes little error will be introduced by considering the system to be in equilibrium. For the sake of simplicity, it will be so considered in the ensuing discussion.

The concept of the water potential of a cell needs refinement because the cell is a multiphasic system. Thus, in equilibrium,

$$\psi^C = \psi^W = \psi^P \tag{76}$$

where the superscripts ^{W} and ^{P} indicate wall and protoplasm, respectively. Both ψ^P and ψ^W require detailed consideration.

6.1.1. Bound Water and the Categorization of Water in Cells

The concept of "bound" water probably arose in connection with the properties of some of the water present in suspensions of macromolecules, such as proteins. A most useful summary of the concept in this context has been provided by Kuntz and Kauzmann (1974), who have noted the difficulty of providing an adequate definition. In essence, they recognize that much of the water present in macromolecular systems, and therefore in membranes and cells, has properties different from that of normal water. Such water is said to be bound in the sense that its molecules exhibit a degree of preferential ordering with respect to the molecules of the protein or biopolymer. This usage of the term "bound water" circumscribes it adequately, emphasizing the macromolecule–water interrelationship but placing no stress on the magnitude of the binding force.

If emphasis is mistakenly reversed and placed on the magnitude of the forces operating, the concept of bound water can be most misleading, particularly if it is assumed that water bound or absorbed by macromolecules is water bound by some extremely (or uniquely) powerful force. It is probably worth pursuing this issue in more detail by considering how much water will be held by a system when it is exposed to a given external water potential. For a soil, this is given by the water sorption isotherm (Fig. 3), the amount of water held being clearly dependent on the applied potential over a very wide range. Comparable water sorption isotherms could be produced for the walls of microorganisms. With these, however, desorption would not commence until the matric potential fell to below -30 to -40 MPa *(Bacillus megaterium)* or -80 MPa *(Neurospora crassa, Saccharomyces cerevisiae)* and might well not be significant until far lower potentials were reached (Mitchell and Moyle, 1956; Gerhardt and Judge, 1964; Trevithick and Metzenberg, 1966; Burnett, 1976) [see equation (72) for the relationship between pore size and matric potential; rigidity of wall is assumed]. Osmotic systems could be considered similarly, although it is not conventional to do so. Thus, in a system consisting of a solution of two phases, A and B, separated by a semipermeable membrane, B having a higher solute potential than A, the water in phase A is "unavailable" to phase B, and therefore in one sense, from the standpoint of B, might well be considered to be bound.

Water sorption isotherms for macromolecules can also be produced. In their normal form, with water absorbed plotted against relative humidity, they are S-shaped and show hysteresis. As relative humidity increases from 0 to 10–20%, water content of the macromolecule–water system increases rapidly. There is then a region, up to about 60% relative humidity, during which the water content of the system increases more slowly. Finally, above 60% and especially above 85% relative humidity, great increases in water content with increase in relative humidity occur. Such curves, which vary in detail with the macromolecular system studied, indicate that water is progressively absorbed by macromolecules over a range of water potential similar to that affecting the water contents of many other natural systems in which quite different mechanisms are operating. Clearly, there is nothing unusual about the magnitudes of potential of the water associated with macromolecules. To avoid confusion, I shall therefore not use the term "bound water" subsequently unless the meaning is quite clear. Following the lead of Kuntz and Kauzmann (1974), I shall instead refer to the water associated with macromolecules as "absorbed." The potential arising from such absorption is conventionally thought of in matric terms, the same mechanisms existing for both macromolecules and their insoluble aggregates (e.g., membranes).

Cooke and Kuntz (1974) and Kuntz and Kauzmann (1974) have critically reviewed the evidence concerning the properties of water resulting from interactions with proteins, membranes, and other macromolecular cellular constituents. The systems are extremely complex, and the water sorption isotherms indicate that multisite–multilayer interactions occur between the macromolecule and the water. A number of treatments of such isotherms exist, of which the Brunauer, Emmett, and Teller (BET) isotherm is probably the best known. Treatments of this kind, e.g., Schneider and Schneider (1972), assume that adsorption is taking place onto a solid surface, first as a monolayer of tightly held water and then as successive layers of more loosely held water. The formation of the monolayer is assumed to be complete at the first discontinuity of the isotherm at 10–20% relative humidity. Such theoretical treatments do not, however, predict very successfully the sorption isotherms that are obtained experimentally, and other theories have been proposed that take account of the lattice structure of many biopolymers and of their partial solubility. Kuntz and Kauzmann (1974) and Cooke and Kuntz (1974), after a detailed consideration of the various approaches, have concluded that none is satisfactory. Instead, they have adopted an empirical approach and, on the basis of the measurement of many physical parameters, have divided water in biological systems into three types, categorized thus:

Type I ("Bulk water"). Properties do not differ significantly from those of pure, free water.

Type II ("Bound water"). Molecules of water of this type form a physically distinct water environment, one to a few molecules thick, adjacent to the macromolecular surfaces. Many physical properties of the water are significantly altered, and it cannot be made to freeze despite any lowering of the temperature. Approximately 0.3–0.6 g H_2O/g dry weight macromolecule is probably of type II.

Type III ("Irrotationally bound water"). The properties of these water molecules are the most altered compared with those of free water, and the molecules are probably bound to specific sites on or within the macromolecule. Perhaps 1 to 10% of the total of types II and III belong in this grouping.

From published water sorption isotherms and other related data, it is possible to estimate, within very broad limits of accuracy, that type III water will be retained by macromolecules and cells against external potentials above *ca.* −150 MPa. Type II water, however, will be released progressively with every diminution in potential below *ca.* −4 MPa.

The preceding paragraphs have concerned binary systems consisting of macromolecule and water only. The system is considerably complicated by the presence of a third component, a solute. Considerable evidence indicates that *ca.* 25% of cellular water is inaccessible to potassium ions and possibly to other solutes, although a much larger proportion appears to be accessible to nonelectrolytes such as glucose (Acker, 1969; Cooke and Kuntz, 1974). It is tempting to equate this water possessing modified solvent properties with absorbed water. Even if this is not strictly true, any such association has interesting consequences.

Consider a ternary system consisting of water, a solute, and a biopolymer. If there is a volume of absorbed water that does not act as a solvent, then its matric potential (ψ_m^A) must be equal to the solute potential of the bulk solution when the system is in equilibrium. If the concentration of solute is then increased, the solute and matric potentials must fall and, from the water sorption isotherm of the biopolymer, so must the volume of absorbed water. This may have deleterious effects on the conformation and hence the properties of the biopolymer.

At the other extreme, a system can be envisaged in which the concentrations of molecules (or ions) of solute and water are the same in the bulk solution as adjacent to the biopolymer. Then the biopolymer obviously causes no preferential arrangement of the components of the solution and so is absorbing neither solute nor solvent. ψ_m^A will be zero. In practice it has been found that, if the solute is a salt, each salt has a different effect on the degree of water absorption by the biopolymer, compared with that in the binary system. Thus, Bull and Breese (1968, 1970) have investigated the absorption of electrolytes and water by proteins and have shown that the moles of water absorbed per mole of protein is very sensitive to the presence of salts. Of the salts tested, sodium sulfate greatly increased water absorption, but all other salts decreased it, the effect being more marked with NaCl than with KCl. The presence of salts may thus have a marked effect, differing in magnitude with each salt, on the denaturation of proteins in addition to the effect noted in the previous paragraph. Comparable data for nonelectrolyte solutes appear not to be available.

6.1.2. Water Potential of Protoplasmic Water

The protoplasm is bounded by the plasma membrane and the cell wall. If the cell is turgid, these structures through their pressure increase the water potential of the water in the protoplasm upon it by a component denoted by ψ_p^P (a pressure potential). This component is identical to the turgor pressure of classical biology. If the cell is not turgid, or is killed, the value of turgor potential is zero. The further analysis of the components of the water potential of protoplasm is hindered by the complexity of cellular contents. Most work has utilized higher plants or giant-celled algae, but the interpretation of data and the hypotheses derived from them are not fully agreed upon by the various authors. Further, these cells are all highly vacuolate, whereas bacteria have no vacuoles and vacuoles are very small and dispersed in the active apical portions of fungal hyphae. It is doubtful, therefore, if the "sap" of higher plants and algae, which must consist largely of vacuolar solution, has any close equivalent in most microorganisms.

The bulk (type I) water of the protoplasm contains solutes and so will have a solute potential (ψ_s^P). The potential of that water will also have a matric component (ψ_m^{PI}), originating *inter alia* from the repulsion between charged particles and from the free energy of the elastic polymer network within the protoplasm (ψ_m^P). Hence,

$$\psi^P = \psi_p^P + \psi_s^P + \psi_m^{PI} \tag{77}$$

Such a relationship has experimental support, particularly from the study of leaves. Thus, if sap is expressed from leaves, the water potential of that sap can be measured (Shepherd, 1973). On the supposition that most of the depression in potential is due to solutes, this sap potential is usually referred to as the solute potential of the cell. If, however, the water potential of the whole, newly killed leaf tissue is measured by thermocouple psychrometry, it is found to be lower than that of the sap (Shepherd, 1975). The extra depression is attributed to matric potential (ψ_m^{PI}). Although the detailed interpretation of such data is open to argument (Noy-Meir and Ginzburg, 1967; Warren Wilson, 1967; Acock, 1975), it seems clear that the water potential of some water molecules in the protoplasm has both solute and matric components and that these are additive. There is no reason to believe that microorganisms are different from leaves in this respect.

In naturally occurring cells, 20–25% of the total cell water is absorbed (types II and III taken together—see previous section) and is unable to freeze. A smaller proportion does not even act as a solvent. By definition, therefore, this latter water cannot have a solute potential. As noted earlier, it possesses a matric potential through absorption (ψ_m^A). In addition, the potential of the absorbed water may be affected, although not necessarily identically, by the factors giving rise to ψ_m^{P1}. The sum of these two matric potentials of the absorbed water will be denoted by ψ_m^{P2}.

In the cell (neglecting turgor at this stage), the bulk solvent water of the protoplasm thus has a potential of ($\psi_s^P + \psi_m^{P1}$), the absorbed, nonsolvent water a potential of ψ_m^{P2}. These two volumes of water are in thermodynamic equilibrium, however, so that

$$\psi_s^P + \psi_m^{P1} = \psi_m^{P2} \tag{78}$$

The sum ($\psi_s^P + \psi_m^{P1}$) will be more recognizable as the osmotic potential (or, with change of sign, the osmotic pressure) of classical biology. Lying between the extreme states of bulk solvent water and absorbed nonsolvent water are zones of intergradation, composed partly of the absorbed water of modified solvency. The water in these zones will, however, have the same final potential as that of the bulk and absorbed water, with components of modified ψ_s^P, ψ_m^A, and ψ_m^P. As discussed earlier, the ratio of absorbed to bulk water will vary with ψ^P.

In summary, then the water potential of the protoplasm can be expressed by

$$\psi^P = \psi_p^P + \psi_s^P + \psi_m^{P1} \quad \text{(for bulk solvent water)} \tag{79}$$
$$= \psi_p^P + \psi_m^{P2} \quad \text{(for absorbed nonsolvent water)} \tag{80}$$

Such equations make it apparent, in my view, that the *full* matric potential arising from absorption of water by macromolecules is not additive to the *full* solute potential because these two potentials are not components of the final potential of any given water molecule. In this, I differ from authors such as Warren Wilson (1967). Rather, I regard the absorbed water as being in thermodynamic equilibrium with the solvent water, in general accord with Noy-Meir and Ginzburg (1967).

6.1.3. Water Potential of Wall Water

Bacterial and fungal walls with pores smaller than *ca.* 5 and 1.8 nm radius, respectively, will be impermeable to molecules of greater than about 57,000 and 4500 g molecular weight, respectively (Gerhardt and Judge, 1964;

Trevithick and Metzenberg, 1966). If ψ_s^E is produced by such large molecules, then

$$\psi^W = \psi_m^W = \psi^E \tag{81}$$

where ψ_m^W is the matric potential of the water within the wall. The same equation will be valid if $\psi_m^E \ll \psi_s^E$, i.e., if the solute potential of the environment external to the wall is near zero. Equation (81) therefore applies to cells either immersed in solutions of dextran or of polyethylene glycol of high molecular weight or growing in a number of natural environments, such as soil. Conversely, if ψ_s^E is produced by smaller molecules so that they can pass into the cell wall,

$$\psi^W = \psi_m^W + \psi_s^W = \psi^E \tag{82}$$

where ψ_s^W is the solute potential of the water within the wall.

6.2. Magnitude of Cellular Potentials

Equations (74) to (76) permit estimates of cellular potentials to be made. The experimental results on growth summarized in Section 4 show that ψ^P must vary with ψ^E through a range characteristic of the species. In xerotolerant species, ψ^P will be as low as -40 MPa.

Richter (1976) has gathered data on the minimum values for water potential found in higher plants. These range to -16.3 MPa for desert plants but only to -1.5 to -2.6 MPa for plants of mesic sites. Many plants thus develop potentials similar to microorganisms, although none are comparable to the extreme microbial xerotolerant and xerophilic species. The effects of water deficits (reduced water potential) on enzymatic activity has been reviewed by Todd (1972).

Traditionally, turgor pressure in the cells of higher plants has been measured by observing the point of incipient plasmolysis. Such a method assumes *inter alia* that the cell wall is rigid and that the forces of adhesion between the plasma membrane and the wall are small. In bacteria (Mitchell and Moyle, 1956; Alemohammad and Knowles, 1974) these suppositions are not valid, and they are most doubtful for fungi. Certainly plasmolysis in all these organisms is neither a clear-cut nor easily observed phenomenon. Alternative methods for measuring ψ_p^P have therefore to be adopted. The value of $(\psi_s^P + \psi_m^{P1})$ can be measured using thermocouple psychrometry on killed hyphae. As ψ^P ($= \psi^E$) is known, ψ_p^P can be calculated using equation (79). Griffin (1978) has collected the data of a number of workers concerning fungi, and Luard and Griffin (1981) have presented many of their own. In general, the data show that ψ_p^P is relatively independent of ψ^E throughout the range of potential permitting growth and has a value characteristic of each species. For most species of fungi, ψ_p^P has a value of about 1 MPa and for xerotolerant and xerophilic species about 2–3 MPa. After due allowance has been made for experimental error, most data show no decline of ψ_p^P with fall in ψ^E. Indeed, the converse is indicated. It is therefore unlikely that the lower limit for external potential permitting growth of microorganisms is associated with maintenance of turgor, as has been suggested for higher plants.

A great deal of research has been conducted on turgor regulation in algae, and particularly on the giant cells of some species. This work has been reviewed by Zimmermann (1978), Zimmermann and Steudle (1978), and Gutknecht *et al.* (1978) and lies largely outside the scope of this article. Reported turgor potentials in marine algae are somewhat less than in fungi, being less than 1 MPa for most species, but, as with fungi, there is usually little correlation with external potential. The giant cells of algae are sufficiently large to permit direct measurement of turgor using a micropressure probe (Hüsken *et al.*, 1978), but this is unfortunately still not possible with fungi and bacteria.

The values of ψ_s^P and ψ_m^P for microorganisms have not been experimentally determined. Normally their sum has been presented, somewhat inaccurately, as the "osmotic" pressure or potential. One experiment using higher plants has shown that the matric component is 14% of the value of -2.13 MPa measured for the "osmotic" pressure (Shepherd, 1975). As the water content of leaves is reduced, however (with consequent decrease in ψ^P), the relative contribution of the matric component increases and may become more than that of the solute component (Warren Wilson, 1967). If my argument presented earlier in this section is accepted, the matric component referred to here is ψ_m^{Pl}, not ψ_m^A. Further, there is a great difference between a wilting leaf with highly vacuolate cells and a growing microorganism so that similar changes in the relationship between the matric and solute components of the potential of bulk water within the microbial protoplasm cannot be assumed.

If, for living cells, the masses of individual solutes and of water per unit dry weight is obtained, it is at first sight an easy matter to calculate the solute potential so generated. In practice, the exercise is not so simple because of the plethora of solutes but more importantly because of uncertainty concerning the proportion of total cellular water that should be regarded as a solvent. Wall water should clearly be excluded: in bacteria, *ca.* 8% of total water is in dextran-impermeable space and has been considered to be wall water (Mitchell and Moyle, 1956). Absorbed, nonsolvent water must also be excluded, but this volume probably varies with cell type and with solute. As noted elsewhere (Section 6.1.1), up to 25% of cellular water has modified properties as a solvent. In total, therefore, perhaps *ca.* 30% of the volume of cellular water should be subtracted from the total when estimates of amount of solvent, and hence of solute potential, are being made.

6.3. Compatible Solutes and Osmoregulation

For growth, a positive turgor potential appears to be a necessary though not a sufficient requirement. This requirement necessitates the presence of a semipermeable membrane that permits a difference of solute composition between the protoplasm and the environment and a lowered protoplasmic solute potential compared to that of the environment. If a cell is immersed in a hypertonic solution, it is therefore unlikely to restore thermodynamic equilibrium by simply allowing the ingress of the external solutes. Such a process would indeed restore equilibrium at zero turgor potential, but the restoration of turgor, at least, must depend on metabolic processes. Further, the external solute, on admission to the cell, is likely to be toxic or at least deleterious to cellular functioning. In the presence of hypertonic solutions of most solutes, therefore, organisms generate organic molecules internally to effect the major reduction in solute potential required to restore equilibrium. Such a strategy is, of course, essential for organisms exposed to a reduced external matric potential.

Brown (1976, 1978, 1979) has discussed at length the nature and significance of compatible solutes. These are intracellular solutes that both act as osmoregulators and protect enzyme activity at low water potentials. The commonest extracellular solute in nature is NaCl, but it seems not to be a compatible solute for any organism. In higher plant halophytes, the high NaCl content is localized in the vacuole and is thus largely separated from the enzyme systems of the protoplasm (Hellebust, 1976). Bull and Breese (1968, 1970) have shown that the sodium ion greatly reduces the solvation of proteins and so affects their configuration and activity. Despite the toxicity of sodium ions for general metabolism, NaCl is necessary for stabilizing the envelope of species of the halophilic bacterial genus *Halobacterium*. This genus, however, largely excludes sodium from its protoplasm and selectively absorbs, against great concentration gradients, the potassium ion. In *Halobacterium,* the potassium ion (accompanied by chloride ions or organic acids) acts as a compatible solute, and indeed many enzymes of this genus have a requirement for high concentrations of the ion for maximum activity. The various aspects of the water relations of the halophilic bacteria have been reviewed by Larsen (1967), Kushner (1968), Brown (1976, 1978), Lanyi (1979), and Bayley and Morton (1979).

In most higher plants and marine algae, KCl is the most important solute in effecting adjustment of solute, and hence turgor, potential (Hellebust, 1976; Kauss, 1977; Kirst, 1977). In the giant cells of the genus *Valonia,* turgor potential controls the action of a potassium ion pump involved in turgor regulation. The kinetics of osmoregulation have been extensively studied in giant algal cells, and reference should be made to recent reviews by Gutknecht *et al.* (1978), Zimmermann (1978), and Zimmermann and Steudle (1978) for information in this regard and on turgor-dependent processes. Jennings (1979) has reviewed evidence concerning the transport of water and ions through membranes and the relation of this to hyphal growth.

Despite the demonstrated importance of KCl in the adjustment of solute potential in some organisms, it now seems clear that this salt is not the solute of primary importance in most microorganisms. In fungi and many algae, a variety of polyols are predominant or important protoplasmic solutes. These include mannitol (*Platymonas, Dendryphiella,* and other marine fungi), cyclohexanetetrol *(Monochrysis),* arabitol *(Dendryphiella, Saccharomyces),* sorbitol *(Stichococcus),* α-galactosyl-(1,1)-glycerol *(Ochromonas),* and especially glycerol (*Chlamydomonas, Dunaliella, Debaryomyces, Saccharomyces,* and many filamentous fungi) (Brown, 1976, 1978; Kauss, 1977; Brown and Hellebust, 1978; Avron and Ben-Amotz, 1979; E. J. Luard, unpublished data). The demonstrated presence of many polyols in fungi and lichens also suggests the importance of these chemicals as osmoregulators, as well as food reserves, in these organisms (Lewis and Smith, 1967). It is interesting to note that the predominant polyol in the same species e.g., *Saccharomyces rouxii,* can vary with the external solute and with the phase of growth (Brown, 1978).

In bacteria (Brown 1976), the diatom *Cyclotella cryptica* (Liu and Hellebust, 1976), the green alga *Stichococcus bacillaris* (Brown and Hellebust, 1978), and the oomycetous fungus *Phytophthora cinnamomi* (E. J. Luard, unpublished data), the amino acid proline is the main or at least an important, compatible solute. Proline and the quaternary ammonium compound betaine may also be significant in halophytic higher plants (Hellebust, 1976). Other amino compounds (glutamic acid and α-aminobutyric acid) appear to be significant in some bacteria (Measures, 1975; Unemoto and Hayashi, 1979).

The processes associated with the regulation of polyol concentration, and especially that of glycerol, have been reviewed by Brown (1976, 1978, 1979). *Saccharomyces cerevisiae* and *S. rouxii* adopt different strategies in responding to lowered external potentials, the former leaking to the exterior a constant proportion of the glycerol produced and increasing the rate of production, the latter maintaining a constant rate of production but reducing the rate of leakage (Edgley and Brown, 1978). The latter strategy is clearly the most conservative of energy, and it is therefore not surprising that *S. rouxii* is far more xerotolerant than *S. cerevisiae.*

The production of large amounts of polyols or amino compounds within the stressed cell must be demanding of energy. This is shown by increased rates of specific respiration and enhanced heat production (Wilson and Griffin, 1975; Gustafsson and Norkrans, 1976; Gustafsson, 1979), both being associated with decreased efficiency of growth at low potentials. Indeed, it is likely that the minimal potential permitting growth of a given species is set when the species can divert no more energy to internal solute production.

As noted earlier, a compatible solute has the dual role of adjusting solute potential while protecting, or at least not severely inhibiting, enzyme activity. The concentrations of KCl and glycerol measured in some cells (*ca.* 5 molal) leave no room to doubt their effects on ψ_s^p. Brown (1978) has also discussed the evidence showing that glycerol to an unusual degree neither inhibits enzyme action nor inactivates ("denatures") the enzyme itself.

Schobert (1977) and Schobert and Tschesche (1978) have recently proposed additional roles for such compatible solutes as glycerol and proline. They have argued that the weak points in water-stress regulatory mechanisms are the hydrophobic side-chains of biopolymers. If reduction in ψ^p occurs, with consequent reduction in the moles of water absorbed per mole of biopolymer, the first positions where water structure is altered are expected to be associated with hydrophobic groups. If the disturbance is great enough, the configuration of the biopolymer will change, and it will enter an inactive state. Protection might occur in two ways. The first is by substitution for water in the absorbed layer at the hydrophobic sites by "water-like" molecules of the general formula R–OH. Polyols are such molecules. By substituting for water, they will not only stabilize the biopolymer but release water into the bulk solution. The second hypothesis is based on the fact that more water molecules per site are associated with hydrophobic as compared to hydrophilic sites. Under stress, water might then be conserved if hydrophobic surfaces were transformed into hydrophilic ones. Such has been shown to occur in the case of the amphiphile proline and is likely to occur with another amphiphile, betaine. By either mechanism, the hydration of the biopolymers and hence their structural integrity and activity would be preserved.

The complex, multifaceted role of compatible solutes no doubt explains why so few solutes are able to act in this way.

ACKNOWLEDGMENT. I thank my colleague Elizabeth Luard for the help she has given in many ways, but particularly through discussion and by allowing me to refer to her unpublished experimental data.

References

Acker, L. W., 1962, Enzymic reactions in foods of low moisture content, *Adv. Food. Res.* **11**:263–330.

Acker, L. W., 1969, Water activity and enzyme activity, *Food Technol. (Chicago)* **23**:27–40.

Acock, B., 1975, An equilibrium model of leaf water potentials which separates intra- and extra-cellular potentials, *Aust. J. Plant Physiol.* **2**:253–263.

Alemohammad, M. M., and Knowles, C. J., 1974, Osmotically induced volume and turbidity changes of *Escherichia coli* due to salts, sucrose and glycerol, with particular reference to the rapid permeation of glycerol into the cell, *J. Gen. Microbiol.* **82**:125–142.

Allen, R. N., and Newhook, F. J., 1973, Chemotaxis of zoospores of *Phytophthora cinnamomi* to ethanol in capillaries of spore dimensions, *Trans. Br. Mycol. Soc.* **61**:287–302.

Anand, J. C., and Brown, A. D., 1968, Growth rate patterns of the so-called osmophilic and non-osmophilic yeasts in solutions of polyethylene glycol, *J. Gen. Microbiol.* **52:**205–212.

Aslyng, H. C., 1963, Soil physics terminology, *Int. Soc. Soil Sci. Bull.* **23:**1–4.

Avron, M., and Ben-Amotz, A., 1979, Metabolic adaptation of the alga *Dunaliella* to low water activity, in: *Strategies of Microbial Life in Extreme Environments* (M. Shilo, ed.), pp. 83–91, Verlag Chemie, Weinheim.

Ayerst, G., 1968, Prevention of biodeterioration by control of environmental conditions, in: *Biodeterioration of Materials* (A. H. Walters and J. J. Elphick, eds.), pp. 223–241, Elsevier, Amsterdam.

Ayerst, G., 1969, The effects of water activity and temperature on spore germination and growth in some mould fungi, *J. Stored Prod. Res.* **5:**127–141.

Ayres, P. G., 1978, Water relations of diseased plants, in: *Water and Plant Disease* (T. T. Kozlowski, ed.), *Water Deficits and Plant Growth,* Vol. 5, pp. 1–60, Academic Press, New York.

Baker, K. F., and Cook, R. J., 1974, *Biological Control of Plant Pathogens,* W. H. Freeman and Company, San Francisco.

Bayley, S. T., and Morton, R. A., 1979, Biochemical evolution of halobacteria, in: *Strategies of Microbial Life in Extreme Environments* (M. Shilo, ed.), pp. 109–124, Verlag Chemie, Weinheim.

Brock, T. D., 1979, Ecology of saline lakes, in: *Strategies of Microbial Life in Extreme Environments* (M. Shilo, ed.), pp. 29–47, Verlag Chemie, Weinheim.

Brown, A. D., 1976, Microbial water stress, *Bacteriol. Rev.* **40:**803–846.

Brown, A. D., 1978, Compatible solutes and extreme water stress in eukaryotic micro-organisms, *Adv. Microb. Physiol.* **17:**181–242.

Brown, A. D., 1979, Physiological problems of water stress, in: *Strategies of Microbial Life in Extreme Environments* (M. Shilo, ed.), pp. 65–81, Verlag Chemie, Weinheim.

Brown, L. M., and Hellebust, J. A., 1978, Sorbitol and proline as intracellular osmotic solutes in the green alga *Stichococcus bacillaris, Can. J. Bot.* **56:**676–679.

Brown, R. W., and van Haveren, B. P., 1972, *Psychrometry in Water Relations Research* (eds.), Utah Agricultural Experiment Station, Logan, Utah.

Bull, H. B., and Breese, K., 1968, Protein hydration, I. Binding sites, *Arch. Biochem. Biophys.* **128:**488–496.

Bull, H. B., and Breese, K., 1970, Water and solute binding by proteins, I. Electrolytes, *Arch. Biochem. Biophys.* **137:**299–305.

Burnett, J. H., 1976, *Fundamentals of Mycology* (2nd ed.), Edward Arnold, London.

Byrne, P., and Jones, E. B. G., 1975, Effect of salinity on spore germination of terrestrial and marine fungi, *Trans. Br. Mycol. Soc.* **64:**497–503.

Chen, A. W., and Griffin, D. M., 1966, Soil physical factors and the ecology of fungi, V. Further studies in relatively dry soils, *Trans. Br. Mycol. Soc.* **49:**419–426.

Cook, R. J., 1973, Influence of low plant and soil water potentials on diseases caused by soilborne fungi, *Phytopathology* **63:**451–458.

Cook, R. J., and Duniway, J. M., 1981, Water relations in the life–cycles of soil borne plant pathogens, in: *Water Potential Relations in Soil Microbiology* (J. F. Parr, W. R. Gardner, and L. F. Elliott, eds.), pp. 119–139, Soil Science Society of America, Special Publication No. 9, Madison, Wisconsin.

Cook, R. J., and Papendick, R. I., 1972, Influence of water potential of soils and plants on root disease, *Annu. Rev. Phytopathol.* **10:**349–374.

Cooke, R., and Kuntz, I. D., 1974, The properties of water in biological systems, *Annu. Rev. Biophys. Bioeng.* **3:**95–126.

Cother, E. J., and Griffin, D. M., 1974, Chlamydospore germination in *Phytophthora drechsleri, Trans. Br. Mycol. Soc.* **63:**273–279.

Duniway, J. M., 1976, Movement of zoospores of *Phytophthora cryptogea* in soils of various textures and matric potentials, *Phytopathology* **66**:877–882.

Duniway, J. M., 1979, Water relations of water molds, *Annu. Rev. Phytopathol.* **17**:431–460.

Edgley, M., and Brown, A. D., 1978, Response of xerotolerant and nontolerant yeasts to water stress, *J. Gen. Microbiol.* **104**:343–345.

Fisher, L. R., and Israelachvili, J. N., 1979, Direct experimental verification of the Kelvin equation for capillary condensation, *Nature (London)* **277**:548–549.

Gerhardt, P., and Judge, J. A., 1964, Porosity of isolated cell walls of *Saccharomyces cerevisiae* and *Bacillus megaterium, J. Bacteriol.* **87**:945–951.

Gould, G. W., and Measures, J. C., 1977, Water relations of single cells, *Philos. Trans. R. Soc. London Ser. B:* **278**:151–166.

Griffin, D. M., 1963, Soil moisture and the ecology of fungi, *Biol. Rev. Cambridge Philos. Soc.* **38**:141–166.

Griffin, D. M., 1968, A theoretical study relating to concentration and diffusion of oxygen to the biology of organisms in soil, *New Phytol.* **67**:561–577.

Griffin, D. M., 1972, *Ecology of Soil Fungi,* Chapman and Hall, London.

Griffin, D. M., 1977, Water potential and wood-decay fungi, *Annu. Rev. Phytophathol.* **15**:319–329.

Griffin, D. M., 1978, Effect of soil moisture on survival and spread of pathogens, in: *Water and Plant Disease* (T. T. Kozlowski, ed.), *Water Deficits and Plant Growth,* Vol. 5, pp. 175–197, Academic Press, New York.

Griffin, D. M., 1981, Water potential as a selective factor in the microbial ecology of soils, in: *Water Potential Relations in Soil Microbiology* (J. F. Parr, W. R. Gardner, and L. F. Elliott, eds.), pp. 141–151, Soil Science Society of America, Special Publication No. 9, Madison, Wisconsin.

Griffin, D. M., and Luard, E. J., 1979, Water stress and microbial ecology, in: *Strategies of Microbial Life in Extreme Environments* (M. Shilo, ed.), pp. 49–63, Verlag Chemie, Weinheim.

Gustafsson, L., 1979, The ATP pool in relation to the production of glycerol and heat during growth of the halotolerant yeast *Debaryomyces hansenii, Arch. Microbiol.* **120**:15–23.

Gustafsson, L., and Norkrans, B., 1976, On the mechanisms of salt tolerance. Production of glycerol and heat during growth of *Debaryomyces hansenii, Arch. Microbiol.* **110**:177–183.

Gutknecht, J., Hastings, D. F., and Bisson, M. A., 1978, Ion transport and turgor pressure regulation in giant algal cells, in: *Transport across Multi-Membrane Systems* (G. Giesbisch, D. C. Tosteson and H. H. Ussing, eds.), *Membrane Transport in Biology,* Vol. 3, pp. 125–174, Springer Verlag, Berlin.

Harris, R. F., 1981, Effect of water potential on microbial growth and metabolism, in: *Water Potential Relations in Soil Microbiology* (J. F. Parr, W. R. Gardner, and L. F. Elliott, eds.), pp. 23–95, Soil Science Society of America, Special Publication No. 9, Madison, Wisconsin.

Harris, R. F., Gardner, W. R., Adebayo, A. A., and Sommers, L. E., 1970, Agar dish isopiestic equilibration method for controlling the water potential of solid substrates, *Appl. Microbiol.* **19**:536–537.

Harrower, K. M., and Nagy, L. A., 1979, Effects of nutrients and water stress on growth and sporulation of coprophilous fungi, *Trans. Br. Mycol. Soc.* **72**:459–462.

Hellebust, J. A., 1976, Osmoregulation, *Annu. Rev. Plant Physiol.* **27**:485–505.

Hocking, A. D., and Pitt, J. I., 1979, Water relations of some *Penicillium* species at 25°C, *Trans. Br. Mycol. Soc.* **73**:141–145.

Hüsken, D., Steudle, E., and Zimmermann, U., 1978, Pressure probe technique for measuring water relations of cells in higher plants, *Plant Physiol.* **61**:158–163.

Jennings, D. H., 1973, Cations and filamentous fungi: Invasion of the sea and hyphal functioning, in: *Ion Transport in Plants* (W. P. Anderson, ed.), pp. 323–335, Academic Press, London.

Jennings, D. H., 1979, Membrane transport and hyphal growth, in: *Fungal Walls and Hyphal Growth* (J. H. Burnett and A. P. J. Trinci, eds.), pp. 279–294, Cambridge University Press, Cambridge.

Jones, E. B. G., Byrne, P., and Alderman, D. J., 1971, The response of fungi to salinity, *Vie Milieu* (Suppl. No. 22), pp. 265–280.

Kauss, H., 1977, Biochemistry of osmotic regulation, in: *International Review of Biochemistry— Plant Biochemistry II,* Vol. 13 (D. H. Northcote, ed.), pp. 119–140, University Park Press, Baltimore.

Kirst, G. O., 1977, Ion composition of unicellular marine and fresh-water algae, with special reference to *Platymonas subcordiformis* cultivated in media with different osmotic strengths, *Oecologia (Berlin)* 28:177–189.

Klotz, I. M., 1950, *Chemical Thermodynamics,* Prentice-Hall, Englewood Cliffs, New Jersey.

Kuntz, I. D., and Kauzmann, W., 1974, Hydration of proteins and polypeptides, *Adv. Protein Chem.* 28:239–345.

Kushner, D. J., 1968, Halophilic bacteria, *Adv. Appl. Microbiol.* 10:73–99.

Lanyi, J. K., 1979, Physicochemical aspects of salt-dependence in halobacteria, in: *Strategies of Microbial Life in Extreme Environments* (M. Shilo, ed.), pp. 93–107, Verlag Chemie, Weinheim.

Larsen, H., 1967, Biochemical aspects of extreme halophilism, *Adv. Microb. Physiol.* 1:97–132.

Lewis, D. H., and Smith, D. C., 1967, Sugar alcohols (polyols) in fungi and green plants, I. Distribution, physiology and metabolism, *New Phytol.* 66:143–184.

Lewis, G. N., and Randall, M., 1961, *Thermodynamics* (revised by K. S. Pitzer and L. Brewer, 2nd ed.), McGraw-Hill, New York.

Liu, M. S., and Hellebust, J. A., 1976, Effects of salinity and osmolarity of the medium on amino acid metabolism in *Cyclotella cryptica, Can. J. Bot.* 54:938–948.

Luard, E. J., and Griffin, D. M., 1981, Effect of water potential on fungal growth and turgor, *Trans. Br. Mycol. Soc.* 76: 33–40.

Marshall, T. J., 1959, *Relations between Water and Soil,* Commonwealth Bureau Agricultural Bureaux, Farnham Royal.

Measures, J. C., 1975, Role of amino acids in osmoregulation of non-halophilic bacteria, *Nature (London)* 257:398–400.

Mitchell, P., and Moyle, J., 1956, Osmotic structure and function in bacteria, *Symp. Soc. Gen. Microbiol.* 6:150–180.

Moustafa, A. F., 1975, Osmophilous fungi in the salt marshes of Kuwait, *Can. J. Microbiol.* 21:1573–1580.

Noy-Meir, I., and Ginzburg, B. Z., 1967, An analysis of the water potential isotherm in plant tissue, I. The theory, *Aust. J. Biol. Sci.* 20:695–721.

Onishi, H., 1963, Osmophilic yeasts, *Adv. Food Res.* 12:53–94.

Papendick, R. I., and Campbell, G. S., 1975, Water potential in the rhizosphere and plant and methods of measurement and experimental control, in: *Biology and Control of Soil-borne Plant Pathogens* (G. W. Bruehl, ed.), pp. 39–49, American Phytopathological Society, St. Paul, Minnesota.

Papendick, R. I., and Campbell, G. S., 1981, Theory and measurement of water potential, in: *Water Potential Relations in Soil Microbiology* (J. F. Parr, W. R. Gardner, and L. F. Elliott, eds.), pp. 1–22, Soil Science Society of America, Special Publication No. 9, Madison, Wisconsin.

Pitt, J. I., 1975, Xerophilic fungi and the spoilage of foods of plant origin, in: *Water Relations of Foods* (R. B. Duckworth, ed.), pp. 273–307, Academic Press, London.

Pitt, J. I., and Hocking, A. D., 1977, Influence of solute and hydrogen ion concentration on the water relations of some xerophilic fungi, *J. Gen. Microbiol.* 101:35–40.

Richter, H., 1976, The water status in the plant—experimental evidence, in: *Water and Plant Life* (O. L. Lange, L. Kappen and E. D. Schulze, eds.), pp. 42–58, Springer Verlag, Berlin.

Robinson, R. A., and Sinclair, D. A., 1934, The activity coefficients of the alkali chlorides and of lithium iodide in aqueous solution from vapour pressure measurements, *J. Am. Chem. Soc.* **56:**1830–1835.

Robinson, R. A., and Stokes, R. H., 1968, *Electrolyte Solutions* (2nd ed.), Academic Press, New York.

Rose, C. W., 1966, *Agricultural Physics,* Pergamon Press, Oxford.

Schneider, M. J., and Schneider, A. S., 1972, Water in biological membranes: Adsorption isotherms and circular dichroism as a function of hydration, *J. Membr. Biol.* **9:**127–140.

Schobert, B., 1977, Is there an osmotic regulatory mechanism in algae and higher plants? *J. Theor. Biol.* **68:**17–26.

Schobert, B., and Tschesche, H., 1978, Unusual solution properties of proline and its interaction with proteins, *Biochim. Biophys. Acta* **541:**270–277.

Schoeneweiss, D. F., 1978, Water stress as a predisposing factor in plant disease, in: *Water and Plant Disease* (T. T. Kozlowski, ed.), *Water Deficits and Plant Growth,* Vol. 5, pp. 61–99, Academic Press, New York.

Scott, W. J., 1957, Water relations of food spoilage microorganisms, *Adv. Food Res.* **7:**83–127.

Shepherd, W., 1973, A simple thermocouple psychrometer for determining tissue water potential and some observed leaf-maturity effects, *J. Exp. Bot.* **24:**1003–1013.

Shepherd, W., 1975, Matric water potential of leaf tissue—measurement and significance, *J. Exp. Bot.* **26:**465–468.

Shilo, M., 1979 (ed.), *Strategies of Microbial Life in Extreme Environments,* Verlag Chemie, Weinheim.

Slatyer, R. O., 1967, *Plant–Water Relationships,* Academic Press, London.

Spanner, D. C., 1964, *Introduction to Thermodynamics,* Academic Press, London.

Spanner, D. C., 1972, Plants, water, and some other topics, in: *Psychrometry in Water Relations Research* (R. W. Brown and B. P. van Haveren, eds.), pp. 29–39, Utah Agricultural Experiment Station, Logan, Utah.

Todd, G. W., 1972, Water deficits and enzymatic activity, in: *Plant Responses and Control of Water Balance* (T. T. Kozlowski, ed.), *Water Deficits and Plant Growth,* Vol. 3, pp. 177–216, Academic Press, New York.

Tresner, H. D., and Hayes, J. A., 1971, Sodium chloride tolerance of terrestrial fungi, *Appl. Microbiol.* **22:**210–213.

Trevithick, J. R., and Metzenberg, R. L., 1966, Genetic alteration of pore size and other properties of the *Neurospora* cell wall, *J. Bacteriol.* **92:**1016–1020.

Troller, J. A., 1977, Statistical analysis of a_w measurements obtained with a Sina scope, *J. Food Sci.* **42:**86–90.

Unemoto, T., and Hayashi, M., 1979, Regulation of internal solute concentrations of marine *Vibrio alginolyticus* in response to external NaCl concentrations, *Can. J. Microbiol.* **25:**922–926.

Warren Wilson, J., 1967, The components of leaf water potential, I. Osmotic and matric potentials, *Aust. J. Biol. Sci.* **20:**329–347.

Wearing, A. H., and Burgess, L. W., 1979, Water potential and the saprophytic growth of *Fusarium roseum "Graminearum,"* *Soil Biol. Biochem.* **11:**661–667.

Wilson, J. M., and Griffin, D. M., 1975, Respiration and radial growth of soil fungi at two osmotic potentials, *Soil Biol. Biochem.* **7:**269–274.

Wilson, J. M., and Griffin, D. M., 1979, The effect of water potential on the growth of some soil basidiomycetes, *Soil Biol. Biochem.* **11:**211–212.

Wong, P. T. W., and Griffin, D. M., 1974, Effect of osmotic potentials on streptomycete growth, antibiotic production and antagonism to fungi, *Soil Biol. Biochem.* **6:**319–325.

Wong, P. T. W., and Griffin, D. M., 1976a, Bacterial movement at high matric potentials, I. In artificial and natural soils, *Soil Biol. Biochem.* **8:**215–218.

Wong, P. T. W., and Griffin, D. M., 1976b, Bacterial movement at high matric potentials, II. In fungal colonies, *Soil Biol. Biochem.* **8:**219–223.

Young, B. R., Newhook, F. J., and Allen, R. N., 1979, Motility and chemotactic response of *Phytophthora cinnamomi* zoospores in an "ideal soil," *Trans. Br. Mycol. Soc.* **72:**395–401.

Zimmermann, U., 1978, Physics of turgor- and osmoregulation, *Annu. Rev. Plant Physiol.* **29:**121–148.

Zimmermann, U., and Steudle, E., 1978, Physical aspects of water relations of plant cells, *Adv. Bot. Res.* **6:**45–117.

4

Ecology of Mycorrhizae and Mycorrhizal Fungi

B. MOSSE, D. P. STRIBLEY, AND F. LeTACON

1. Introduction

It is normal for the roots of most plant species, including Angiosperms, Gymnosperms, Pteridophytes, and some Thallophytes, to form mutualistic associations with different groups of soil fungi. Charged to investigate the culture of truffles in the Kingdom of Prussia, Frank (1885) instead discovered and described such mutualistic associations in temperate forest trees and named them *mycorrhiza* (fungus root). He believed that the fungi performed the function of root hairs, which were lacking in these much modified dual structures.

The term *mycorrhiza* has since been adopted for many other mutualistic associations between soil fungi and below-ground plant organs, some of which, notably the protocorms of orchids, are not even roots. The only common feature of these diverse, sometimes highly specialized, mycorrhizal associations is the low degree of pathogenicity exhibited by the fungus towards its host. In the vesicular-arbuscular mycorrhiza—the most widespread and probably the least specialized of all mycorrhizal association—this has led to complete dependence of the fungus on a living host and inability of the fungus to grow on synthetic media. Conversely, some hosts lacking chlorophyll depend largely on their associated fungi for reproduction and growth.

B. MOSSE • Formerly in Department of Soil Microbiology, Rothamsted Experimental Station, Harpenden, Herts, England (Retired). **D. P. STRIBLEY** • Department of Soils and Plant Nutrition, Rothamsted Experimental Station, Harpenden, Herts, England. **F. LeTACON** • Institut National de la Recherche Agronomique CNRF, Champenoux 54280, Seichamps, France.

It is probable that the mycorrhizal habit has a long evolutionary history (Nicolson, 1975), possibly associated with the evolution of land plants (Pirozynski and Malloch, 1975). Lewis (1973) considers that an evolution towards a biotrophic nutrition has occurred in all the fungi capable of forming mycorrhizal associations. He attributes this to parallel or convergent evolutionary trends in fungi of diverse phylogeny rather than to a common ancestor. The higher plants can generally live without their fungal associate if the nutritional status of the environment is suitably adjusted, but in undisturbed ecosystems they rarely do so. The fungi on the other hand are rarely free-living in the soil.

The diversity of mycorrhizal structures applies not only to the plants and fungi involved, but also to the structural and functional details of the association. In this bewildering assembly of disparate, often only superficially connected phenomena loosely called "mycorrhizal associations," Harley's (1969) book *The Biology of Mycorrhiza* has been a beacon of factual information to specialists as well as ecologists, soil microbiologists, and others with peripheral interest in the subject. To create some order, mycorrhizal associations are usually divided into four main groups: ectotrophic or sheathing mycorrhizae occurring chiefly in temperate tree species and in Eucalypts; vesicular-arbuscular (VA) mycorrhizae occurring in most annuals, herbaceous perennials, temperate and tropical fruit trees, and most tropical timber trees; mycorrhizae of the Ericales; and mycorrhizae of the Orchidaceae. A range of other "mycorrhizal" associations have been described, notably in the Pteridophytes (Boullard, 1958) and Hepaticae (Stahl, 1949), but there is little information on these beyond their description.

With the exception of certain families and some particular species and habitats, nearly all plants are mycorrhizal in natural ecosystems, though the extent of infection may vary. It is only when man has made introductions of plant species not indigenous to particular regions, or created soil disturbances by such practices as fumigation or mining and agricultural operations, that conditions arise where the necessary mycorrhizal fungi may be lacking. Although a few mycorrhizal associations are highly specific, the majority are not, and usually a range of fungal species can act as mycorrhizal associates of a particular host. To what extent the availability of mycorrhizal fungi may have determined the worldwide distribution of plant species is difficult to assess. On the whole, it seems more probable that this is chiefly determined by climatic and edaphic factors affecting the host. Appropriate mycorrhizal fungi must therefore have arrived simultaneously with the plant species.

Infection is never systemic or seed transmitted, and there are no mixed propagules as in the lichens. Furthermore, many mycorrhizal fungi, except in the Orchidaceae, have poor saprophytic ability or are even obligate symbionts. Methods of overcoming this handicap have varied with the different kinds of mycorrhizal associations, and we have given some consideration to this aspect of mycorrhizal associations in the current review.

Finally, it is clear that in the large majority of cases the host plant determines which of the main types of mycorrhizal association shall be formed. A few plant species, like *Corylus* and *Populus,* can form either ecto- or VA mycorrhiza seemingly simultaneously on the same plant; even fewer form one after the other as the plant matures. But as a working hypothesis, it can be assumed that the host determines its mycorrhizal associate. Some families, like Pinaceae and Fagaceae, are exclusively ectomycorrhizal, others like the Rosaceae are exclusively vesicular-arbuscular, and still others like Leguminosae can be either but are mostly vesicular-arbuscular. Even now, the reasons for this selection are quite unknown. The result, however, is that the large-scale distribution of the different types of mycorrhizae, and therefore to some extent that of their associated fungi, is determined by the distribution of the host plant. This aspect is further discussed below.

The ecto- and VA mycorrhizal associations are the most widespread and include the large majority of economically interesting plants. But mycorrhizal studies have in the past been rather neglected because many of the fungi can not be grown on synthetic media, and this, together with the general inaccessibility of roots, has greatly restricted their study, particularly in natural ecosystems. As a result, their significance is such systems still remains largely a matter of conjecture. Available information is fragmentary and is often based on pot experiments in sterilized soils with doubtful applicability to plants in the field. Yet practical ways of exploiting mycorrhizal systems for improved crop production are being sought, and the implications of this and possible means by which such efforts might be furthered, or impeded, are also discussed in this review.

2. The Hosts: Types of Mycorrhizal Structures and Large-Scale Distribution of Hosts

Because it is much easier to say something precise about the distribution of the two relatively uncommon types, the usual order will be reversed, and the mycorrhizae of orchids and Ericales will be considered first.

2.1. Mycorrhizae of Orchidaceae

Mycorrhizae formed by this family involve a variety of organs in which there is extensive and repeated penetration of cells by septate hyphae that often bear clamp connections. The hyphae form intracellular loops and coils known as *pelotons*. Degradation of the intracellular fungal structures by host reaction is a characteristic feature. Host plants include a range of types, from species containing abundant chlorophyll at maturity to completely achlorophyllous and presumably saprophytic forms.

Orchid mycorrhizae are formed at a very early stage of plant development. Bernard (1909) showed that the embryo resulting from germination of the characteristically minute seeds is very soon colonized by a mycorrhizal fungus. Subsequently, an achlorophyllous protocorm develops in which the fungus becomes confined to the cortical cells. The protocorm may either be killed by the fungus or form a stable relation with it (Harvais and Hadley, 1967), the frequency of parasitism being determined by the strain of fungus and the nutritional environment (Hadley, 1969). Frequent reinvasion of the protocorm may occur, but in those species which become autotrophic, the development of autotrophy seems to preclude further parasitic attack. In the adult saprophyte *Vanilla,* however, pathogenic and mutualistic phases of the mycorrhizal association have been observed adjacent to each other in the roots (Alconero, 1969).

In adult orchids, many different structures are infected. Roots of all types of orchids are infected, as are underground rhizomes in complete saprophytes such as *Corallorhiza.* Tubers of the Ophrydeae are commonly regarded as lacking infection, but Warcup (1971) has reported the occurrence of *Tulasnella calospora* in tubers of several species of *Diurus.* Montfort and Küsters (1940) arranged the orchids in a series of forms of increasing saprotrophy and showed that autotrophic forms tended to be lightly infected at maturity, with the mycorrhizae restricted to roots, whereas the complete saprophytes showed heavy infection in a variety of underground structures.

The Orchidaceae is a large and diverse family of some 12,000–30,000 species, and is most richly represented in the tropics where epiphytic forms are common (Sanford, 1974). Only a small fraction of the mycorrhizal associations have been studied, and it is not possible to generalize about them. Northern temperate forms such as *Dactylorchis, Goodyera,* and *Corallorhiza* have been extensively studied by Dr. G. Hadley and colleagues at Aberdeen University, U.K. They have found adult plants to be strongly mycorrhizal in most cases. Hadley and Williamson (1972) reported that infection in green terrestrial and epiphytic Malayan orchids was sparse and suggested that poor infection may be typical in the tropics. This does not apply to achlorophyllous orchids, which depend on their mycorrhizae for a supply of carbon. Achlorophyllous angiosperms, including *Monotropa,* which bears arbutoid mycorrhizae, may generally be obligately mycorrhizal, and for a detailed discussion of this topic the reader is referred to the excellent review by Furman and Trappe (1971).

2.2. Mycorrhizae of Ericales

Mycorrhizae of the plant order Ericales are characterized by a penetration of the root cortex by septate hyphae which form intracellular coils. They are divided into two morphological groups. In the *arbutoid* type, the mycorrhizae are of the ect-endo type, in which there is an organized fungal sheath as well as inter- and intracellular penetration of the cortex (Lutz and Sjolund, 1973). The hosts of this type include autotrophic plants (the Arbutae section of the Ericaceae, e.g., *Arbutus* and *Arctostaphylos*), partial saprophytes (Pyrolaceae), and complete saprophytes (Monotropaceae). In the second group, the *ericoid* mycorrhizae, there is no distinct sheath, but there may be considerable external mycelium. The fungus forms loops and coils within cells which appear to be degraded by the host (Bonfante-Fasolo and Gianinazzi-Pearson, 1979), and there is little lateral spread within the cortex. Host plants include family members of the Ericaceae (except the Arbutae section), Empetraceae, and Epacridaceae, and the important genera are *Calluna, Erica, Vaccinium,* and *Epacris.*

Ericoid mycorrhizae are formed only by the "hair-roots" typical of their hosts. These, the ultimate branches of the root system, lack root hairs, possess only 1–3 layers of cortical cells, and are only about 20–50 μm in diameter.

The hosts of arbutoid mycorrhizae have very diverse mycorrhizal structures. The root system of *Arbutus* trees is differentiated into long and short roots, and only the latter are extensively infected. *Arctostaphylos* has undifferentiated and evenly infected root systems, and finally the complete saprophyte *Monotropa* has short, much-branched absorbing organs which may be heavily mycorrhizal.

The hosts of ericoid mycorrhizae typically occur as dominant and codominant members of heathland communities. Gimingham (1972) has pointed out that heathlands occur where conditions favor dominance of a dwarf-shrub type vegetation. These conditions are summarized in Table I together with reports

Table I. Distribution of Ericoid Mycorrhizae[a]

Conditions favoring heathland vegetation	Geographical location	Reports of ericoid mycorrhizae
Forest excluded in oceanic regions of north temperate belt	Western Europe	Pearson and Read (1975a), Burgeff (1961), Vegh *et al.* (1979), Largent *et al.* (1980b)
High altitudes with adequate humidity	South Africa (*Erica*-dominated communities)	Robinson (1973), Read (1978)
	Austrian Tyrol	Haselwandter (1979)
	Pakistan	Khan (1972)
	Australia (Epacridaceae-dominated)	McLennan (1935), Burges (1936)
Subarctic	Alaskan tundra	O. K. Miller (1981)

[a]Partly after Gimingham (1972).

of ericoid mycorrhizae from each environment. Ericoid mycorrhizae are clearly very widespread and are found over large areas of the world's surface.

2.3. Ectotrophic Mycorrhizae

Ectotrophic mycorrhizae often cause marked changes in gross root morphology that serve as the basis for some classification systems. Essentially, the ectotrophic mycorrhizal root consists of a more or less well-developed fungal sheath or mantle surrounding the root proper. On the inside, the mantle is connected to a *Hartig net*—a network of hyphae extending between the first layer of epidermal cells and sometimes up to the endodermis. On the outside, the sheath is connected to an external network of hyphae that extends into the soil, sometimes several centimeters from the root surface. The sheath may be smooth with few radiating hyphae, as in mycorrhizae formed by many species of *Lactarius* and *Russula* (Voiry, 1980), but more usually there are outgrowths which may be either simple hyphae, formed in particular by the fungi *Cenococcum graniforme* and *Hebeloma crustuliniforme,* or aggregates forming strands and highly specialized rhizomorphs. This external mycelium is often neglected because it is difficult to observe and easily destroyed when roots are taken from the soil.

In general, ectotrophic mycorrhizae are formed only on roots with primary tissues, and secondary thickened roots remain uninfected. Mycorrhizal tree roots often show a division, most pronounced in *Pinus* spp., into long roots that grow continuously, and short roots that branch repeatedly and grow little. Mycorrhizae are chiefly formed on the short roots but the long roots can also by colonized, and here a loose sheath, lacking a Hartig net, may be formed (Harley, 1969). It is still controversial whether formation of short roots precedes, or is a result of, infection.

Ectotrophic mycorrhizae have their greatest host range among temperate forest trees. A few alpine herbaceous plants—species of *Dryas, Helianthemum, Salix, Polygonum,* and *Kobresia*—form ectomycorrhiza with *Cenococcum graniforme* at high altitudes (Haselwandter and Read, 1979), and Warcup (1980) found many Australian nonleguminous herbs, members of Fabaceae, to be ectomycorrhizal.

Boreal and high-elevation forests in the northern hemisphere are exclusively ectomycorrhizal, as are also forests on very poor acid soils (podzols) containing such species as pine, spruce, birch, oak, and beech. Only at lower altitudes, further south, on mull soils below pH 5, can such VA mycorrhizal species as *Acer, Taxus, Sorbus,* and others compete with the ectomycorrhizal species. Eucalypt species, with a very wide distribution in Australia, are also ectomycorrhizal.

There is some uncertainty about the mycorrhizal status of trees in the tropics. While most Nigerian timber trees, including mahogany and also

Araucaria, form VA mycorrhiza (Redhead, 1968), St. John (1979) reported that many trees in the Amazonian basin were nonmycorrhizal. Singer (1979) noted that ectomycorrhizal trees occurred not only at high elevations in the tropics but also in periodically inundated forests and white sand podzols of the Amazonian basin. He connected occurrence of ectotrophic forests in this region with a reduction in litter decomposing fungi. Ectomycorrhizae occur habitually in the important subtropical and tropical tree families Dipterocarpaceae and Caesalpinioideae, a subfamily of the Leguminosae (Alexander, 1979).

2.4. VA Mycorrhizae

VA mycorrhizae (VAM) consist of roots little changed in gross root morphology and are in fact not recognizable without staining, except for an occasional yellow tinge that occurs in some host species. The primary cortex is invaded both inter- and intracellularly by aseptate fungi, though occasional irregular septa may occur. As in the orchidaceous mycorrhizae, the intracellular stage (arbuscule) is short-lived and disintegrates while the host cell maintains its integrity. As in the ectotrophic mycorrhizae, the root is organically connected to hyphae ramifying on and around the root surface and extending some centimeters into the surrounding soil.

VAM are usually confined to the primary cortex of roots. They are not found in such underground overwintering structures as bulbs (Daft *et al.,* 1980), tubers, or stolons, but they have been reported in such diverse structures as the modified leaves of the water fern *Salvinia cucullata* (Bagyaraj *et al.,* 1979), gametophytes (Rabatin, 1980) and rhizoids (O. K. Miller and Laursen, 1978) of mosses, and the fruiting peg of the peanut (Graw and Rehm, 1977). They are generally excluded from nitrogen-fixing nodules and proteoid roots (Lamont, 1972).

VAM have the widest host range and distribution of all the mycorrhizal associations. Briefly, they occur in all the plant species not infected by any of the other types, although it does not follow that every specimen of every species will show infection, or that levels of infection will necessarily be the same in all species and under different environmental conditions. Hosts include the economically important families Rosaceae, Gramineae, and Leguminosae, all grain crops, all temperate fruit trees and shrubs, and also citrus, coffee, tea, rubber, and sugar cane among others. They also occur in the roots of ferns (Cooper, 1976). Families essentially immune to VA infection are Cruciferae, Chenopodiaceae, Proteaceae, and most species of *Carex*. Even among these families, regularly infected species can occur; e.g., *Atriplex canescens* (Williams *et al.,* 1974) and *Lunaria annua* (Medve, 1979). Also, much-attenuated forms of infection can sometimes be induced in essentially immune species in the presence of a normal host (see Section 6.2.2). Occasionally, immune species may occur within a normally susceptible family; e.g., *Lupinus consentinii*

(Trinick, 1977), which has a thick fringe of long root hairs resembling the proteoid roots of members of the Proteaceae that are also immune. The only habitats free from infection were once thought to be permanently inundated sites. Recently, however, infection has been reported in some temperate aquatic species lacking root hairs (Søndergaard and Laegaard, 1977) and in some tropical aquatic plants lacking true roots (Bagyaraj *et al.,* 1979).

3. The Fungi

3.1. Classification

Many mycorrhizal fungi, notably VA endophytes, do not grow on synthetic media, and many ectomycorrhizal fungi are difficult to isolate. Still others, notably among ericaceous and orchidaceous endophytes, are known only in their imperfect forms. To meet practical needs, some systems of classification have therefore been based on descriptive, rather than orthodox taxonomic criteria. Ectomycorrhizae have been classified according to gross root morphology, branching pattern, color, and surface structure of the mycorrhizal root. But these classifications do not necessarily give any indication of the fungi involved. VA endophytes have been classified according to morphological characteristics of the resting spores formed in the soil. Ericaceous mycorrhizae are divided into two groups according to the type of infection (arbutoid or ericoid), and orchidaceous mycorrhizal fungi are classified according to orthodox systems of fungal classification.

Unless the purpose of such unorthodox classification systems is clearly defined, these systems run the risk of becoming unmanageable because of the proliferation of new mycorrhizal types and species. Three objectives of a mycorrhizal classification system might be: (a) identification of fungi by morphological or anatomical characteristics of the mycorrhizal structure to which they give rise; (b) correlation of mycorrhizal characteristics with symbiotic effectiveness (for instance, yellow mycorrhizal roots of *Abies balsamea* took up 3, 10, and 16 times more phosphate than brown or black ones as reported by Langlois and Fortin, 1978); or simply (c) ease of communication. As will be seen, none of these objectives is particularly well served by some of the systems currently used, perhaps because of insurmountable difficulties in the classification of relatively unspecifically associated dual organisms that require the soil as an inseparable third component for their symbiotic performance.

3.1.1. Ectotrophic Fungi

Two systems of classification (Björkman, 1941; Harley, 1969) are based on the gross morphology of the root, and one (Dominik, 1969), in the form of a binomial key, is based on the anatomy of the root surface in transverse section.

The main morphological types of beech mycorrhizae (Harley, 1969) are listed in Table II. In addition to these types, typically black mycorrhizae with stiff bristle-like hyphae projecting from a sheath are formed by *Cenococcum graniforme*. Björkman's classification, which was designed mainly for *Pinus* spp., also contains this type as a distinct form. Some typical surface structures of ectomycorrhizae described by Dominik are shown in Fig. 1, as well as three types of cystidia diagnostic for different fungal species. But the surface structure of mycorrhizal roots is diagnostic for only a small number of species.

Table III shows the main groups and genera of fungi that form ectomycorrhizae. Most occur in the order Agaricales. Lycoperdales contain fewer ectomycorrhizal species, but some are very important. Most ascomycetes that are ectomycorrhizal form hypogeous sporophores and are relatively poorly known. *Elaphomyces* and the imperfect fungus *Cenococcum graniforme*, both in the Eurotiales, form ectomycorrhizae with many tree species. The genera *Terfezia* and *Tirmania* form ecto- or endomycorrhizae with Cistaceae (e.g., *Helianthemum* and *Cistus* spp.), and *Terfezia* produces the desert truffle, which was probably the "manna" of the Bible. The most famous of the Tuberaceae is *Tuber melanosporum*, the black truffle of Perigord, which forms mycorrhizal associations with *Quercus lanuginosus, Q. petraea, Q. ilex, Q. coccifera, Corylus avellana*, and probably other deciduous trees; e.g., *Carpinus* and *Betulus* spp., *Fagus sylvatica*, etc. *Tuber melanosporum* is also associated with *Pinus nigra, P. halepensis, Picea abies*, and others and possibly with *Helianthemum*. Only one phycomycete genus, *Endogone*, forms ectomycorrhizal associations.

Table II. Types of Mycorrhizal Roots in Beech[a]

1. *Uninfected roots*
2. *Superficial ectotrophic mycorrhiza*. These roots are morphologically similar to uninfected roots except that an easily separated fungal sheath encloses the apex.
3. *The diffuse infected system*. This system consists of lateral root branches composed of fully infected roots and uninfected or superficially infected roots.
4. *The pyramidal infected system*. The lateral roots take the form of a Christmas tree. Uninfected apices are very rare.
5. *The coralloid infection*. The branches are less regularly arranged and the main axes show secondary thickening.
6. *The nodular infection*. The racemosely branched mycorrhizal axes are bound together in a nodule by mycelium.
7. *Apically infected roots*. In this system, the roots are infected at their apices only, so that the root axis shows the unusual structure of an uninfected root with root hairs.
8. *Loose weft infection*. Associated with roots of the last type may be found small lateral systems that branch racemosely and have an abundant weft of hyphae around them.

[a]After Harley (1969).

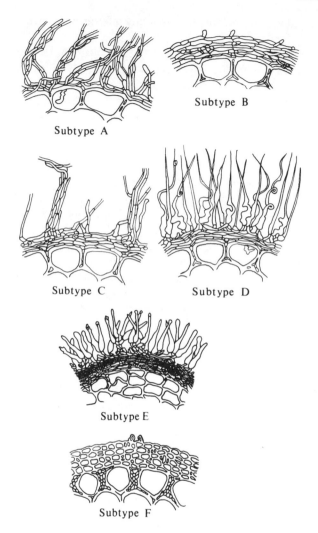

Figure 1. Some types of ectomycorrhizal structure (in transverse section) from the classification of Dominik (1969).

3.1.2. *VA Endophytes*

The identity of VA endophytes was very imperfectly known before the 1950s, although there was a consensus of opinion (Peyronel, 1923; Butler, 1939) that these fungi belonged to the Endogonaceae, a family in the Mucorales. The subdivision of the Endogonaceae into zygosporic, chlamydosporic, and sporangiosporic species according to spore structure within a usually sub-

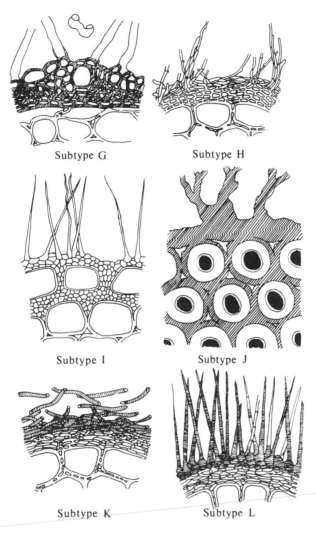

Subtype G   Subtype H

Subtype I   Subtype J

Subtype K   Subtype L

Figure 1. *(Cont.)*

terranean fruit body or sporocarp was generally accepted until a revision of the Endogonaceae was proposed by Gerdemann and Trappe (1974). This revision seemed appropriate because a range of *Endogone*-like spores had been discovered borne singly on the soil mycelium without an enclosing sporocarp, which, till then, had been regarded as an essential diagnostic feature of Endogonaceae. Other reasons for the revision are given by Gerdemann and Trappe (1975). A revision also seemed timely because of the growing interest shown in VA mycorrhizal associations and the need for communication between workers.

Table III. Genera of Ectomycorrhizal Fungi

Class	Order	Family	Genus
Basidiomycetes	Agaricales	Thelephoraceae	*Thelephora*
			Corticium
		Boletaceae	*Boletus*
		(Polysporaceae)	*Leccinum*
			Phlebopus
			Poria
			Suillus
			Xerocomus
		Agaricaceae	
		Leucosporeae	*Amanita*
			Cantharellus
			Clitocybe
			Hygrophorus
			Laccaria
			Lactarius
			Russula
			Tricholoma
		Ochrosporeae	*Cortinarius*
			Hebeloma
			Inocybe
			Paxillus
		Melanosporeae	*Gomphidius*
		Rhodosporeae	*Entoloma*
			Clitopilus
		Hydnaceae	*Hydnum*
	Lycoperdales	Lycoperdaceae	*Astraeus*
			Pisolithus
			Scleroderma
			Archangeliela
		Hymenogastraceae	*Hysterangium*
			Hymenogaster
			Melanogaster
			Rhizopogon
		Phallaceae	*Phallus*
Ascomycetes	Eurotiales	Elaphiomycetaceae	*Elaphomyces*
			Cenococcum
	Tuberales	Balsamiaceae	*Balsamia*
		Geneaceae	*Genea*
			Myrmecocystis
		Hydnotryaceae	*Barssia*
			Choiromyces
			Hydnotrya
		Terfeziaceae	*Mykagomyces*
			Picoa
			Terfezia
			Tirmania
		Tuberaceae	*Tuber*

Table III. *(Cont.)*

Class	Order	Family	Genus
	Helotiales	Geoglossaceae	*Circinans*
			Spathularia
	Pezizales	Hevellaceae	*Helvella*
		Otideaceae	*Otidea*
		Pyronemaceae	*Geospora*
		Rhizinaceae	*Geospora*
		Sarcoscyphaceae	*Sarcoscypha*
			Sarcosoma
Phycomycetes		Endogonaceae	*Endogone*

The revised classification of the Endogonaceae (Gerdemann and Trappe, 1974) proposes seven genera, of which four are known to form VAM. The type genus *Endogone*, reserved for the truly zygosporic species like *E. lactiflua* (Bucholtz, 1912), apparently does not form VAM but can form ectomycorrhizae with *Pinus* and *Pseudotsuga* spp. (Fassi and Palenzona, 1969). A more recently described species *E. eucalypti* (Warcup, 1975) forms ectomycorrhizae with species of *Eucalyptus*.

The four genera known to form VAM are *Glomus, Gigaspora, Acaulospora,* and *Sclerocystis*. They are distinguished by the morphology of their resting spores (see Fig. 2). *Gigaspora* has a single gametangialike suspensor, which leads to the assumption that the resting spore, termed an *azygospore*, may resemble a zygospore formed without prior sexual stage. *Acaulospora* resting spores are formed by migration of spore contents from a previously developed, short-lived mother spore (Mosse, 1970) and are usually sessile on the stalk of the mother spore. In *Sclerocystis*, a number of chlamydospores are arranged peripherally around a central plexus in a sporocarp. The genus *Glomus* contains both sporocarpic and nonsporocarpic species with chlamydospores borne on a more or less straight subtending hypha. Species are differentiated mainly by wall structure, morphology of secondary spores (vesicles), presence or

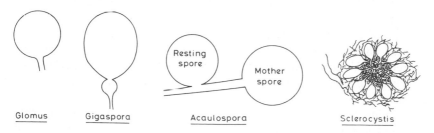

Figure 2. Spore morphology of four VA mycorrhizal genera of the Endogonaceae.

absence of sporocarps, size and color of spores, or a combination of these features, including also cytoplasmic structure and method of spore germination. Criteria employed in this classification are neither those of ontogeny nor related to any particular property of the fungi as mycorrhizal symbionts, with the possible exception of *Gigaspora* spp., which rarely form vesicles in the root. Predictably, a range of additional new species and genera have been described since 1974 with ever-increasing frequency (e.g., Tandy, 1975; Ames and Lindermann, 1976; Becker and Hall, 1976; Gerdemann and Bakshi, 1976; Becker and Gerdemann, 1977; Hall, 1977; Trappe, 1977a, 1979; Nicolson and Schenck, 1979; Rose *et al.,* 1978; Rothwell and Trappe, 1979; Rose and Daniels, 1979; Rose and Trappe, 1980), sometimes on very limited material and of questionable validity. Even two new genera, *Entrophospora* (Ames and Schneider, 1979a) and *Complexipes* (Walker, 1979), have been suggested. Spores of the single-type species *Complexipes moniliformis* have previously been described by Mosse and Bowen (1968), Wilcox *et al.* (1974), and Thomas and Jackson (1979). The fungus forms ect-endomycorrhiza with spruce and was thought to belong to the Ascomycetes. As the hyphae contain septal pores and Woronin bodies (G. Thomas, personal communication), the wisdom of this transfer seems very doubtful. Since the relative taxonomic importance of any particular morphological feature is unknown, it is very tempting to describe a new species on what may turn out to be small, relatively unimportant, and possibly unstable variations in structure. At least a description of the range of variation in single spore progenies should be mandatory before a new species description can be accepted. Species like *Glomus infrequens* (Hall, 1977; renamed *Entrophospora infrequens* by Ames and Schneider, 1979a), whose classifications are based on random collections of spores obtained from soil sievings, are questionable both on taxonomic and practical grounds because no mycorrhizal associations have been demonstrated. Furthermore, it is doubtful that even the practical objective of identifying spores obtained in soil sievings is aided by the indiscriminate multiplication of species names. Rarely are more than 50% of the spores in any soil sieving sufficiently well preserved to identify them with any certainty, and correct identification from written descriptions is difficult, particularly for people working in isolation. Two binomial keys (Hall and Fish, 1979; Nicolson and Schenck, 1979) are now available to help in such identification.

One of the more interesting new species described is *Glomus epigaeus* (Daniels and Trappe, 1979), which forms sporocarps containing many hundreds of chlamydospores on the soil surface. Only one other species, *Endogone sphagnophila,* is known to form aboveground sporocarps (at the apex of sphagnum shoots).

In addition to the VA endophytes distinguished by their spore morphology, there is a group of endophytes with a much finer mycelium (usually between 2 and 5 μm) and other distinguishing features. These endophytes, which do not form identifiable resting spores, are common in many grasslands and undisturbed ecosystems. They were first named *Rhizophagus tenuis* (Greenall, 1963), now changed to *Glomus tenuis* (Hall, 1977), but they are frequently referred to as "fine endophyte." Such a descriptive name, based simply on hyphal diameter and other anatomical features, may be preferable to their controversial inclusion in the genus *Glomus,* from which they differ markedly. Also, it is quite uncertain whether there are different fine endophytes and, if so, how similar they may be taxonomically or in symbiotic potential. Naming new species on inadequate premises makes it very probable that such names also will shortly be superseded, adding to the ever-increasing confusion of nomenclature. It is becoming increasingly evident that strains with different tolerances to temperature, pH (Abbott and Robson, 1978), heavy metals (Gildon, 1981), germination requirements (Daniels and Duff, 1978), and probably different symbiotic efficiency exist within species. The coenocytic character of the mycelium and the propensity to anastomose make this very probable. Clearly the classification of VA endophytes needs reexamination.

3.1.3. Ericalean Endophytes

The fungi so far isolated from the family Ericaceae have all been septate, slow growing, and sterile. In culture, they have dark pigmentation and lack clamp connections. Read (1974) induced one isolate to fruit and proved it to be an ascomycete because it formed small apothecia. It was named *Pezizella ericae* and was subsequently shown by Webster (1976) to be homothallic. Electron microscopy of intracellular hyphae of *Calluna* mycorrhizae from northern Italy revealed simple septa with Woronin bodies, typical of ascomycetes (Bonfante-Fasolo and Gianinazzi-Pearson, 1979), and *P. ericae* has been isolated from ericaceous plants growing in France (Vegh *et al.,* 1979). It seems that *Pezizella* and related fungi are the most common ericoid endophytes. The claims by Seviour *et al.* (1973), Peterson *et al.* (1980), and Bonfante-Fasolo (1980) that basidiomycetes, probably *Clavaria* spp., are mycorrhizal with ericaceous plants await confirmation by isolation of the putative endophyte and reinoculation in axenic culture.

The fungal partners of arbutoid mycorrhizae are probably all basidiomycetes (Francke, 1934; Khan, 1972; Lutz and Sjolund, 1973). Two have been positively identified: a species of *Cortinarius* is mycorrhizal with *Arbutus menziesii* (Zak, 1974) and *Amanita gemmata* is a non-host-specific mycosymbiont of *Arctostaphylos marzanita* (Largent *et al.,* 1980a).

3.1.4. Orchid Endophytes

The commonest fungi included in this group are various species and strains of the form genus *Rhizoctonia,* including *R. solani,* which is a ubiquitous plant parasite. Warcup and Talbot (1967) identified the perfect stage of many rhizoctonias from orchids as species of the basidiomycete genera *Ceratobasidium, Sebacium, Thanatephorus,* and *Tulasnella.* Certain rhizoctonias isolated from orchids are difficult to preserve in culture and have proved to be nutritionally exacting, requiring certain vitamins and organic N sources (Hijner and Arditti, 1973; Hadley and Ong, 1978). These may be specialized orchid root-infecting symbionts.

It has long been known that the aggressive parasite *Armillaria mellea* is mycorrhizal with the saprophytic orchid *Gastrodia elata* (Kusano, 1911), and there is growing evidence of the importance of similar wood-destroying hymenomycetes as mycorrhiza formers (for examples see Furman and Trappe, 1971). The basidiomycete affinity of other endophytes is often inferred from the presence of clamp connections. Ascomycetes may also be mycorrhizal with orchids.

3.2. Dispersal of Mycorrhizal Fungi

Except for the sporophores of basidiomycetes and ascomycetes, all mycorrhizal fungi are strictly soil- or root-inhabiting and have no aerial stage. Systemic infection, once thought to occur in the seeds of *Lolium* and ericaceous plants, has now been discounted. Acorns may, however, provide a source of ectomycorrhizal inoculum from fungal spores, not basidiomycetes, lodged on their seed-coat covering (Chastukhin, 1955). Compared to air dispersal, dissemination of propagules in the soil is relatively inefficient. Except for that of epigaeous ectomycorrhizal fungi, therefore, the dissemination of mycorrhizal fungi is quite limited, confined to spread from one living root to another, spread of spores and other propagules in drainage water and wind-blown particles, and spread by animal vectors. These vectors can include rodents (Maser *et al.,* 1978), earthworms, slugs, beetles, ants, flies, and small soil fauna as well as birds and even man. Gerdemann and Trappe (1974) found that the stomach contents of three rodent species contained several sporocarpic but no chlamydosporic species, suggesting that the former are preferentially eaten. Fructifications of some of the hypogaeous mycorrhizal fungi produce distinctive odors attractive to mammals, including pigs and dogs, and to insects. Even fungal species poisonous to man can be eaten with impunity by slugs. Spores of some ectomycorrhizal fungi have already germinated when the fecal pellets are voided. Others, like *Glomus fasciculatus,* remain viable after passage through the alimentary canal (Rothwell and Holt, 1978). Apparently viable endogonaceous spores have also been found in the digestive tracts of rodents (Dowding, 1955; Bakerspiegel, 1956) and embedded in the cementing matrix of wasps' and birds' nests (McIlveen and Cole, 1976). Indeed, germination of some spores may be assisted by digestive enzymes, but endogonaceous zygospores, whose germination has never yet been observed, were not induced to germinate by treatment with snail enzymes (Godfrey, 1957 and others). It is possible that some nematodes ingest spores of VA endophytes, thus assisting their dispersal, but this view rests on circumstantial evidence. Sporulation of endophytes was reduced by fungivorous nematodes (Salawu and Estey, 1979), and sometimes nematode cysts may be full of endogonaceous spores. Movement of endogonaceous spores in drainage water is probably restricted because the resting spores are large, often greater than 100 μm, and generally remain attached to the soil mycelium. Spread can occur from detached and senescent roots (Schrader, 1958; Powell, 1976a), and in an experimental system, limited saprophytic spread ($<$ 2 cm) occurred in soil without any intervention of living host roots (Warner and Mosse, 1980). The relative importance of interroot spread and spread by animal vectors in the dissemination of hypogaeous mycorrhizal fungi is difficult to assess. The much faster spread of VA endophytes in the field (Mosse *et al.*, 1981a) and in unsterilized soil (Powell, 1979) compared with spread rates in sterilized soils (generally less than 1 cm/week; Sparling and Tinker, 1978b; Powell, 1979; Warner, 1980) could be due to dissemination by animals.

The sporophores of epigaeous basidiomycetes and ascomycetes have, by contrast, a seemingly limitless profligacy in the production of air-borne spores, which can be carried over very long distances. Similar spores of pathogenic fungi have been entrapped at altitudes of 5000 m and can be carried from one continent to another. One mycorrhizal sporophore can produce several billions of spores, and *Boletus chrysenteron*, for example, can produce more than 100 sporophores/ha per year in a beech forest. Other species may be less prolific. The spores of many species are pulverulent so that dissemination is mainly in air currents, but others accumulate in a mucilagenous matrix and require rain for successful dissemination. Spores of some species, e.g., *Laccaria, Lactarius, Paxillus,* and *Leccinum,* are sensitive to organic inhibitors, and others require special organic stimulators for germination (Fries, 1978, 1979). Sporophore production is much affected by climatic and edaphic factors. A wet autumn favors fructification, but water stress during the summer increases the number of sporophores. Fertilization also affects sporophore production (Menge and Grand, 1978; Garbaye *et al.*, 1979). Last *et al.* (1979) showed that sporophore production is disrupted when leaf function is impaired and suggest that, in addition to carbon compounds, other substances such as auxins may be needed for sporophore enlargement.

The means of dissemination and the nature of the inoculum of ericoid mycorrhizae are completely unknown. The fungus could be disseminated by ascospores and possibly by cells resulting from segmentation of hyphae (Pearson and Read, 1975a). The endophyte has little ability to degrade complex polysaccharides (Pearson and Read, 1975b) and may therefore have a limited independent existence in the soil. Infection of seedlings from mycorrhizal roots already present undoubtedly occurs in heathlands and may be very efficient.

Singh (1974) observed that *Calluna* seedlings became infected only five to six weeks after germination under such conditions.

The rhizoctonias of orchid mycorrhizae do not have asexual spores, and dissemination may be by basidiospores or by oidial spores formed on vegetative mycelium. Sclerotia are formed by many species and may act as inoculum and as a mechanism for long-term survival. Free-living mycelia of the more parasitic strains of rhizoctonias may act as inoculum in the soil. Rhizomorphs and hyphal strands growing from food bases are probably sources of infection by mycorrhizal hymenomycetes.

3.3. Distribution of Mycorrhizal Fungi

A study of the distribution of mycorrhizal fungi presents some interesting and unique problems, partly because of the inability to or difficulty of isolating the fungi and partly because the presence of a fungus able to form mycorrhizae can only be demonstrated in the presence of the appropriate host. Absence of sporophore or spore type does not necessarily indicate absence of the fungus.

3.3.1. Ectotrophic Fungi

Most ectotrophic mycorrhizal fungi are unable to decompose cellulose or lignin (Melin, 1948) and require simple sugars as a carbon source. They are therefore poor competitors with saprophytic fungi that decompose cellulose and often also lignin. Some of these ectotrophic fungi are difficult to isolate and they grow very slowly; others, like *Russula* spp., have not been cultured. Nevertheless, some ectomycorrhizal fungi, e.g., *Tricholoma fumosum,* can use cellobiose or lignin (Norkrans, 1950), and *Boletus luteus, B. variegatus,* and *Amanita citrina* use cellulose (Lyr, 1963). *Boletus subtomentosus* can cause rapid breakdown of leaf litter (Lindeberg, 1948) and can probably live as a saprophyte and form sporophores in the absence of a host.

In general, epigaeous mycorrhizal fungi are more widespread than the hypogaeous. Lamb (1979), investigating the spread of mycorrhizal fungi from a young pine plantation in Australia, found that species occurring more than 800 m from plantations were either *Rhizopogon roseolus* or *Suillus luteus*. A fungus with a worldwide distribution is *Pisolithus tinctorius;* Marx (1977) lists 33 countries and 38 states in the U.S.A. where it has been found. It has no host specificity and can find hosts in all the world, except in equatorial forests, where ectomycorrhizal trees—other than in the Caesalpineae and some other families—are rare. It is possible to introduce *Pisolithus* in such conditions by also introducing suitable hosts, e.g., *Pinus* or *Eucalyptus* spp. But the occurrence of *P. tinctorius* can also be controlled by other factors; it is a thermophilic fungus unable to fructify or survive in cool climates and it does not occur in calcareous soils. In France, *P. tinctorius* occurs along the Atlantic and North Sea coasts, in the southwest and southeast, but not in the center or the northeast where the climate is more continental. Nevertheless, in mountainous areas of North America and Canada, cold-adapted strains apparently exist. Equally worldwide in distribution, although apparently discontinuous in its occurrence, is *Cenococcum graniforme* (Trappe, 1964), which has no host specificity (associating also with some herbaceous hosts), no climatic or edaphic exigencies, and can be found beyond the polar Arctic circle as well as in equatorial and tropical regions. Nevertheless, Dominik (1961) failed to isolate it from cultivated soils. It is difficult to explain how a hypogaeous fungus, like *C. graniforme,* which does not fructify and lacks an aerial dispersal phase, achieves such worldwide distribution, being associated with *Eucalyptus gigantea* in Australia and *Nothofagus* in South America. One hypothesis is that this fungus has never evolved since the separation of the continents. By contrast, another hypogaeous fungus, *Tuber melanosporum,* has a very limited distribution and is confined to southwestern Europe (France, Spain, Italy, and Yugoslavia). It seems that it has not spread across the Mediterranean into North Africa, although ecological conditions (calcareous soils, climate, and hosts) appear to be suitable.

In spite of the very efficient aerial dispersal of many ectomycorrhizal fungi, there is some doubt about their presence in steppe and prairie soils. These are grassland areas that occur extensively in Russia and the midwestern United States; they are normally devoid of trees except along stream banks. It is widely agreed that oaks and beeches in the steppe soils and pines in the prairies succeed only if the seedlings become mycorrhizal. Whereas the older literature (e.g., Vysotskii 1902) stressed the need for inoculation of oak and beech seedlings in steppe soils, Vlasov (1952), and Potebnja (1952) reported mycorrhizal development in oak seedlings without inoculation. Shemakhanova (1962), summing up Russian experience, concluded that young oak seedlings in Russia became mycorrhizal not only in soil long denuded of forests but also in soil far removed from forests. This may be explained by Chastukhin's (1955) observation that inoculum could be carried on the acorn seed coat. Mishustin (1955) observed that in the chernozem zone, oak seedlings formed mycorrhizae naturally with indigenous fungi, whereas they failed to do so in the more southerly chestnut soils further removed from the inoculum of the temperate forest belt. One reason for the confusing and divergent opinions may be that indigenous mycorrhizal fungi are present but relatively inefficient, and thus foresters insist on the need for inoculation, while researchers simply note the presence of mycorrhizae in test seedlings. Dominik (1961, 1963) in Poland and Fassi and De Vecchi (1962) in Italy showed that young seedlings became ectomycorrhizal in cultivated soils, but the number of fungal species is smaller than in normal forest soils and the fungi are often relatively inefficient. Bowen (1963) made a somewhat analogous observation for *Pinus radiata* in Australia.

Most prairie soils of midwestern United States lack the mycorrhizal fungi necessary for the successful introduction of conifers (Hatch, 1936; Mikola, 1953, 1970).

3.3.2. VA Endophytes

VA endophytes are among the commonest soil fungi, although recognition of this fact rests on visual identification of the characteristic soil mycelia and, more convincingly, on formation of VAM in indicator or naturally occurring plants. This indicates that few, if any, natural soils are entirely free from infection, except possibly at depths beyond the normal rooting range of plants. Such ubiquitous occurrence, in spite of relatively poor dispersal mechanisms and total dependence on a living host, illustrates the value of an extremely wide host range, combined with the virtual absence of host specificity. Nevertheless, soils can differ considerably both in the number of propagules and in the composition of the endophyte population.

3.3.2a. Techniques. Among the techniques used to obtain qualitative and quantitative information on endophyte populations, sampling soil populations of endogonaceous spores has been the most popular. Techniques used for spore recovery are often adapted from nematological techniques. All depend on differential sedimentation to separate spores and organic debris from the heavier particles in a soil suspension (Gerdemann, 1955a). Variations of this are a floatation–adhesion technique (Sutton and Barron, 1972), a countercurrent floatation–bubbling system (Furlan and Fortin, 1975), and a separation by centrifugation on discontinuous sucrose gradients (Mertz *et al.,* 1979) that has been proposed for large-scale spore separation. Unless exposure to high osmotic pressures is very short, spores can be damaged. Having obtained a suspension of spores mixed with organic debris, there remains the problem of counting and identification. Hayman (1970) filtered suspensions through discs made of a loosely woven synthetic fiber (Dicel) which allowed good visibility and orientation under a dissecting microscope. Probably the best device is a Petri dish with concentric ridges (Doncaster, 1962), for which a mathematical formula exists giving the sampling error of counting only some of the rings.

Smith and Skipper (1979) found severalfold differences in spore recovery using different methods of extraction. Numbers usually reported are of the order of 20–2000 in 100 g of soil. Sutton and Barron (1972), however, reported figures of 2–92 and up to 132 in 1 g of soil. Their figures are likely to have included empty, nonviable spores, which are usually excluded from most surveys.

Although identification of spore types is at present the most practicable way of obtaining qualitative information on endophyte populations, it is open to certain criticisms. Not all endophytes produce large resting spores that can be identified. The "fine endophyte" is the obvious example of a nonsporing type, but there may be others. Some endophytes sporulate much more profusely than others, and spore production is differentially affected by environmental conditions. Probably spores of some species are more susceptible to microbial degradation than others. Alternative methods of species differentiation have therefore been sought, but they are in their infancy. There are consistent differences in the anatomy of infection caused by different endophytes (Mosse and Smith, 1976; Abbott and Robson, 1979; Mosse *et al.*, 1981b). In experimental systems where fungal species with clearly differentiated anatomical characters have been selected, anatomy can be used to examine spread and relative aggressiveness of two species. The technique is, however, subjective and requires specialized skills. Another technique successfully used for identification of ectomycorrhizal species is immunofluorescence (Schmidt *et al.*, 1974).

Although spore counts have been and are often used as a measure of soil infectivity, doubts have gradually arisen about this, as more has been learned about the factors affecting sporulation (see Section 3.3.2b). Alternative, more direct methods of assessing infectivity are therefore being tried. These are generally based on soil dilution and the most-probable-number (MPN) principle. Porter (1979) has described a method combining this with a baiting technique, inducing plants to grow through a suitably diluted soil core. A similar technique has been used in studies of "take-all" disease (Hornby, 1969), but the processes of dilution and thorough mixing of the soil introduce certain artifacts (Hornby, 1979), such as damage of propagules. Clark (1964) completely destroyed infectivity by passing a VA-infested soil through a 2-mm sieve. The MPN technique is laborious and requires greenhouse space protected against chance infection. It is only useful in application to homogeneous soils, as are used in agriculture, and would need to be carefully standardized if it were to be used as a universal measure of infectivity. For instance, degree of compaction of the diluted soil and date of harvest will affect root development, which will in turn affect the likelihood of a root making contact with a fungal propagule. Nevertheless, the method has obvious value for certain investigations.

3.3.2b. Distribution of VA Endophyte Species as Indicated by Spore Counts. Spore surveys have been conducted in particular regions (Gerdemann and Nicolson, 1963; Mosse and Bowen, 1968; Gilmore, 1968; Khan, 1971; Saif and Iffat, 1976; Hayman *et al.*, 1976; Redhead, 1977; Johnson, 1977; Hall, 1977; Nicolson and Schenck, 1979; Ames and Schneider, 1979b); at particular sites, e.g., dunes (Koske *et al.*, 1975; Koske, 1975, 1981b); in xerophytic environments (Khan, 1974), heathlands (Sward *et al.*, 1978), a Pennine grassland (Sparling and Tinker, 1975a), and coal spoils (Daft *et al.*, 1975; Khan, 1978); and in connection with particular plants, for example, *Araucaria* (Bevege, 1971), *Festuca* spp. (Molina *et al.*, 1978), and soybeans (Schenck and Hinson, 1971). As a result, it is clear that all four mycorrhizal genera, viz. *Glomus, Gigaspora, Acaulospora,* and *Sclerocystis,* are distributed throughout the world. *Glomus* spp. may have the widest distribution, but they are also the least well-defined genus. *Gigaspora* and *Sclerocystis* spp. appear to be more common in tropical soils. *Acaulospora* has been found in nearly every spore survey, though rarely as the dominant species. Experimental evidence indicates that it is better adapted to soils below pH 5, whereas *Glomus mosseae* requires a higher pH and is common in alkaline soils.

At most sites, more than one spore type has been found. Molina *et al.* (1978) give an average of 2.7 spore types per community. In an early survey based on single (rather than pooled) samples that were randomly collected over parts of southern Australia and New Zealand, 17 and 54%, respectively, contained no spores.

There is a general consensus of opinion that spore numbers fluctuate seasonally. In temperate climates, they tend to increase towards autumn and decline during winter. It is also widely found that spores tend to be fewer under forest and pasture than in arable land. A generalized conclusion of many surveys is that sporulation may be favored by intermittent root growth. This, however, could not be confirmed experimentally (Baylis, 1969; Redhead, 1975).

Soil samples taken no more than 20 cm apart can yield very different numbers of spores (Redhead, 1974; Mosse *et al.*, 1981a). Walker and McNabb (1979) found frequency distributions highly variable and usually nonnormal. Hayman and Stovold (1979) found a divergence between mean and maximum spore numbers ranging from 12- to 44-fold for eight spore types recorded. Such variability raises the question of whether sporulation may occur in discrete microsites and emphasizes the importance of technique of sample collection. Maximum sporulation is usually close to the densest part of the root system, near the collar roots.

At specific experimental sites (Hayman, 1970; Hayman *et al.*, 1975) and in sterilized soils in pots (Daft and Nicolson, 1969), spore numbers have been directly proportional to infection levels in the plants, but there is now much doubt that they are always so related in natural sites. Apart from the widespread occurrence of the nonsporing "fine endophyte," sporulation can be markedly affected by interaction between endophyte species (Ross and Ruttencutter, 1977; Warner and Mosse, 1979; Nemec, 1979). It can also be depressed by hyperparasitic fungi (Ross and Ruttencutter, 1977; Schenck and Nicolson, 1977), by nematodes, and by fertilizer treatments. Ross (1980), in a very interesting paper, has drawn attention to a nondiffusible, heat-labile factor present in natural soil which reduced sporulation of *Glomus* and *Gigaspora* spp. This factor appears to be associated with living organisms because it did not pass a bacteriological filter, and it may be vital in controlling inoculum density in soils.

One of the most thorough and informative surveys (Hayman and Stovold, 1979) covered 73 sites in New South Wales. The authors concluded, "the survey illustrates the variability of VA mycorrhizal populations in New South

Wales, which is largely unrelated to geographical features or crop. Only at specific localities could variation between sites be accounted for by definite factors such as fertilizer application. Populations varied considerably for the same crop at different locations." Hayman (1978) also concluded from a general spore survey of pastures in Otago that there was a large measure of ecological equivalence between Endogonaccae species. General surveys then are only useful for providing qualitative information. Ideally they should be well illustrated, like the survey of heathland spores in southeastern Australia (Sward et al., 1978), rather than containing simply a collection of Latin names.

To yield useful information about particular associations of fungal species and particular hosts, surveys need to be much more detailed. Koske (1981b) examined the interactions between five species of VA endophytes sporulating in association with the dominant species on a single barrier dune. Two fungal species were absent or rare on two hosts, and spore production of *Acaulospora scrobiculata* was significantly greater in association with *Ammophila breviligulata* than with *Lathyrus japonicus*. He concluded that the interaction between VA species is less important in determining spore density than are the species of host plant and other environmental factors. Rabatin and Wicklow (1979) studied plant diversity in relation to VA mycorrhizal diversity and concluded that "where the majority of plant taxa present were VA hosts, diversity of Endogonaceae was relatable to plant diversity." These findings may provide a clue to the inexplicable diversity of spore populations over small distances coupled to the seemingly worldwide distribution of most of the known species.

3.3.3. Ericaceous Endophytes

Inoculum has been reported in soils lacking ericoid hosts (Pearson and Read, 1975a).

4. The Association

The most obvious prerequisite for mycorrhiza formation is the presence of the appropriate host and fungus. When both are present, infection depends on the proximity of root and inoculum and on the ability of the infection to spread. This in particular can be much influenced by environmental conditions. These points are considered at greater length in the following two sections.

4.1. Host–Fungus Specificity

Specificity in mycorrhizae may arise from of either the host or the fungus being restricted to one particular partner. Thus, a fungus may be restricted to one host which can accept many other fungi, or, on the other hand, a host may only accept a single fungal species which, however, has many alternative hosts. The determinant for specificity may therefore reside either in the host or in the fungus.

Most trees which form ectotrophic mycorrhizae will do so with many different fungal species (for example, Trappe [1977b] considers that Douglas fir *(Pseudotsuga menziesii)* can probably form mycorrhizae with 2000 species), while a host accepting only one species is unknown. Conversely, while some fungal species, such as *Pisolithus tinctorius,* will infect many hosts (Marx, 1977), others are restricted to single hosts, e.g., *Boletus elegans* with larch *(Larix europaeus),* and *Boletus carpini* with hornbeam *(Carpinus betulus).*

VA mycorrhizae are characterized by an almost complete lack of either host- or fungus-determined specificity. A single root system may harbor more than one species of fungus, and hyphal links between roots of two different hosts can also occur (Heap and Newman, 1980). Graw *et al.* (1979) have reported a notable exception to the general lack of VA specificity—a species of *Glomus gerdemannii* from Brazil which would only infect one of their test plants, *Eupatorium odoratum.* Other similar cases may yet be found.

Orchids also show little evidence of specific host–fungus relations. Hadley (1970) tested protocorms of 10 orchids against 32 *Rhizoctonia* isolates from orchids, nonorchid hosts, and soils, all of worldwide distribution, and found no evidence of specificity. Similarly, Jonsson and Nylund (1979) found that the hymenomycete *Favolaschia dybowskyana* from Africa would infect protocorms of *Bletilla stricta* from Japan. By contrast, Warcup (1971) claimed that protocorms of certain orchids of south Australia formed symbiotic associations only with certain fungi. There is too little evidence about adult orchids to make generalizations, but these too seem to form mycorrhizae with many different fungi (Gäumann *et al.,* 1961; Alconero, 1969). Hadley (1970) suggested that apparently specific associations in the field, e.g., *Rhizoctonia goodyeras* with *Goodyera repens,* may be the result of selection of a particular fungus by the prevailing soil environment.

Absence of specificity in the ericoid mycorrhizae is well documented. For example, Pearson and Read (1975a) showed that endophytes from each of the genera *Calluna, Erica, Vaccinium,* and *Rhododendron* would infect the other three genera, regardless of the genus of origin. Little is known about relations in arbutoid mycorrhizae, but nonspecificity may characterize these also, for Zak (1976) reported that seven different genera of basidiomycetes could form mycorrhizae with *Arctostaphylos* in the laboratory.

Not only have most mycorrhizal fungi a range of hosts, but all host plants are not necessarily restricted to one type of mycorrhiza. *Populus* may bear ectotrophic (Levisohn, 1954) or VAM (Dangeard, 1900), but it is uncommon to find a species such as juniper *(Juniperus communis)* with both types of infection in one root system. Further, typical VAM have been reported in the orchid

Corybas macranthus (Hall, 1976) and in the ericoid host *Empetrum nigrum* (Miller, 1981). McNabb (1961) claimed that ericoid mycorrhizae occurred in four different nonericaceous genera, but the fungi were not isolated. Largent *et al.* (1980b) found that large ericaceous shrubs in northern California possessed either ecto- or arbutoid mycorrhizae, and they also observed rare associations of *Vaccinium* with arbutoid mycorrhizae and *Arctostaphylos* with ericoid mycorrhizae. Although such examples could probably be multiplied, they are nevertheless exceptional and only illustrate the great diversity of mycorrhizal associations.

It is a general conclusion then that mycorrhizae largely lack specificity. This feature has enormous ecological implications and has led to much speculation about evolutionary origins.

4.2. The Infection Process

The factors which may determine the establishment of mycorrhizal infections in developing root systems are summarized in Fig. 3, which is intended not as a complete model but to illustrate the complexity of the processes involved and to act as a guide for this section. It is difficult to represent the

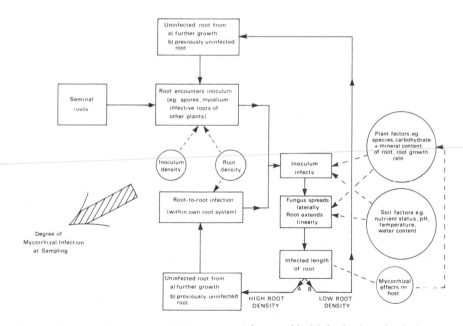

Figure 3. Factors that may control the progress of mycorrhizal infection in a developing root system.

dynamic process of infection using a static model; hence, Fig. 3 should be interpreted as a representation of a continuous process rather than a series of discrete steps.

Infection in a seminal root originating from a seed or specialized perennating organ will occur when that root encounters a suitable propagule in the soil. The chance of this happening is a function of both root and inoculum density in the soil; the critical distance between root and propagule will depend largely on the growth potential of that propagule. A quantitative model embodying these principles has been proposed by Gilligan (1979) for rhizosphere infection by propagules of pathogenic fungi.

The only information on the quantitative relation between inoculum density and mycorrhizal infection concerns VAM. Carling *et al.* (1979) found that infection was directly proportional to inoculum density when propagule numbers were low, but as the inoculum increased, infection progressively decreased to a plateau where further inoculum density had no effect. A similar relationship exists for other host-parasite systems (Curve B of Van der Plank, 1975) and indicates that propagules of low density act independently but compete for available sites when inoculum is high.

Information on the distances over which the various forms of mycorrhizal inocula can infect is scarce, but a little is known about VA chlamydospores. These are not stimulated to germinate by host roots, neither do their germ tubes show tropism (Hepper and Mosse, 1975), but they have a considerable capacity for production of germ-mycelium (Hepper, 1979) and can infect roots 11–15 cm away (Powell, 1976a). This distance, however, is likely to be a maximum estimate and probably resulted from the experimental conditions used whereby two-dimensional growth of germ tubes along slides was observed, rather than the natural three-dimensional growth. Warner (1980) quotes a maximum inter-root spread of 2 cm for three fungal species. The ability of higher fungi to form mycelial aggregates (strands and rhizomorphs) capable of infection over long distances must be important in the initiation of ectotrophic, orchidaceous, and Ericalean mycorrhizae.

When a root has encountered a propagule, the process of infection will be influenced both by the soil environment and by the properties of the host root, as shown in Fig. 3. In VAM, the fungus, after infection, extends within the cortex. The relationship between the rate of root extension and the rate of spread of the infection from the initial entry point—which is influenced by the development of neighboring infections—to the rate of root extension will have a large effect on the percentage infection recorded at harvest.

Further root growth will now produce uninfected roots, which may either encounter further propagules in the soil (route B, Fig. 3) or become infected from the mycorrhizae already formed. This could occur either (a) by external mycelium traversing the soil between roots or (b) by infective hyphae spreading over the root surface and setting up new infections on the same root and on

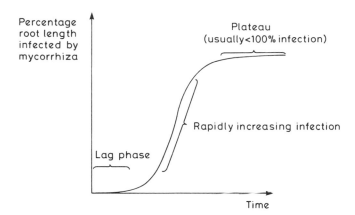

Figure 4. Model of change with time in percentage root infection by vesicular-arbuscular mycorrhizae in a developing root system.

new, emerging laterals. The relative frequency of root colonization from
already infected roots, compared to that from new propagules encountered in
the soil (routes A and B, Fig. 3) is largely unknown for all types of mycorrhizae
and warrants immediate attention. In VAM, spread of infection within a root
system is undoubtedly efficient, as shown by Daft and Nicolson (1969), who
obtained considerable infection at harvest from an inoculum of only a few
spores. Also little understood are the relative rates of fungal spread within and
between roots as described above. Study of this problem may explain why
spread of VAM from one root system to another, where a continuous surface
is lacking, is very much slower than spread within a single root system. This
problem is also highly relevant to ectomycorrhizae.

The progress of VA infection in a single root system commonly follows a
curve of the type shown in Fig. 4. These sigmoid progressions have been
observed in controlled environments as well as in the field (Sutton, 1973; Saif,
1977), and show them to be an intrinsic property of root systems and not solely
the result of seasonal changes. The lag phase presumably reflects the slow initial establishment of infection, but the sharp rise in infection is not understood.
It could be due to the attainment of a critical root density allowing inter-root
infection (Black and Tinker, 1979); however, the sigmoid curve has been successfully reproduced in a model by Tinker (1976) based on the simple assumptions of exponential growth of a root system where the rate of formation of new
infected material is proportional to the amount of uninfected but receptive root
and to the amount of infected root, all per unit soil volume. This model warrants further testing for it implies that a change in the relative growth rate of
the root could influence mycorrhizal development, irrespective of concomitant
changes in root biochemistry. Other authors have related plant morphogenesis

to the phases of mycorrhizal development (Sutton, 1973; Furlan and Fortin, 1977; Saif, 1977), but this is only a partial explanation for the sigmoid curve that has also been observed in plants growing exponentially (Stribley, unpublished data). In such cases, the attainment of a plateau in percent infection is intriguing, as it suggests a balanced relationship between root growth and mycorrhizal spread. The reasons for the existence of this plateau, why it is nearly always less than 100%, and what determines the stable value reached are not understood. Warner (1980) has clearly shown, however, that the maximum infection attained is highly species-dependent.

It is very likely, based on current evidence, that increased host growth and mineral content resulting from infection will themselves influence the further development of the mycorrhizae, as implied in Fig. 3. This point has not yet been tested, possibly because of the difficulty of separating cause and effect in mycorrhizal systems (Stribley *et al.,* 1980).

Studies on infection dynamics of mycorrhizae other than VA are rare, but the data of Laiho and Mikola (1964) on ectomycorrhizal development in *Pinus sylvestris* show that the rate of increase in the roots colonized by ectomycorrhizae is roughly comparable to that of VAM in other species.

The mechanisms discussed so far refer to development of mycorrhizae in a new root system. In perennial root systems where new roots grow seasonally, the mechanisms may be quite different. Sparling and Tinker (1978a) reported that infection levels of VAM in a perennial grassland were relatively constant despite considerable root turnover, and a stable percentage of infection in a grassland ecosystem was also reported by Read *et al.,* (1976). These observations suggest a tight feedback control between host and fungus in perennial roots.

The complexity of the infection processes discussed in this section shows that the effects of plant and soil factors are not easy to predict or to interpret. They may operate by affecting (a) the germination of fungal propagules and their subsequent growth in the rhizosphere, (b) the infection process, (c) the postinfection spread of the fungus, and (d) the inoculum density of the fungus in the soil.

Although Fig. 4 shows the degree of mycorrhizal infection at sampling to be the result of root and inoculum density, root density is in fact very rarely recorded. In most papers, measurements of infection are expressed as percent root length infected, based on the proportion of infected to uninfected root in a sample. Giovanetti and Mosse (1980) give some information on the sampling error of different methods and recommend a grid sampling technique. In many situations, such a method is the only practicable measurement. But percent length infected gives, by itself, no indication of root density and may lead to some curious distortions. Some figures (Table IV) obtained by Warner (1980) show total root length, root length infected, and percent infection 20 and 40 days after inoculation in five host species. The figures show that, under com-

Table IV. Progressive Changes in Root Length and Percentage Infection in Five Different Hosts[a]

Host	Increase in root length between 20- and 40-day harvest	Percent infection		Increase in root length infected between days 20 and 40 (cm)
		Day 20	Day 40	
Onion	× 2.5	31	48	33
Fescue	× 33	1.6	2	152
Lettuce	× 5	1.6	10	192
Bean (*Phaseolus* sp.)	× 4.5	3	10	330
Clover (*Trifolium* sp.)	× 18	11	17	676

[a]Warner (1980).

parable conditions, increases of 6 and 7% may represent, respectively, 330 and 676 cm of new infected root. A virtually negligible increase of 0.4% in fescue represented an additional 152 cm of infected root, whereas a 17% increase in onions represented only 33 cm of new infected root. A change in infection level in fescue, and perhaps by implication in natural swards, would need to be very large before it registered as an increase in percentage infection. Clearly, root density is a very important parameter in any comparative studies. Table IV also shows a marked effect of host species on the development of infection. It would be interesting to know where in Fig. 3 the host species exerts its influence.

The questions of whether certain plant species are habitually more strongly infected than others, and what may be the underlying causes, have attracted a certain amount of speculation. Boullard (1979) made the interesting point that the intensity of infection in roots of Pteridophyte gametophytes that harbor endophytes possibly related to VA species is inversely related to their degree of evolutionary development. Baylis (1975) proposed that, through evolution, species with primitive, coarse "magnolioid" roots may have become *physiologically dependent* on mycorrhizae for nutrient uptake, whereas those with fine, highly branched "graminoid" roots rely mainly on root hairs for absorptive function. This hypothesis, for which there is a good deal of support, has been taken to imply that the *degree of infection* is somehow inversely related to the graminoid nature of the root system, and evidence for this has been offered by St. John (1980). Nevertheless, there are exceptions; for instance, the tropical grasses *Paspalum notatum* and *Digitaria procumbens* have long and abundant root hairs, high levels of infection, and good responses to the mycorrhizal association. Also, slight infection need not necessarily imply intrinsic barriers to infection. For example, the low percentage infection generally reported in potato *(Solanum tuberosum)* does not extend to roots grown from true seed (Stribley, unpublished data). Recently, Bertheau *et al.* (1980) reported large differences in VAM development in different wheat cultivars, which suggests some kind of genetic control of infection levels.

5. Factors Affecting the Development of Mycorrhizae

5.1. Light

Pot experiments have shown that mycorrhizal development is greatly affected by the light received by the shoot of the host plant. Hatch (1937) first showed that the number of ectomycorrhizal roots formed was greatest at high light intensities, and this was confirmed by Björkman (1942). Boullard (1961) found that increasing the photoperiod from 8 to 24 hr increased the number of short roots in *Pinus pinaster* by sevenfold, and the percentage which were infected rose from zero to over 50%.

Similarly, low irradiation and short day lengths can reduce the development of VAM (Hayman, 1974; Daft and El-Giahmi, 1978), but they also reduce plant growth markedly in controls. Furlan and Fortin (1977) have reported the converse—that infection developed faster under low light intensities but that spore production increased with light intensity. Johnson (1976) has made the interesting observation that shading reduced development of infection in shade-tolerant species but not in light-demanding species.

It is not clear whether the observed effects are determined by the total light energy received, i.e., whether the effect of light is simply cumulative, or whether photoperiod and intensity have separate effects. More work on light effects is clearly necessary, and it is worth reiterating Lewis's (1975) plea for studies on *established* mycorrhizae because of the possibility of pathogenicity of the fungi in light-starved plants.

The mechanism of light effects is difficult to investigate because of the manifold effects of shoot photosynthesis on root growth and biochemistry. Björkman (1942) first proposed that there were positive correlations between the concentration of soluble sugars in the root, light, and mycorrhizal development. The theory that sugar levels control infection has been controversial for several reasons (see remarks by Lewis, 1975) but has recently received strong support from Marx *et al.* (1977b).

The effects of grazing on VA mycorrhizal infection in pastures are of considerable ecological interest and may be similar to those of light, since defoliation greatly changes the soluble carbohydrate content of grass roots (Kigel, 1980). Daft and El-Giahmi (1978) defoliated pot-grown grasses and found a marked reduction in mycorrhizal development. Again, it would be interesting to know the effects of grazing on established mycorrhizae. Ericaceous shrubs, notably heath *(Calluna vulgaris),* also provide grazing material, and it would be of interest to know the effect of this activity on the ericoid mycorrhizae.

5.2. Soil Factors

5.2.1. Chemical

5.2.1a. Nutrients. Numerous pot experiments have demonstrated beyond doubt that formation of ectotrophic, VA, and ericoid mycorrhizae is reduced by added nutrients, in particular P and N. The pioneering work of Hatch (1937) on ectotrophic infections, which suggested that intensity of infection was greater under conditions of low or unbalanced nutrient supply than when nutrients were ample or balanced, has been largely confirmed. Björkman (1942) pointed out that *very* severe deficiencies of P or N had an adverse effect. Fowells and Krauss (1959) showed further development of established mycorrhizae to be also sensitive to the nutrient environment. They transplanted mycorrhizal *Pinus* seedlings into sand culture, where they observed that new roots were heavily mycorrhizal when the plants were supplied with 1 ppm P and 5 ppm N but remained uninfected if the nutrient solution contained 5 ppm P and 25 ppm N. These results were confirmed by Marx *et al.* (1977b), who found a negative correlation between high levels of N and P and susceptibility to ectomycorrhizal infection by *Pisolithus tinctorius* of short roots of loblolly pine.

It is well established that formation and continued development of VAM are inversely related to P supply. At high levels of P, VAM may be eliminated, and Johnson (1976) found this effect was greater in fine rooted species than in those with coarse roots. At very low levels of P, as in the tree mycorrhizae, infection may be poor.

Ericoid mycorrhizae also form less readily when P and N are abundantly supplied (Morrison, 1957; Burgeff, 1961; Stribley and Read, 1976), but Morrison reported that very high applications of P to *Pernettya macrostigma* stunted root growth, resulting in atypical but intense infection.

Field work has mostly confirmed these pot trials. Boullard (1960) reported that in *Pinus pinaster* growing on poor, sandy soil, the number of ectomycorrhizal root tips was notably reduced by N additions and by complete NPK fertilizer. Menge *et al.* (1977) found the density of mycorrhizal tips in an 11-year-old stand of *Pinus taeda* was reduced by added N, but the effect did not persist for more than 2 years. Work on VAM has shown that P addition reduces field infection in wheat (Hayman *et al.,* 1975) but that the effect of N is greater (Hayman, 1975). Sparling and Tinker (1978a) added various fertilizers to a grassland in the Pennine hills of England and found that only P reduced infection. They observed a 25% reduction which lasted one year only.

Hatch (1937) and Björkman (1942) considered that nutrients did not affect ectotrophic mycorrhizae by acting on the fungi directly but by altering the composition of the root, and there is strong evidence that the internal P content of the root controls VA infection (Sanders, 1975; Azcón *et al.*, 1978b; Menge *et al.*, 1978). There is evidence that for ectomycorrhizae (Marx *et al.*, 1977b) and VAM (Jasper *et al.*, 1979), nutrients act by controlling the soluble sugar content of the root, but it has also been proposed that, in VAM at least, internal P may work by controlling the permeability of root cell membranes (Ratnayake *et al.*, 1978).

Experiments such as these have unfortunately prompted the general belief that mycorrhizae are found only in nutrient-poor soils. This notion must be dispelled. LeTacon and Valdenaire (1980) report that *Hebeloma cylindrosporum* forms abundant mycorrhizae with fir and spruce trees in nurseries given 150 kg N, 150 kg P_2O_5, and 150 kg K_2O/ha. In a careful study of the Granada region of Spain, Hayman *et al.*, (1976) found relatively constant levels of VA infections over a considerable range of fertility. For example, grapevine bore 37% infection at 2 mg/kg of $NaHCO_3$-soluble P and 22% in a soil containing 171 mg P/kg.

A paradox exists here. Many pot studies and field trials purport infection to be sensitive to added nutrients, yet surveys such as that of Hayman *et al.* (1976) suggest nutrient levels to be unimportant. Different host species will differ in their reaction, but most probably, the answer lies in the existence of fungal species and strains adapted to particular levels of soil nutrients. Some strains, for example, may be superior to others in spreading within a root cortex low in carbohydrate. If this is so, the ecological relevance of many nutrient experiments becomes questionable when the levels of added nutrient go far beyond the range naturally encountered by a particular fungal isolate.

Several authors report that fertilizers change the type of ectomycorrhiza formed on trees. For example, Menge *et al.* (1977) showed that P addition selected for mycorrhizae formed by *Cenococcum graniforme*. Endophyte species of VAM appear to differ in their tolerance for it has often been found that introduced endophytes infect better at high levels of P than endophytes native to the relatively infertile experimental soil (Mosse, 1977a,b; Abbott and Robson, 1978; Powell and Daniel, 1978a,b; Jasper *et al.*, 1979; Clarke and Mosse, 1981). Porter *et al.* (1978), however, found no evidence of endophytes adapted to high P levels in field plots which had received different rates of P for 11 years. Further work on the nature and extent of adaptation to nutrient levels by mycorrhizal fungi is badly needed.

In orchid mycorrhizae, the external supply of organic rather than inorganic nutrients may control mycorrhizal development. Hadley (1969) found that rhizoctonias were less prone to parasitize protocorms when cellulose rather than soluble sugars were supplied to the fungus. The forms of carbohydrate available to epiphytic orchids could conceivably control the fate of protocorms and select for particular strains of endophyte.

5.2.1b. pH. Effects of pH are particularly difficult to evaluate since many chemical properties of the soil vary with changes in pH. In acid soils, much soluble Al, Mn, and Fe are released, and these metals, rather than H^+, are probably responsible for many observed effects of low pH. Also, plant roots excrete bicarbonate ions (as a mechanism for maintenance of electrical neutrality during ion uptake), making the rhizosphere alkaline (Nye and Ramzan, 1979). Consequently, the pH measurements commonly made on bulk soil are but a poor measure of the conditions encountered by the mycorrhizae.

Ectotrophic mycorrhizae are present over a wide pH range but are more abundant on raw humus soils (pH 3.5–4.0) than on mull soils (pH 5.0–5.5). This wide tolerance appears to result from the presence of various fungal species, as shown in Fig. 5, and perhaps various strains. The reason for these differences in pH optima and latitude of tolerance in these fungi is obscure, since they all show rather similar pH optima of around 3.5–6.0 in pure culture.

On the whole, VAM resemble ectomycorrhizae in their wide tolerance to pH. Read *et al.,* (1976) found a remarkable consistency in VA infection of *Festuca ovina* growing in natural communities over the pH range 4.2–7.0, and Sparling and Tinker (1978a) found no obvious effect of pH on infection in three grassland sites at pH 4.9, 5.9, and 6.2; however, such infection may have been due to different endophyte species. There is some evidence for differences in adaptation of strains and species of VA fungi to pH. Lambert and Cole (1980) reported that six isolates of *Glomus tenuis* differed in their ability to form mycorrhizae at low pH and that an isolate of *Gigaspora gigantea* failed

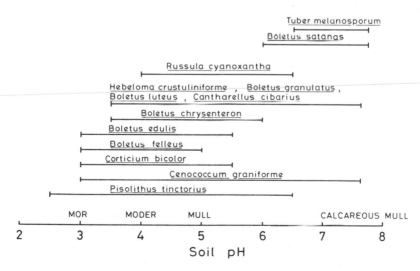

Figure 5. pH ranges over which different fungi form ectotrophic mycorrhizae (based on field observations by several authors).

to infect at low pH, even though this species had previously been found in mine spoils more acid than pH 4.8. In western Australia, the yellow vacuolate spore type *(Glomus mosseae)* is absent from acid soils, where it is replaced by honey-colored spores (*Acaulospora* sp.) (Abbott and Robson, 1977), and in pot experiments *G. mosseae* has often failed to infect at a low pH (Hayman and Mosse, 1971; Mosse, 1972), whereas *Acaulospora* has been successful (Mosse, 1972). Green *et al.* (1976) showed that spore germination of *G. mosseae* on agar was best under neutral and alkaline conditions, whereas two *Gigaspora* spp. were favored by an acid pH, indicating that the ability of spores to germinate may be a factor limiting species or strains to particular soils.

Production of VA spores may also be affected by pH. Read *et al.* (1976) found that spore numbers and viability in natural soils declined with increasing acidity.

On highly acidic natural soils, VAM are often excluded because of a lack of suitable hosts but may be found on man-made habitats. Daft *et al.* (1975) reported considerable VA infection in plants growing in a bituminous mine spoil of pH 2.7. Natural soils of the world cover the pH range 2.8 to greater than 10.0 (Baas Becking *et al.*, 1960). The acid group comprises acid sulfate soils, where acidity is generated through oxidation of pyrite following drainage (Bloomfield and Coulter, 1973), and highly leached soils such as podzols. However, nothing is known about mycorrhizae on acid sulfate soils. Ectotrophic and Ericalean mycorrhizae seem in general to be well adapted to acidic natural soils. Podzols are often dominated by calcifuge ericaceous plants and hence by ericoid mycorrhizae, which have been observed (in *Calluna*) at a pH as low as 2.9 (Stribley, unpublished data), but Pearson and Read (1975b) reported a pH optimum of 6.6 for growth *in vitro* of a *Calluna* endophyte. Information on the mycorrhizae of ericaceous plants which grow on alkaline soils (e.g., species of *Erica*) is absent. In fact, at the alkaline pH extreme—which occurs in natural soils of arid regions where upward movement of water precipitates mineral salts—reports of mycorrhizae are confined to the single discovery by Bowen (1980a) of typical VAM at pH 9.2.

Finally, soil pH may affect the distribution of mycorrhizae in a subtle way. Some workers (Mosse, 1975; Graw, 1979) have noted that pH can affect the extent to which a certain species of mycorrhizal fungus increases the P supply to its host, and hence increases host vigor, in a way unrelated to the extent of infection. In this way, the given pH could select for particular mycorrhizal symbionts.

5.2.1c. Metals. The effect of metals on mycorrhizae is of considerable interest because the metal contents of soils can vary widely. Acid soils contain much soluble Al, Mn, and Fe, and in neutral soils Mn and Fe can be released in large quantities when reducing conditions prevail after waterlogging (Ponnamperuma, 1972). Geering *et al.* (1969) found that soil solution Mn concentrations in a silt loam varied 100-fold during one year as a result of seasonal influences. Many soils are grossly polluted with heavy metals as a result of man's activities of mining and smelting.

Studies of the tolerance of ectomycorrhizae to metals are few, and remain limited to the effects of heavy metals. Harris and Jurgensen (1977) found Cu in mine tailings was very inhibitory to formation of *Populus* mycorrhizae, and Göbl and Pümpel (1973) showed that Cu applied to seedlings as fungicide selected preferentially for certain fungal symbionts. Berry and Marx (1976) have reported that mycorrhizae formed by *Pisolithus tinctorius* could partly inactivate high concentrations of toxic metals, such as Zn, resulting from application of sewage sludge.

The germination of spores of VA endophytes is sensitive to metals. On agar, germination of chlamydospores of *G. mosseae* and *G. caledonius* was severely inhibited by Mn and Zn at concentrations commonly found in soil solutions (Hepper and Smith, 1976; Hepper, 1979). However, the metals used in these tests were in ionic form, whereas in soil solutions they are mainly complexed (see Table I in Loneragan, 1975). These results indicate that metal content of soils could greatly influence the distribution and abundance of VAM by affecting the infectivity of the propagule.

The effects of metal pollutants on VAM is interesting. Trappe *et al.* (1973) found that apple trees on soils with an As content of ca 50 ppm formed very few mycorrhizae, but apparently healthy *Glomus* spores were present. Gildon (1981) investigated the reclaimed site of an old lead mine at Shipham, England and found considerable infection in white clover growing on soil heavily contaminated with Zn and Cd. Mycorrhizae were even formed by plants with foliar concentrations of 3750 ppm Zn and 2.1 ppm Cd. A strain of *G. mosseae* isolated from the site was shown to form mycorrhizae in pot-grown clover at concentrations of Zn which completely prevented infection by *G. mosseae* from an agriculture field at Rothamsted. Previous pot work had also demonstrated that the latter agricultural endophyte was incapable of forming mycorrhizae with onion and maize when low concentrations of Cu, Ni, and Cd were added to the soil. This work is an unequivocal demonstration that VA endophytes can develop strains adapted to particular soil conditions.

5.2.1d. Salinity. Effects of salts such as NaCl on mycorrhizae have not been studied in a systematic way. There are several reports of VAM in maritime salt marshes, where hosts of other types of mycorrhizae are absent. Here, about half the host species present are mycorrhizal, but certain genera, notably *Salicornia* and *Triglochin,* always seem to be uninfected (Mason, 1928; Fries, 1944; Wojciechowska, 1966). It is not known whether the endophytes involved show a special adaptation to the saline conditions.

Chloride ions inhibit germination of spores of *Gigaspora margarita* (Hirrel and Gerdemann, 1980). Bowen (1980a) has reported typical VA infections and spores in a soil with > 5000 ppm Cl$^-$.

5.2.1e. Organic Matter and Distribution in Depth. Ectomycorrhizae as well as endomycorrhizae can develop in plants reared on completely inorganic media. But soil organic matter will undoubtedly affect the incidence of mycorrhizae indirectly through its effects on soil structure, water-holding capacity, nutrient mineralization, etc.

Fassi *et al.* (1972) considered highly humified organic matter to be beneficial to ectomycorrhizal establishment in nurseries, and Minchink (in Shemakhonova, 1962) reported that pine seedlings did not form mycorrhizae in soils with less than 2% humus. No direct chemical effects of organic matter have been clearly shown, but Melin (1953) demonstrated that organic matter may contain substances which stimulate the growth of both fungus and host and the establishment of the symbiosis. Furthermore, organic matter may play an important role in providing the substrate for the saprophytic existence of ectomycorrhizal fungi (Boullard, 1964; Dimbleby, 1953) and of orchidaceous mycosymbionts.

Together with the physical and chemical effects imparted by it, organic matter may control the distribution in depth of the fine roots that can become mycorrhizal. Some plants, like apple and strawberry, develop a special system of fine, fibrous collar roots, and if the surface soil remains undisturbed and does not dry out too much, such roots are generally strongly mycorrhizal (Mosse, 1956). The relationship between root depth, fresh weight, percent mycorrhizae, and available phosphate in a Pennine grassland is well illustrated by some figures shown in Table V (Sparling and Tinker, 1975). The data show that in a Pennine grassland, it was primarily the availability of roots that governed mycorrhizal distribution at different soil depths. St. John and Machado (1978) found that in cacao the number of roots and the proportion that were mycorrhizal were closely related to depth. Marked decreases in spore numbers were recorded by Redhead (1977) and Sutton and Barron (1972) between 15 and 30 cm. St. John (1979) also found that in a tropical rain forest, nonmycorrhizal species were more likely to have deep root systems.

Many authors have observed that ectotrophic mycorrhizae are much more abundant in the upper horizons of the soil and have remarked on the concentration of mycorrhizal roots in the litter and raw humus layers (Mikola and Laiho, 1962). Mikola *et al.* (1966) showed that the distribution of mycorrhizal

Table V. Root Fresh Weight, Percent Mycorrhizal Root Length, and Soil Phosphate at Different Depths

Depth (cm)	Root fresh wt. (g/liter)	Mycorrhizal root (%)	Olsen P (mg/kg)
5	91	30	15.5
10	25	42	4.5
25	7	43	3.4

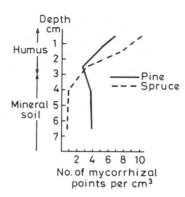

Figure 6. Distribution of ectomycorrhizal roots of Norway spruce and Scots pine with depth (after Mikola *et al.*, 1966).

roots in a mixed stand of pine and spruce differed markedly according to the host species (Fig. 6). Spruce had 84% of its mycorrhizal roots in the humus layer, but pine only had 42%. Göttsche (1972) made a detailed study of the distribution of fine roots and mycorrhizae in a mixed beech and spruce stand. Marks *et al.* (1968) studied the vertical distribution of the mycorrhizae in a good and a poor site of *Pinus radiata* in Australia. On the poorer site, the majority of mycorrhizae were restricted to the top 10 cm. On the good site, mycorrhizal numbers were also greatest near the surface, but there were in addition appreciable numbers of mycorrhizae at a depth of 35–40 cm.

The apparent concentration of mycorrhizal roots in the litter layer, particularly in tropical forests, has led to much speculation about their possible importance in nutrient cycling and preserving nutrients from leaching and mineralization in this environment (Went and Stark, 1968; Stark, 1977). Bowen (1980b) has reviewed the possible role of mycorrhizae in nutrient conservation. Fogel (1979) stressed the importance of the organic input resulting from decomposition processes of fine roots themselves, which he considered to be five times greater than the nutrient release from litter fall and decomposition in a Douglas fir ecosystem. Blaschke (1979) found that short root formation in silver fir was stimulated by the organic soil fraction, namely decayed wood, litter, and humus.

5.2.2. Physical Factors

5.2.2a. Water. The effects of water on mycorrhizae are complex. In very dry or very wet soil, a particular type of mycorrhiza may be absent because its host does not exist, having been replaced by specialized xerophytes or hydrophytes. Waterlogging may inhibit mycorrhizal formation through lack of O_2 since this gas is slow to diffuse through water, but as Scott-Russell (1977) has pointed out, soil water will often contain sufficient dissolved O_2 to prevent anaerobiosis. If reducing conditions do prevail, a complex array of toxic substances such as H_2S, Mn, and various organic acids are liberated. The action of these toxins on mycorrhizae is, apart from the effects of Mn already discussed, unknown and may be at least as important as lack of O_2. Unfortunately, papers on mycorrhizae and soil water have not presented data on redox potentials.

Ectomycorrhizae are usually near the soil surface and may easily be killed by drought (Redmond, 1955). Palmer (1954) showed that periods of drought caused cortical degeneration of mycorrhizae, but restoration of normal moisture levels regenerated roots which rapidly became reinfected. Excessive water killed mycorrhizae of spruce *(Picea excelsa)* and prevented infection of new, short roots, but intensive infection occurred again when the watertable dropped (Wojciechowska, 1960). Under continuously wet conditions, ectomycorrhizae are rare but are formed by older trees *(Alnus)* with the fungi *Hymenogaster alnicola* and *Lactarius obscuratus* (Trappe, 1977b).

Changes in soil water were shown by Worley and Hacskaylo (1959) to alter the species of fungi mycorrhizal with *Pinus virginiana;* under dry conditions, *Cenococcum graniforme* replaced white fungi. Wet soils will generally suppress ectomycorrhizal fungi, which require a good O_2 supply. Miller and Laursen (1978), for example, showed that mycorrhizal fungi of an Alaskan tundra were much less abundant in the bottom of wet sloughs and basins than in their drier rims.

VAM occur over a wide range of soil water contents. Infection has been found in arid regions of the U.S. (R. M. Miller, 1979; Reeves *et al.,* 1979), in xerophytic plants in Pakistan (Khan, 1974), and in Libyan soils (El-Giahmi *et al.* 1976). At the other extreme, VAM occur in the very wet soils of marshes (Dowding, 1959) but are comparatively rare under such conditions; this is because the vegetation may be dominated by intrinsically nonmycorrhizal hosts of the Cyperaceae and Juncaceae and other hydrophytes (Khan, 1974) and, most importantly, because the chemical and physical environment is unfavorable. Read *et al.* (1976) reported low levels of infection in an English wetland, and Mejstrik (1965) found that infection intensity varied inversely with the height of the water table in plants of a *Cladietum mariscii* association. Further evidence of the adverse effects of waterlogging is provided by Mmbaga (1975), who found infection in upland rice but not in paddy rice, and by Shuja *et al.* (1971), who noted ecto- and endomycorrhizae only in roots of *Populus americana* growing away from, but not towards, an adjacent canal. The reports of VAM in aquatic environments (Søndergaard and Laegaard, 1977; Bagyaraj *et al.* 1979) do not contradict these observations, since in water there is probably adequate dissolved O_2, and toxic concentrations of metals are probably absent. Typical VAM have been produced in corn and beans *(Phaseolus vulgaris)* in nutrient-flow cultures in which roots were exposed to a shallow layer of circulating nutrient solution (Mosse *et al.,* 1978).

VA infection is usually found to be most rapid at soil water contents less than the maximum water-holding capacity (WHC) of the soil (Wojciechowska, 1966; Reid and Bowen, 1979; Sieverding, 1979). The sensitivity of spore germination to soil water content reported for *G. mosseae* (Reid and Bowen, 1979) may be a general feature of VA endophytes and could account for some of the observed effects of water.

Soil water may select for certain species of VA fungi. Saif *et al.* (1975), working in Pakistan, showed that the bulbous reticulate spore type (*Gigaspora* sp.) was only prevalent in soils which remained at around 50–60% WHC.

Ericoid mycorrhizae also tolerate a wide range of soil water content. They are found in the dry *Erica* heaths of South Africa, the epacridaceous heaths of Australia, and also the wet *Calluna*- and *Vaccinium*-dominated communities of Western Europe. Read (1978) found that mycorrhizae of *Erica bauera* in South Africa degenerated during the dry season, but new roots which grew after heavy rainfall were found to be healthy and well infected. In the pot experiment of Bannister and Norton (1974), formation of *Calluna* mycorrhizae was unaffected by the watering regime.

5.2.2b. Temperature. Pot experiments have shown mycorrhizal development to be strongly temperature-dependent. Where soil temperature varies considerably with the season, as in temperate zones, it may be the major rate-limiting factor in infection spread. Most tropical soils remain at a relatively constant temperature, and here the effects may be less.

The influence of temperature upon *in vitro* growth of ectomycorrhizal fungi has been extensively studied (Melin, 1925; Mikola, 1948; Moser, 1956; Hacskaylo and Vozzo, 1965; Marx *et al.,* 1970). In general, the fungi grow best at around 20°C, that is, at an optimum greater than the temperatures they would encounter in the field. Hacskaylo and Vozzo (1965) reported that the species they tested would not grow at 2 and 35°C. There is little known about the influence of temperature on development of ectomycorrhizae apart from the study of Marx *et al.* (1970), who showed that the temperature optima varied for different fungi. *Pisolithus tinctorius* was notably tolerant of high temperatures, even forming mycorrhizae at 34°C. In the field, certain fungi are associated with extremes of temperature. *Pisolithus* mycorrhizae are found in the very hot substrates of coal mine wastes (Schramm, 1966), whereas *Boletus plorans* forms mycorrhizae in alpine conditions. At relatively high temperatures, ectomycorrhizae can be formed by a very wide range of fungi, as strikingly demonstrated by the successful establishment of mycorrhizae on *Pinus caribea* in Puerto Rico with Scandinavian strains of the fungi *Rhizopogon roseolus* and *Corticium bicolor* (Vozzo and Hacskaylo, 1971).

Increased soil temperatures hasten the development of VA mycorrhizae. Furlan and Fortin (1973) found infection in onion to be markedly faster in a growth cabinet at 21°C night/26°C day than at 11°C night/26°C day, but they did not control root and shoot temperatures separately. Smith and Bowen (1979) reported infection in *Medicago* and *Trifolium* to be most sensitive to increase in temperature in the range 12–16°C. They showed that temperature strongly affected the infection process, as measured by the number of entry points per centimeter of root. This sensitivity of infection to temperature may explain the slow development of infection in agricultural crops on temperate soils (Black and Tinker, 1979), where soil temperatures range from 2 to 15°C (Hay, 1976).

A feature lacking in temperature studies is a comparison of endophytes from different climates, for it is possible that strains and species may be temperature-adapted. This is suggested by the work of Schenck *et al.* (1975), who found that two isolates of *Glomus* from Florida germinated best at 34°C, whereas one from Washington had an optimum of 20°C; additionally, the observation of Daft *et al.* (1980) suggested that new roots growing from a bulb of bluebell *(Endymion nonscriptus)* became well infected during the winter months when soil temperatures were generally as low as 5°C. Since most species of VA fungi are worldwide in their distribution, it seems likely that temperature adaptation is common. This is an area worthy of future study.

5.2.2c. Aeration. The O_2 content of surface layers of soil is relatively high and stable and may be unimportant to mycorrhizae, except when, as discussed above, anaerobiosis is induced by waterlogging. By contrast, the CO_2 concentration in soil air may vary widely, but its effects on mycorrhizae are unknown.

Lack of O_2 is more likely to affect the survival and spread of mycorrhizae in soil than within the root since O_2 is transported down air-filled spaces within the plant from shoot to root (Greenwood, 1967). In agreement with this, Read and Armstrong (1972) showed that O_2 diffusion from roots may provide the initial stimulus for ectomycorrhizal infection and maintain the physiological activity of the mycorrhizae, once formed.

LeTacon, Skinner, and Mosse (unpublished data) found that germination of *Glomus mosseae* spores was inhibited by reduced O_2 tension and that subsequent hyphal growth was poor. When spores were germinated in air, however, their subsequent growth was little affected by lack of O_2 until the O_2 tension was below 3%. These results imply that lack of O_2 could markedly reduce VA infection through its effect on spores, but this idea has not been tested.

5.3. General Observations

The effects of soil and other environmental factors on the development of mycorrhizae are difficult to assess or review. Much of the literature consists of disparate observations, lacking comprehensive data and not undertaken to test a specific hypothesis. One is sometimes reminded of MacFadyen's (1975) warning that "the aimless and disorientated ecological survey is a bottomless pit for precious time, effort and enthusiasm." Perhaps it is possible to itemize some particular points worth considering.

a.   Soil factors rarely vary independently, and for that reason alone, a simple relationship between soil properties and the prevalence of mycorrhizae is unlikely to be found in an ecological survey. Also, some confusion might be avoided if individual effects of environment on infection processes and on subsequent development of mycorrhizae were considered and studied separately.

   b. A striking feature of mycorrhizae in general is their tolerance to a wide range of soil conditions and, especially for VA endophytes, a worldwide distribution of the fungal species. This strongly implies adaptation to a whole range of soil factors. The findings of Lambert et al. (1980) and of Gildon (1981), among others, support this. There will be little consistency in reported findings till this is further investigated and fully appreciated.

   c. It is now clear that the external mycelium is very important in the functioning of mycorrhizae. For obvious reasons of technique, fragility and inaccessibility of the soil mycelium, and inability to distinguish between active and nonfunctional mycelium, very little quantitative information is available and almost none relating to natural soils. In sterilized soils, VA mycorrhizae can produce up to 80 cm hyphae/cm of infected root (Sanders and Tinker, 1973), which can extend up to 8 cm into the soil (Rhodes and Gerdemann, 1975). The mass produced is proportional to root growth (Sanders et al., 1977), and in agar cultures, the mass of internal and external mycelium is roughly equal (Hepper, unpublished data) but can be varied by Ca levels (Elmes and Mosse, 1980) and by phosphate source (Mosse and Phillips, 1971). Skinner and Bowen (1974) showed that the length of ectomycorrhizal strands was strongly influenced by soil type and compaction. Also, other soil microorganisms, such as mycophagous nematodes, Aphelenchoides spp. (Riffle, 1967; Schafer, unpublished data), can affect the soil mycelia of ectotrophic, VA, and ericoid mycorrhizae. Collembola also eat VA mycelia, reducing symbiotic efficiency of the fungi (Warnock, Fitter, and Usher, unpublished data). Hyperparasitic fungi (Ross and Ruttencutter, 1977) and other unspecified biotic factors affect sporulation. Ross (1980) has speculated that such factors may be vital in controlling inoculum density of VA propagules. Integrated studies and also new techniques are needed to obtain more information about the soil mycelia, inoculum density, and the nature and distribution of infective propagules, with the ultimate objective of a better understanding of the significance of mycorrhizae in natural ecosystems.

6. Interactions with Other Components of the Ecosystem

6.1. Microorganisms

   Concepts of rhizosphere microbiology are presently undergoing considerable change (Bowen, 1980b). They are becoming more dynamic, concerned as they are with colonization patterns and with interactions in defined spaces, the rhizosphere, and the rhizoplane. Also, the physiology and distribution of roots are now often included as emphasis is shifting from the mere identification and enumeration of organisms that can be isolated on synthetic media. Therefore, the important role of mycorrhizal fungi as components of the rhizosphere population—previously ignored because they could not be cultured—is receiving more recognition. The new approaches may lead to a better understanding of mycorrhizal systems, but that lies in the future. So far, most observations on relationships between mycorrhizal fungi and other microorganisms have been on a one-to-one basis, though a few may include also the plant. At best, one can pick out some facets, from a wealth of indiscriminate observations, where mycorrhiza may have an impact on the ecology of other soil microorganisms and vice versa. Some of the more interesting are: (a) increased root and therefore organic matter production because of increased plant growth; (b) interactions with other soil microorganisms (fungi and bacteria) that affect the spread of mycorrhizal infection, germination of spores, and establishment and persistence of the other organism; (c) synergistic effects with other microorganisms affecting plant growth, notably symbiotic nitrogen fixation; and (d) interactions with plant pathogens (viruses, fungi, and nematodes).

6.1.1. Factors Arising from Nutritional Effects on Plant Growth

Fogel's (1979) findings that mycorrhizae in a Douglas fir system provided 50% of the annual biomass throughput and 42% of the nitrogen released annually give some measure of the importance of mycorrhiza in providing substrate for other soil microbial processes. Tranquillini (1964) found similar results with *Pinus cembra* at the tree-line in the Alps. About 40% of the carbon fixed was released as CO_2 by the tree, while the rest, about two-thirds, was utilized mainly by the mycorrhiza and by secretion in the rhizosphere. It is likely that, in terms of root exudate, such effects are not only quantitative but also qualitative. Katznelson *et al.* (1962) showed that nonmycorrhizal birch roots harbored mostly bacteria with simple mineral nutrient requirements, whereas those associated with mycorrhizal roots had more complex amino acid requirements and a lower proportion of rapidly growing phosphate-solubilizing bacteria. Oswald and Ferchau (1968) reported a general stimulation of microbial activity in soil surrounding ectomycorrhizae and coined the term *mycorrhizosphere*. This may extend to the rhizomorphs and soil mycelia. Slankis (1974) reviewed such findings in greater detail. The hyphae of some, but not all, VA endophytes appear to be coated with a mucilaginous material (Clough and Sutton, 1978; Tisdall and Oades, 1979) and are often covered with bacteria. Bagyaraj and Menge (1978) found that mycorrhizal tomatoes harbored more bacteria and actinomycetes in their rhizosphere than did nonmycorrhizal plants. Other similar observations exist, but it is difficult to assess their significance. Although one might well surmise that nutritionally richer plants would produce more and richer exudates, Ratnayake *et al.* (1978) made the interesting observation that roots of P-deficient plants were actually more leaky than those adequately supplied with P.

6.1.2. Interactions Affecting Establishment of Other Microorganisms

In a careful study of eight bacteria and eight ectomycorrhizal fungi, Bowen and Theodorou (1979) found that different bacteria could depress, have no effect, or even stimulate root colonization by mycorrhizal fungi and that such effects did not necessarily correspond to their *in vitro* effects on fungal growth. Some bacteria gave protection against the depressive effects of others. Depression effects may be based on antibiosis or competition for nutrients. Interestingly, the strain of *Bacillus* that stimulated growth of *Rhizopogon luteolus* was isolated from the mycelia of that fungus in soil.

Mycorrhizal fungi can inhibit or stimulate each other as well as other soil fungi. Levisohn (1957) found that *Alternaria tenuis,* which is common in arable soils, inhibited growth of *Rhizopogon luteolus* and four *Boletus* species and that in the field only *Leccinum scabrum* was established, but not the other mycorrhizal fungi that were more susceptible to *A. tenuis* in laboratory media.

Spore germination of ectomycorrhizal fungi has been studied extensively by Fries (1978, 1979). There is some evidence that in natural environments other microorganisms, like *Rhodotorula* and *Ceratocystis,* as well as growing roots, may be producers of germination-inducing substances.

Persistence of various bacterial inocula can be greater in the rhizosphere of mycorrhizal than of nonmycorrhizal plants (see Section 6.1.3).

6.1.3. Interactions with Synergistic Effects on Plant Growth

The most important of synergistic effects is probably the interaction between mycorrhizae and rhizobia and possibly other symbionts of nonleguminous, nitrogen-fixing plants. This subject has been quite extensively reviewed by Daft (1978), Mosse (1977b), and Munns and Mosse (1980) and also in a range of research papers. Mycorrhizae markedly improve nodulation and nitrogen fixation by helping to provide the rather high phosphorus requirements for the fixation process. They possibly have some other, as yet undetermined, secondary effect leading to better nitrogen fixation. In return, the improved plant growth due to a better nitrogen supply offers more root for mycorrhizal colonization, which is important for an obligate symbiont. It is not known whether ectomycorrhizal infection in the Caesalpinioideae has similar effects.

Mycorrhizal infection allowed introduced populations of *Azotobacter chroococcum* (Bagyaraj and Menge, 1978), *Azotobacter paspali* (Brown, 1976), an *Azotobacter* species epiphytically associated with *Glomus* spores

(Ho and Trappe, 1979), and phosphorus-solubilizing bacteria (Azcón *et al.*, 1976) to maintain high numbers for longer than nonmycorrhizal plants and to exert synergistic effects on plant growth. In the case of the azotobacter, these effects were thought to be due to hormone production rather than to nitrogen fixation by the free-living azotobacter. In the case of the phosphate-solubilizing bacteria, the benefits were probably due to localized mobilization of added rock phosphate in alkaline soils by the bacteria and the effect of the mycorrhizae in transferring some of this mobilized phosphate to the plant. Azcón *et al.* (1978a,b) also found that cell-free extracts of *Rhizobium, Azotobacter,* and a phosphate-solubilizing *Pseudomonas* sp. increased the growth of three plant species by 40–180% and marginally increased the percentage infection in plants inoculated with VA endophytes. A mixture of plant hormones had a comparable effect. It is unlikely that the small increases in infection would by themselves account for the large growth increases, and one might speculate that the hormones produced by the bacteria could exert some synergistic effect on plant growth or mycorrhizal efficiency.

6.1.4. Interactions with Pathogens

Interactions between mycorrhizae and pathogens are only a special case of interaction with other microorganisms. In spite of numerous investigations, much the same degree of unpredictability and ambivalence prevails. Almost every article on mycorrhizae contains a section on disease resistance. Marx (1973) reviewed the case of ectomycorrhizae and Schenck and Kellam (1978) and Schönbeck and Dehne (1979), that of VA endophytes.

Like most instances of biological control, mycorrhizae never confer complete immunity against any disease or pest. They can impart a degree of resistance or tolerance against root and collar rots, often caused by species of *Phytophthora, Pythium,* or *Rhizoctonia,* to wilt diseases, and to attack by nematodes. Conversely, mycorrhizal infection appears to make many plants more susceptible to leaf pathogens (Schönbeck and Dehne, 1979) and to viruses (Daft and Okusanya, 1973) which multiplied primarily in arbuscule-containing cells (Schönbeck and Spengler, 1979). Both increased tolerance to root pathogens and susceptibility to leaf pathogens may be the result of better plant nutrition. Various direct suppression mechanisms have been postulated and demonstrated, viz. mechanical exclusion of pathogens by the fungal sheath, competition for infection sites (Becker, 1976), production of antibiotics, volatiles (Krupa and Fries, 1971), or polyphenols (Foster and Marks, 1967) by mycorrhizal fungi or roots, production of phytoalexins (Nuesch, 1963), and changes in host metabolism, e.g., higher arginine, chlorophyll, or phosphatase content affecting pathogen metabolism (Baltruschat and Schönbeck, 1975; Dehne and Schönbeck, 1978).

On the whole, mycorrhizal fungi seem relatively resistant to most biocides applied at recommended rates. Benomyl appears to be one of the most toxic. They may even be stimulated by some nematicides, like aldicarb, possibly through elimination of nematodes or other fauna that feed on or otherwise destroy the integrity of the soil mycelium. Ectomycorrhizal fungi seem more sensitive. Calomel and benzene hexachloride severely inhibited mycorrhizal development (Persidsky and Wilde, 1960). Hacskaylo (1961) reported that cycloheximide, which is used against *Cronartium ribicola,* inhibits the growth of different ectomycorrhizal fungi. Most biocides used in forest nurseries affect ectomycorrhizal development, but ethylene dibromide and Nemagon do not suppress mycorrhiza (Hacskaylo and Palmer, 1957). Recolonization after soil fumigation can also be quite rapid, presumably by propagules in the lower soil layers, and there is even some evidence that resistance can build up (Hayman, 1970). Certainly, pineapple plantations in Hawaii that are regularly treated with nematicides have remarkably high rates of mycorrhizal infection. However, after partial soil sterilization by steaming in forest nurseries, pseudomycorrhizal mycelia attributed to *Rhizoctonia* sp. form root infections earlier— or they produce a more limited number of infections—than ectomycorrhizal fungi that are prevented from forming efficient associations (Levisohn, 1965).

One of the most interesting aspects of mycorrhizal systems is the control of the fungal symbiont by the host. This applies particularly where the symbiont is a pathogen on other hosts. The most obvious example of this is *Armillaria mellea,* a virulent pathogen on plantation trees, that is yet apparently widespread and fairly harmless in many forests. It applies to the orchidaceous rhizoctonias, where endogenous antifungal substances of a phytoalexin nature have been demonstrated (Nuesch, 1963). But it applies equally to the much less aggressive VA endophytes, for which we have no knowledge how the fungus is confined to the primary cortex.

6.2. Higher Plants

Mycorrhizae, or lack of mycorrhizal fungi or hosts, can affect the composition of plant communities. Two types of interaction are considered here— the preferential stimulation of one species in a mixed community and the effect of host and nonhost plants growing together.

6.2.1. Competition between Species in Mixed Communities

The ability of plant species to grow at a particular site is governed by climatic and edaphic factors. Among the latter, nutrient availability is one of the most important, and the ability to form mycorrhiza improves the growth potential in nutritionally marginal sites. The ability of a species with a relatively restricted root system, e.g., many trees and legumes (Barley, 1970), to establish in such environments in competition with less demanding species or ones with more extensive root systems, e.g., many grasses, can be contingent on mycorrhizal infection. Bowen (1980a) discusses ways in which competition for scarce nutrients among species in a mixed community can be reduced. He lists three mechanisms, viz., stratification of rooting zones (allowing some surface-rooting species to explore the upper soil layers while others draw nutrients from lower down), staggering the times of major nutrient demand, and preferential stimulation of one species through mycorrhizal association. This can occur, for instance, when legumes are grown in pastures. Only if the legume can obtain sufficient phosphate, by virtue of its mycorrhizal system, to supply the phosphate necessary for symbiotic nitrogen fixation can it compete successfully with the grasses. The greater responsiveness of legumes to mycorrhizal infection has been shown in several pot experiments in which grasses and clover were grown, either singly in the same soil (Sparling, 1976) or together (Crush, 1974; Powell, 1977a; Hall, 1978). Whereas root density of fescue became supra-optimal after ten weeks and rate of spread of the infection thereafter declined, rate of spread and root density increased roughly proportionally up to 15 weeks in clover (Warner, 1980). Thus, in a mixed community, the more sparsely rooted plants may derive more benefit from the mycorrhizal infection and thus improve their competitive ability. Bowen (1980b) postulated that this might occur if infection decreased as nutrient requirements were met in the more efficient species. Such species differences in response to infection also occurred when two grasses, *Lolium perenne* and *Holcus lanatus,* were grown together (Fitter, 1977). Mycorrhizal infection gave considerable advantage to *H. lanatus* in the uptake of both potassium and phosphate and actually decreased the growth of *L. perenne* in the presence of the other grass. But *L. perenne* decreased infection of *Plantago lanceolata* grown with it by 40% (Christie *et al.,* 1978). On the whole, mycorrhizal infection in this investigation was less influenced by host species interactions than was the bacterial population.

Long ago, Stone (1949) showed a similar interaction between ectotrophic Monterey pine seedlings and the normally VA mycorrhizal sorghum. When the two were competing for scarce phosphate, the sorghum in two soils weighed, respectively, 1/20 and 1/3 as much when grown with mycorrhizal than with nonmycorrhizal pine seedlings. Onset of infection in the two species may have influenced this result as the pine, which was mycorrhizal by inoculation, may have become infected sooner than the sorghum, which depended on natural infection in the soil. Once such an advantage is established, it is likely to be self-perpetuating, as the plant with the more extensive root system will have better chances of contacting undepleted soil. Given sufficient phosphate, there was no obvious antagonism between the sorghum and the pine.

Antagonism (allelopathy) can, however, also occur between the mycorrhizal fungi of one species and another, as in the case of *Calluna* heaths, whose unsuitability for afforestation was attributed by Handley (1963) to the effects of the *Calluna* endophyte itself, and by Robinson (1972) to *Calluna* root exudates on ectomycorrhizal fungi. Another interaction between a mycorrhizal fungus and nonhost species occurs with *Tuber melanosporum*. When roots of *Quercus* and *Corylus* are almost totally infected by this fungus, all herbaceous plants around these trees appear "burnt," seemingly by some volatile compound emitted by the fungal mycelia in the soil (Grente and Delmas, 1974).

It is well known that, as well as antagonisms, there are synergistic relationships between different plant species in mixed communities. Occasionally, the suggestion is made that transfer of substances might occur between synergistic plant species through their common mycorrhizal fungi. Such transfer could involve two live plants but may also occur between senescent and live roots of the same plant, different plants of the same species, or plants of different species. Woods and Brock (1964) injected tree stumps with radioactive Ca and P and recovered isotope in nearby vegetation, but this experiment is open to considerable criticism since dying roots attached to the tree stumps would certainly release isotope into the soil, where it could be picked up by the soil mycelium in the normal way. More recently, Read *et al.* (1979) claimed to have demonstrated transfer of ^{32}P between *Plantago* and *Festuca* spp. via mycelial bridges. With one plant acting as source and the other as sink, transfer was greater when both were mycorrhizal. Although mycelial connections between disparate plant species presumably occur, since it is possible to infect one from the other, it is exceedingly difficult to demonstrate this in practice (Mosse, unpublished; Heap and Newman, 1980), and the frequency of such connections remains to be established. Read *et al.* (1979) suggested that seedling survival under a dense and intensely competitive sward may be assisted by transfer of nutrients from mature to developing plants by way of a common mycorrhizal associate.

6.2.2. Effects of Nonhosts

It is evident that soil infectivity will be reduced in sites long inhabited by nonhost plants and that such sites may be difficult to colonize subsequently with highly mycorrhiza-dependent plants. Examples of this are discussed in Section 7. There may also be more immediate effects when host and nonhost plants are grown in close proximity, but the experimental evidence is conflicting. Hayman *et al.* (1975) found that swedes, a nonhost, depressed infection in onions grown together with them in the same pot by 70%, but in later experiments Ocampo *et al.* (1980) found no subsequent adverse effects of four cruciferous and one chenopodiaceous nonhost species. Iqbal and Qureshi (1976) found a reduction of 13–40% infection in eight wheat varieties grown with mustard in the field and attributed this inhibition to sulfur-containing exudates from the crucifer roots. They considered this finding might have relevance to agricultural practices in India. There can also be leachable toxic substances in some seeds or seed coats, as in *Lupinus cosentinii,* a legume normally not or only very slightly infected (Morley and Mosse, 1976; Trinick, 1977).

Conversely, it seems that mycorrhizal hosts can have some effect in allowing VA endophytes to colonize, to a very limited extent, plants normally immune to mycorrhizal infection (Hirrel *et al.,* 1978; Parke and Linderman, 1980; Ocampo *et al.,* 1980). Ocampo and Hayman (1981) believe that even roots of nonhost species may have some stimulatory effect on inoculum placed into a soil as compared with the complete absence of roots such as might be produced by some herbicide treatments.

7. Successions

Four types of successions will be considered:

a. Successions in sites without previous vegetation, e.g., dunes and reclaimed soils or locations such as mine spoils, oil shale wastes, open-cast mining sites, gravel pits, and other rehabilitation sites where top soil has been brought to the surface or stored for prolonged periods with marked reduction in indigenous mycorrhizal fungi.

b. Successions from natural ecosystems to cultivated ecosystems, which can be planted forests or agriculture.

c. Successions within forests as they mature or change species composition, or within agricultural systems in relation to monoculture or crop rotation.

d. Changes in fungal populations after hosts are harvested.

7.1. In Previously Uncolonized Sites

Nicolson (1960), Nicolson and Johnston (1979), Koske (1975), and Koske *et al.* (1975) have studied the occurrence of mycorrhizal infection and spores of VA endophytes in dune successions in Scotland, Australia, and Lake Huron. All have found increasing mycorrhizal development, both in number of infected roots and of spores, with increasing dune stabilization from foredune to first and second dune. Some marine dunes contained endophyte spores, but others did not. They also remarked on the organic particles, which were probably infested root debris that frequently harbor many spores and may function as an important inoculum source in the dunes. Presumably, such particles can be windblown from the more established dunes and inland vegetation.

Koske *et al.* (1975) observed that endophyte mycelia in the soil have the exceptional ability to bind sand and soil particles, and Tisdall and Oades

(1979) have since confirmed that hyphae have an important function in increasing stable soil aggregates greater than 2.0 mm. A layer of amorphous material on the hyphal surface appears to help in the binding process. Mycorrhiza may thus have an important function in soil stabilization and perhaps in erosion control.

Old and young dunes in southwestern France have been successfully colonized by direct seeding with *Pinus pinaster.* Presumably, the spores of ectomycorrhizal fungi were windblown.

Schramm (1966), Marx (1975), Daft *et al.* (1975), Lindsey *et al.* (1977), Harris and Jurgensen (1977), Khan (1978), and others have studied the mycorrhizal status of plants colonizing coal spoils. Shrubby species such as *Robinia hispida, Comptonia perigrina,* and *Alnus glutinosa* that colonized these wastes naturally had bacterial or actinomycete nitrogen-fixing nodules and were also ectomycorrhizal (Schramm, 1966), but according to Daft *et al.* (1975), many herbaceous plants on such sites were endomycorrhizal, including *Robinia hispida* and *Eleagnus umbellata.* Schramm (1966) observed that the normally endotrophic maples and sweet gums adjacent to the coal spoil could not successfully colonize it and assumed that this was due to lack of VA endophytes in the coal spoil. One of the first ectomycorrhizal fungi growing on coal wastes is *Pisolithus tinctorius* (Marx, 1975) because of its tolerance of high temperatures.

Several workers have noted that ephemeral species invading disturbed sites are inherently nonmycorrhizal (R. M. Miller, 1979; Reeves *et al.,* 1979; Allen and Allen, 1980). Any speculation about this should include the point made by Grime (1979) that ephemerals have evolved a strategy whereby resources are quickly diverted from rapid growth to flower and seed production. This *opportunistic* strategy will in itself render such ephemerals less competitive with subsequent colonizers regardless of their mycorrhizal status. Reeves *et al.* (1979) and Allen and Allen (1980) have made the suggestion that nonmycorrhizal ephemerals may delay the establishment of mycorrhizae by maintaining low inoculum density in the soil.

7.2. From Natural to Man-Made Ecosystems

Two examples of this, the afforestation of prairie and steppe soils and the introduction of temperate trees into tropical and equatorial soils, have already been noted.

Under tropical or equatorial conditions, the mycorrhizal fungi are mostly endotrophic, except in marginal conditions at high elevations or on very poor soils (Singer, 1978). Ectotrophic trees cannot establish themselves without inoculation with appropriate fungi. Many relevant observations have been made in Australia, Africa, and some Caribbean islands.

One of the best examples of the introduction of ectomycorrhizal inoculum occurred in Puerto Rico. According to Vozzo and Hacskaylo (1971), the first attempt to establish pines on the island was made in 1928. No success was recorded in attempts with *P. caribea, P. pinaster,* and *P. radiata.* In 1948 superphosphate was suggested as a possibility for improving growth. The result was a complete failure. Inoculation with ectomycorrhizal fungi was then suggested, and in 1955 a humus and A_1 horizon from a U.S. plantation of *Pinus echinata* and *P. taeda* were used. It was a complete success. One year after inoculation, the tallest pine seedlings were approximately 1.5 m high and had developed abundant mycorrhizae, whereas most of the controls had died. In 1963 pure cultures of four ectomycorrhizal fungi were introduced, as well as spores of three other species, and there was marked growth stimulation in the seedlings that became mycorrhizal.

Another type of succession is that from heath and moorland to forest plantation. In Europe, large areas of cultivated land have been abandoned by agriculture for decades or even centuries. Such areas, particularly abundant on acid soils in France and Britain, are always difficult to re-afforest. They are rarely colonized by ectomycorrhizal hosts. Often the first tree to appear when grazing is discontinued in the "Massif central" is an endotrophic species, *Juniperus communis.* Except for birch, which is mycorrhizal with *Boletus scaber* (Dimbleby, 1953), trees planted on *Calluna* moorland grow very slowly and have few or no mycorrhizal roots. Handley (1963) showed that aqueous extracts of *Calluna* roots or humus inhibited growth of many ectomycorrhizal fungi, except *Boletus scaber* and *Amanita muscaria.*

Relatively little is known about the changes in VA endophyte populations when virgin soil is brought into cultivation. Schenck and Kinloch (1976) annually monitored the endophyte population of a woodland area for four years after clearance. Most spores were found after the first year and the maximum number of species after the second. Thereafter, both spore and species numbers fell. Janos (1980) found that a tropical rain forest was established with difficulty in swampy areas that had been colonized mainly by *Carex* (a nonmycorrhizal genus). He attributed the difficulty in establishment to lack of VA endophytes in the soil.

Crush (1978) studied the effectiveness of endomycorrhizal populations during pasture development, and he concluded that there was a rapid increase in the effectiveness of soil endophyte populations as pasture development proceeded. He assumed this to be the result of host selection for effective endophyte strains from a diverse soil population under a changing environment. Others, however, have found that the indigenous endophytes are not always the most efficient, even if soil conditions are not altered (Mosse, 1975, 1977a; Powell, 1976b, 1977b; Powell and Daniel, 1978b; Abbott and Robson, 1978).

The sensitivity of many indigenous endophytes to changes in pH and soil nutrient levels has been demonstrated in a series of pot experiments already discussed. The results clearly show that changes in endophyte populations can be induced by inoculation and probably occur naturally in response to changed soil conditions brought about by agricultural practices.

7.3. Within Established Systems

7.3.1. Monoculture

The cultivation of ectomycorrhizal forest trees in pure stands is a common practice in many European forests. Such stands may be monotonous and uniform with only one or two species, but diversity is present belowground. One can commonly find 50 or more species of ectomycorrhizal fungi in a pure spruce or beech forest. But in mixed forests, there is still greater diversity. In a mixed stand of beech, birch, and hornbeam one finds both the fungi common to all three hosts and those specific for birch (e.g., *Lactarius torminosus* and *Boletus leucophaeus*) or hornbeam (e.g., *Boletus carpini*). Equally, a mixed stand of deciduous trees and conifers is always richer in ectomycorrhizal species than is a pure stand of either. Also, an artificial stand is often richer in ectomycorrhizal species than a natural stand of the same species in the same ecological conditions. When, in the Jura, a natural spruce stand is followed by a planted one, new species of ectomycorrhizal fungi are often introduced with the seedlings from the nursery. Some fungi, like *Thelephora terrestris,* are well adapted to nursery soils and disappear after outplanting in forest conditions. Distinct successions of different sporophores are generally observed as the plantation ages and a closed canopy develops.

Foster and Marks (1967) observed that in a 42-year-old stand of *Pinus radiata,* a new fungus could displace an older one. When a change in mycorrhiza occurred, the new fungus broke the continuity of the old mantle and formed a new mantle on the fresh root surface. Smooth-black, bristly-black, and white-rhizomorphic forms of mycorrhizae frequently replaced the reddish-grey, smooth-white, and reddish-brown types. Foster and Marks considered that this change in association during the life of a root could be related to changes in microenvironment. Read *et al.* (1977) even observed a succession from VA to ectomycorrhiza *(Cenococcum graniforme)* in *Helianthemum chamaecistus.*

Prolonged monoculture of agricultural crops can lead to a buildup of VA infection and spore numbers or to a decline according to host species and fertilizer treatment (Strzemska, 1975; Kruckelmann, 1975). Whereas infection and spore numbers can remain high after continuous wheat (Strzemska, 1975; Kruckelmann, 1975; Hayman, 1975), even after appreciable fertilizer additions, they dropped to very low levels after 16 years of potatoes (Kruckelmann, 1975).

7.3.2. Forest Successions and Crop Rotations

A common forest succession in the planted forests of Europe is coniferous after deciduous, e.g., spruce or Douglas fir after beech or oak. Most of the ectomycorrhizal fungi associated with the deciduous trees, viz. *Amanita muscaria, Boletus edulis, Russula cyanoxantha,* etc., can also form mycorrhiza with the coniferous species, but some, like *Craterellus cornucopioides,* tend to disappear, while new species specific to conifers, e.g., *Lactarius deliciosus,* appear. *Boletus elegans* often develops with *Larix europeus,* and *Boletus amabilis* with *Pseudotsuga douglasii* (Mikola, 1970). In the tropics, many ectomycorrhizal fungi were introduced from Europe or North America into coniferous plantations. According to Mikola (1970), the most common in Africa are *Boletus luteus, B. granulatus, B. edulis, Rhizopogon luteolus, R. roseolus, Amanita muscaria, Thelephora terrestris,* and *Hebeloma crustuliniforme.*

There has been some interest in the effect of crop rotations, particularly those involving nonhost species, on endophyte populations of agricultural soils. Infection and spore numbers in barley declined somewhat after one year fallow, markedly after two years fallow, and even more drastically after a nonmycorrhizal crop, kale (Black and Tinker, 1977, 1979); after a second barley crop, such reductions were no longer apparent. Kruckelmann (1975) also noted interactions between a crop and precrop affecting spore numbers. However, Ocampo and Hayman (1981) found that in four crop pairs tested, infectivity was not depressed in soil previously cropped with a nonhost plant. This conclusion was based on a pot experiment in which the precrop grew for less than three months in the test soil. While these results do not allow a firm conclusion about the effects of crop rotations on mycorrhizal infection, it seems likely that a nonmycorrhizal precrop left in the soil for a whole growing season would reduce soil infectivity, at least during the important stage of early seedling growth of the subsequent crop. Not only nonhosts but also different host species can have differential effects on infectivity.

7.4. Survival of Fungi after Disappearance of the Host

It is difficult to obtain good evidence about the survival of mycorrhizal fungi after disappearance of the host, because fresh spore inoculum of ectomycorrhizal fungi is difficult to exclude from any natural, or even experimental site (Robertson, 1954); furthermore, VA endophytes have such a wide host range, including weeds, that it is difficult to exclude all potential hosts.

Ectomycorrhizal fungi differing in saprophytic ability are also likely to differ in their ability to survive. Dimbleby (1953) assumed that they persisted for several years in a saprophytic state after deforestation until stumps were completely mineralized. Warcup (1959) reported that an ectomycorrhizal fungus could frequently be isolated from dead or decaying grass roots. Most ectomycorrhizal fungi can probably survive for a few months. In forest nurseries of Europe, seedlings are usually transplanted in autumn. The young seedlings sown the following spring probably become infected from hyphae remaining in the soil or in decaying or dead roots, because with the exception of *P. tinctorius, T. terrestris, Laccaria laccata,* and *Tuber albidum,* and some others, ectomy-corrhizal fungi do not fructify in nurseries.

Resting spores of VA endophytes have chitinous walls that are relatively resistant to microbial degradation, and can survive for 2–3 years in the laboratory, although viability gradually declines. They also have the ability to return to dormancy after germination if the germ tube does not encounter a suitable host root. They may then germinate again (Mosse, 1959) up to ten or more times (Koske, 1981a) if the original germ tube is removed. The spores are relatively resistant to desiccation and survive well at low temperatures (Tommerup and Kidhy, 1979; Mertz *et al.,* 1979). The mycelium also has an unusual ability to repair damaged sections (Gerdemann, 1955b) and to move cytoplasmic contents, forming septa to cut off empty, nonfunctional portions.

8. Effects of Man on Distribution

Most of man's activities on the land, ranging from organized agriculture and forestry to such disruptive processes as mining and road building, will have effects on the extent of mycorrhizal infection and on the population of mycorrhizal fungi. As the previous sections show, detailed knowledge of such incidental effects is very limited. Relatively recently, the notion of deliberately changing populations of mycorrhizal fungi or of introducing new ones has received some serious attention. These two aspects will be considered separately.

8.1. Incidental Effects

It is likely that the development of agriculture has greatly modified the distribution of mycorrhizal fungi. The deforestation that occurred in temperate countries has resulted in the disappearance of most ectomycorrhizal fungi from the cultivated soils and has favored the spread of endomycorrhizal fungi. All soil movements on implements or machinery and on man and livestock will assist the spread of mycorrhizal propagules and could, in some instances, be a contributory factor in localized distribution patterns. Nevertheless, such assorted methods of mechanized spread are probably minor influences, especially in view of the long evolutionary history of mycorrhizal associations.

The introduction of monoculture may well have reduced species diversity of mycorrhizal fungi. For instance, the *Calluna* heathlands, which are maintained to provide food for game birds and sheep, are deliberately burned at intervals to discourage reinvasion by trees. This practice has led to dominance of ericoid mycorrhizae in large tracts of land.

Soil sterilization and the use of biocides have increased with increasing monoculture and more intensive crop production. Soil fumigation, if successful, will eliminate the mycorrhizal fungi from the soil. How soon and from where natural reinfestation occurs will depend on the area and depth of soil affected and on the availability of new inoculum. There is also the possibility that breeding for disease resistance may, at the same time, introduce some resistance to mycorrhizal infection. In addition, the development and use of systemic fungicides active against pathogens like *Pythium* and *Phytophthora* and translocated downwards in the plant—though offering great potential as a research tool—may severely affect the mycorrhizal association.

In the transition from natural forest to agriculture, most soils have been limed, and it is likely that the change in soil pH has favored the spread of "neutrophile" fungal species or strains. The liberal application of fertilizers for maximum crop production will certainly change mycorrhizal populations and may, in extreme cases, lead to decreased mycorrhizal development. Some plant breeding programs of the past decades have produced cultivars dependent on high fertilizer input for good performance. Such endeavors may select for plants with increasing immunity to mycorrhizal infection. New practices such as minimal cultivation may affect mycorrhizal populations, possibly favorably.

Major disturbances of the surface soil involving removal and storage, as in open cast mining, gravel extraction, and road building, will reduce mycorrhizal populations at least temporarily. Reeves *et al.* (1979) and Moorman and Reeves (1979) have shown that, after soil disturbance, infection levels of natural colonizers or introduced plant species were 1–2% compared with 99% and 77%, respectively, in undisturbed soils. Top-soil storage for 3 years had a detrimental effect on mycorrhizal populations and particularly on the infectivity of root fragments in the soil (Rives *et al.,* 1979). The extent of such soil disturbance is considerable. In Britain alone, 40,000 acres left from open cast mining requires rehabilitation.

The extent and diversity of orchid mycorrhizae are being reduced as man destroys the host plants. Development of agriculture has probably eliminated many species, leaving mainly those that are successful colonizers of disturbed land (Sanford, 1974). The habitats of many orchids, particularly epiphytic forms, are being destroyed as rain forests are felled.

8.2. Deliberate Effects

The deliberate inoculation with specific mycorrhizal fungi may have one of the following objectives: (a) the increase of total inoculum potential or (b) the establishment of more efficient fungal symbionts than those already present. Two general principles apply. First, for strongly mycorrhiza-dependent plant species, any mycorrhizal fungus is better than none. Second, indigenous mycorrhizal fungi are not necessarily the most efficient, especially when soil conditions are being changed or new plant species are introduced.

Relevant to the first situation are soils that have been fumigated or treated with soil sterilants or in which, for other reasons, no appropriate mycorrhizal fungi exist. Under such conditions, inoculation is very likely to succeed because the introduced fungi will have little competition from other mycorrhiza formers. Examples of such situations are the introduction of temperate timber trees, chiefly *Pinus* spp., into many African and Australasian countries and of *Eucalyptus* species into South America and Africa. A very recent, less well-known example is the inoculation of the Chinese gooseberry *(Physalis peruviana),* which is being planted in New Zealand in plowed up pasture land. With no reservoir of suitable endophytes in the pasture soil, inoculation is producing large crop increases (Powell, personal communication). Into this category also falls the inoculation of plantation crops (primarily trees like citrus, avocado, and coffee), some endomycorrhizal hardwoods (like sweet gum, yellow poplar, sycamore, dogwood), and some ornamentals, which are raised in sterilized soil in the nursery or in containers. It has been shown that such seedlings transplant better and suffer less from drought during the establishment stages, when they are mycorrhizal.

The successful introduction of specific, more beneficial ectomycorrhizal symbionts was the outcome of several experiments in Australia, England, and Africa (Samuel, 1926; Rayner, 1938; Chilvers and Pryor, 1965; Mikola, 1970; Bowen *et al.,* 1971; Momoh and Gbadegesin, 1980), where the soils already contained some relatively inefficient mycorrhizal fungi. It is the normal situation with VA endophytes, because these are present in nearly all soils. To find particular, more beneficial endophyte species here is much more problematical than with the more host-specific ectomycorrhizal fungi. Not only the efficiency of the introduced mycosymbiont but also its establishment and spread are important factors in determining success of any inoculation. The importance of persistence has not really been studied adequately. It may be that, in tree species in particular, the most mycorrhiza-dependent phase is during early establishment, and normal successions of mycorrhizal fungi will inevitably take over after the first 1–2 years. This does not necessarily invalidate the benefits of inoculation. Even for annuals with VAM, increased soil infectivity during early seedling growth may lead to crop increases, but inoculation costs may require the persistence and spread of the inoculum during subsequent years. The greatest question mark in the successful use of introduced mycorrhizal fungi is the cost and feasibility of producing and marketing the inoculum.

8.2.1. Inoculum Production

Four types of inocula are available: infested soil, infected roots, pure cultures of fungi, and spores. The earliest attempts to introduce mycorrhizal fungi relied on infested soil, and often this was successful. The main objections to this method of inoculation are the risks of inadvertently introducing pathogens and the possible heterogeneity of inoculum; for example, *Amanita phalloides,* a fungus poisonous to man. was apparently introduced into South America as a mycorrhizal fungus on oak seedlings imported from Europe and in South Africa (Mikola, 1970). The first use of a pure inoculum was in a nursery at Wareham Heath, England, where sporophores of *Rhizopogon luteolus* were introduced into a nursery soil naturally infested with *Boletus bovinus;* the inoculum produced marked growth improvements in Sitka spruce seedlings (Levisohn, 1956). Moser (1958, 1959) inoculated *Pinus cembra* with pure cultures of *Boletus plorans* and greatly improved growth of the seedlings at high altitudes. Inoculation of nurseries with a pure culture of *Hebeloma cylindrosporum* has substantially improved nursery performance of Norway spruce and Douglas fir, but subsequent growth in the forest has yet to be observed (Le Tacon and Valdenaire, 1980).

The largest single effort at commercial inoculum production is currently being made with *Pisolithus tinctorius* (Marx, 1978), and there is an extensive literature much of which describes the successful introduction of this fungus in difficult or routine sites (Marx *et al.,* 1976, 1977a). Some of the practical difficulties of large-scale production of mycelial inoculum are being discovered. One is the need to wash the inoculum free from the culture medium, as the latter offers a rich substrate to a range of soil saprophytes that will otherwise swamp the inoculum.

The relative merits of spore versus mycelial inoculum are not yet fully resolved. Shemakhanova (1962) reported variable results with inoculated seed in Russia. Theodorou (1971) obtained good mycorrhiza formation in *Pinus radiata* with seed inoculated with spores of *Rhizopogon luteolus* in a fumigated nursery and in nonsterile soil. Seeds encapsulated with *P. tinctorius* can be used to inoculate conifers (Jarl and Marx, 1979). Oak and hazelnut seedlings inoculated with spores of, or whose roots already contained, *Tuber melanosporum* are now produced commercially in France. Truffles can be obtained five years after planting, even in sites where truffles did not previously exist.

Some efforts are being made to culture VA endophytes, but at present, hope of a clean inoculum rests on the controlled production of infected roots, preferably using a host species different from that for which the inoculum is intended. Resting spores are another possibility. With all forms of inoculum, the practical difficulties of storage, reactivation, and application on a large scale have hardly been touched. Mosse and Hayman (1980) have summarized some of these difficulties.

A large question in all mycorrhizal inoculation attempts is whether to aim for a single or a mixed species inoculum. Probably for a given host, at a given site, and maybe even in a given year, one fungal species or even isolate will be the best. But the particular one is neither predictable nor feasible to determine in each instance. Should efforts therefore be directed toward producing a limited number of good inocula well adapted to particular hosts, soils, or climatic conditions—or should the objective be the production of a mixed inoculum of several fungal species in the hope that natural selection will operate fast enough to outweigh the disadvantages of competition between the more and less efficient species?

Future mycorrhizal studies should not, and probably will not, depend on success or failure of inoculation programs. Mycorrhizal associations are clearly very widespread and important in natural ecosystems and merit study toward a better understanding in this context alone.

References

Abbott, L. K., and Robson, A. D., 1977, The distribution and abundance of vesicular arbuscular endophytes in some western Australian soils, *Aust. J. Bot.* **25**:515–522.

Abbott, L. K., and Robson, A. D., 1978, Growth of subterranean clover in relation to the formation of endomycorrhizas by introduced and indigenous fungi in a field soil, *New Phytol.* **81**:575–587.

Abbott, L. K., and Robson, A. D., 1979, A quantitative study of the spores and anatomy of mycorrhizas formed by a species of *Glomus,* with reference to its taxonomy, *Aust. J. Bot.* **27**:363–375.

Alconero, R., 1969, Mycorrhizal synthesis and pathology of *Rhizoctonia solani* in *Vanilla* orchid roots, *Phytopathology* **59**:426–430.

Alexander, I. J., 1979, Ectomycorrhizas on tropical trees, in: *Abstracts of the 4th North American Conference on Mycorrhizae,* Fort Collins, Colorado.

Allen, E. B., and Allen, M. F., 1980, Natural re-establishment of vesicular-arbuscular mycorrhizae following stripmine reclamation in Wyoming, *J. Appl. Ecol.* **17**:139–147.

Ames, R. N., and Lindermann, R. G., 1976, *Acaulospora trappei* sp. nov. *Mycotaxon,* **3**:565–569.

Ames, R. N., and Schneider, R. W., 1979a, *Entrophospora:* A new genus in the Endogonaceae, *Mycotaxon,* **8**:347–352.

Ames, R. N., and Schneider, R. W., 1979b, Vesicular-arbuscular mycorrhizal fungi of California, *Abstracts of the 4th North American Conference on Mycorrhizae,* Fort Collins, Colorado.

Azcón, R., Barea, J. M., and Hayman, D. S., 1976, Utilization of rock phosphate in alkaline soils by plants inoculated with mycorrhizal fungi and phosphate-solubilizing bacteria, *Soil Biol. Biochem.* **8**:135–138.

Azcón, R., Azcón de Aguilar, C., and Barea, J. M., 1978a, Effects of plant hormones present in bacterial cultures on the formation and responses to VA endomycorrhiza, *New Phytol.* **80**:359–364.

Azcón, R., Marin, A. D., and Barea, J. M., 1978b, Comparative role of phosphate in soil or inside the host on the formation and effects of endomycorrhiza, *Plant Soil* **49**:561–567.

Baas Becking, L. G. M., Kaplan, I. R., and Moore, D., 1960, Limits of the natural environment in terms of pH and oxidation-reduction potentials, *J. Geol.* **68**:243–284.

Bagyaraj, D. J., and Menge, J. A., 1978, Interaction between a VA mycorrhiza and *Azotobacter* and their effects on the rhizosphere microflora and plant growth, *New Phytol.* **80**:567–573.

Bagyaraj, D. J., Manjunath, A., and Patil, R. B., 1979, Occurrence of vesicular-arbuscular mycorrhizas in some tropical aquatic plants, *Trans. Br. Mycol. Soc.* **72**:164–167.

Bakerspiegel, A., 1956, *Endogone* in Saskatchewan and Manitoba, *Am. J. Bot.* **43**:471–475.

Baltruschat, H., and Schönbeck, F., 1975, Untersuchungen über den Einfluss der endotrophen Mykorrhiza auf den Befall von Tabak mit *Thielaviopsis basicola, Phytopathol. Z.* **84**:172–188.

Bannister, P., and Norton, W. M., 1974, The response of mycorrhizal and non-mycorrhizal rooted cuttings of heather (*Calluna vulgaris* [L.] Hull) to variations in nutrient and water regimes, *New Phytol.* **73**:81–89.

Barley, K. P., 1970, The configuration of the root system in relation to nutrient uptake, *Adv. Agron.* **22**:159–201.

Baylis, G. T. S., 1969, Host treatment and spore production by *Endogone, N. Z. J. Bot.* **7**:173–174.

Baylis, G. T. S., 1975, The magnolioid mycorrhiza and mycotrophy in root systems derived from it, in: *Endomycorrhizas* (F. E. Sanders, B. Mosse, and P. B. Tinker, eds.), pp. 373–389, Academic Press, New York and London.

Becker, W. N., 1976, Quantification of Onion Vesicular Arbuscular Mycorrhizae and Their Resistance to *Pyrenochaeta terrestris,* Ph.D. Thesis, University of Illinois, Urbana.

Becker, W. N., and Gerdemann, J. W., 1977, *Glomus etunicatus* sp. nov., *Mycotaxon* **6**:29–32.

Becker, W. N., and Hall, I. R., 1976, *Gigaspora margarita,* a new species in the Endogonaceae, *Mycotaxon* **4**:155–160.

Bernard, N., 1909, L'évolution dans la symbiose, *Ann. Sci. Nat. Bot.* **14**:235–258.

Berry, C. R., and Marx, D. H., 1976, Sewage sludge and *Pisolithus tinctorius* ectomycorrhizae; their effect on growth of pine seedlings, *For. Sci.* **22**:351–358.

Bertheau, Y., Gianinazzi-Pearson, S., and Gianinazzi, S., 1980, Développement et expression de l'association endomycorrhizienne chez le blé. 1. Mise en évidence d'un effet variétal, *Ann. Amélior. Plantes* **30**:77–79.

Bevege, D. I., 1971, Mycorrhizas of *Araucaria.* Aspects of their Ecology, Physiology and Role in Nitrogen Fixation, Ph.D. Thesis, University of New England, Armidale, N.S.W.

Björkman, E., 1941, Die Ausbildung und Frequenz der Mykorrhiza in mit Asche gedüngten und ungedüngten Teilen von entwässertem Moor, *Medd. Skogsförsöksanst* **32**:255–296.

Björkman, E., 1942, Conditions favouring the formation of mycorrhizae in pine and spruce, *Symb. Bot. Ups.* **6**:1–191.

Black, R. L. B., and Tinker, P. B., 1977, Interaction between effects of vesicular-arbuscular mycorrhiza and fertilizer phosphorus on yields of potatoes in the field, *Nature* **267**:510–511.

Black, R., and Tinker, P. B., 1979, The development of endomycorrhizal root systems II. Effect of agronomic factors and soil conditions on the development of vesicular-arbuscular mycorrhizal infection in barley and on the endophyte spore density, *New Phytol.* **83**:401–413.

Blaschke, H., 1979, Observations on the ectomycorrhizae of silver fir (*Abies alba* Miel.), in: *Abstracts of the 4th North American Conference on Mycorrhizae,* Fort Collins, Colorado.

Bloomfield, C., and Coulter, J. K., 1973, Genesis and management of acid sulphate soils, *Adv. Agron.* **25**:265–326.

Bonfante-Fasolo, P., 1980, Occurrence of a basidiomycete in living cells of mycorrhizal hair roots of Calluna vulgaris, *Trans. Br. Mycol. Soc.* **75**:320–325.

Bonfante-Fasolo, P., and Gianinazzi-Pearson, V., 1979, Ultrastructural aspects of endomycorrhiza in the Ericaceae. 1. Naturally infected hair roots of *Calluna vulgaris* L. Hull, *New Phytol.* **83**:739–744.

Boullard, B., 1958, La mycotrophie chez les Ptéridophytes, *Botaniste* **XL1**:5–185.

Boullard, B., 1960, Influence des engrais sur la constitution des mycorrhizes par le Pin maritime (*Pinus pinaster* Sol.), *Rev. Gen. Bot.* **67**:47–59.

Boullard, B., 1961, Influence du photopériodisme sur la mycorhization de jeunes conifères, *Bull. Soc. Linn. Normandie* **2**:30–46.

Boullard, B., 1964, Acquisitions récentes relatives à l'écologie des mycorrhizes, *J. Agric. Trop. Bot. Appl.* **11**:515–542.

Boullard, B., 1979, Considération sur la symbiose fongique chez les Ptéridophytes, *Syllogeus* **19**:1–58.

Bowen, G. D., 1963, The natural occurrence of mycorrhizal fungi for *Pinus radiata* in south Australian soils, *Commonw. Sci. Ind. Res. Org., Div. Rep., Div. Soils, Adelaide* **6**:1–11.

Bowen, G. D., 1980a, Mycorrhizal roles in tropical plants and ecosystems, in: *Tropical Mycorrhiza Research* (P. Mikola, ed.), pp. 165–190, Oxford University Press, Oxford.

Bowen, G. D., 1980b, Misconceptions, concepts and approaches in rhizosphere biology, in: *Contemporary Microbial Ecology* (D. C. Ellwood, J. N. Hedger, M. J. Latham, J. M. Lynch, and J. H. Slater, eds.), pp. 283–304, Academic Press, London.

Bowen, G. D., and Theodorou, C., 1979, Interactions between bacteria and ectomycorrhizal fungi, *Soil Biol. Biochem.* **11:**119–126.

Bowen, G. D., Theodorou, C., and Skinner, M. F., 1971, Towards a mycorrhizal inoculation programme, in: *American Australian Forest Nutrition Meeting,* p. 13, Canberra.

Brown, M. E., 1976, Role of *Azotobacter paspali* in association with *Paspalum notatum, J. Appl. Bact.* **40:**341–348.

Bucholtz, F., 1912, Beiträge zur Kenntniss der Gattung *Endogone* Link., *Beih. Bot. ZbL.* **29:**147–225.

Burgeff, H., 1961, *Mikrobiologie des Hochmoores,* Gustav Fischer Verlag, Stuttgart.

Burges, A., 1936, On the significance of mycorrhiza, *New Phytol.* **35:**117–131.

Butler, E. J., 1939, The occurrences and systematic position of the vesicular-arbuscular mycorrhizal fungi, *Trans. Br. Mycol. Soc.* **22:**274–307.

Carling, D. E., Brown, M. F., and Brown, R. A., 1979, Colonization rates and growth responses of soybean plants infected by vesicular-arbuscular mycorrhizal fungi, *Can. J. Bot.* **57:**1769–1772.

Chastukhin, V. Ya., 1955, Biology and ecology of mycorrhiza-forming fungi in: *Mycotrophy in Plants,* pp. 87–100, published for the U.S. Department of Agriculture by the Israel Program for Scientific Translations (1967), Jerusalem.

Chilvers, G. A., and Pryor, C. D., 1965, The structure of Eucalypt mycorrhizas, *Aust. J. Bot.* **13:**245–259.

Christie, P., Newman, E. I., and Campbell, R., 1978, The influence of neighbouring grassland plants on each others endomycorrhizas, *Soil Biol. Biochem.* **10:**521–527.

Clark, F. B., 1964, Micro-organisms and soil structure affect yellow-poplar growth, U.S. Forestry Service Research Paper CS-9, 12 pp.

Clarke, C., and Mosse, B., 1981, Plant growth response to vesicular-arbuscular mycorrhiza. XII. Field inoculation responses of barley at two soil levels, *New Phytol.* **87:**695–703.

Clough, K. S., and Sutton, J. C., 1978, Direct observations of fungal aggregates in sand dune soil, *Can. J. Microbiol.* **24:**333–335.

Cooper, K. M., 1976, A field survey of mycorrhizas in New Zealand ferns, *N. Z. J. Bot.* **14:**169–181.

Crush, J., 1974, Plant growth responses to vesicular-arbuscular mycorrhiza. VII. Growth and nodulation of some herbage legumes, *New Phytol.* **73:**743–749.

Crush, J. R., 1978, Changes in effectiveness of soil endomycorrhizal fungal populations during pasture development, *N. Z. J. Agric. Res.* **21:**683–685.

Daft, M. J., 1978, Nitrogen fixation in nodulated and mycorrhizal crop plants, *Ann. Appl. Biol.* **88:**461–465.

Daft, M. J., and El-Giahmi, A. A., 1978, Effect of arbuscular mycorrhiza on plant growth. VIII. Effects of defoliation and light on selected hosts, *New Phytol.* **80:**365–372.

Daft, M. J., and Nicolson, T. H., 1969, Effect of *Endogone* mycorrhiza on plant growth. III. Influence of inoculum concentration on growth and infection in tomato, *New Phytol.* **68:**953–963.

Daft, M. J., and Okusanya, B. O., 1973, Effect of *Endogone* mycorrhiza on plant growth. V. Influence of infection on the multiplication of viruses in tomato, petunia and strawberry, *New Phytol.* **72:**975–983.

Daft, M. J., Hacskaylo, E., and Nicolson, T. M., 1975, Arbuscular mycorrhizas in plants colonising coal spoils in Scotland and Pennsylvania, in: *Endomycorrhizas* (F. E. Sanders, B. Mosse, and P. B. Tinker, eds.), pp. 561–580, Academic Press, London and New York.

Daft, M. J., Chilvers, M. T., and Nicolson, T. H., 1980, Mycorrhizas of the Liliflorae 1. Morphogenesis of *Endymion non-scriptus* (L.) Garke and its mycorrhizas in nature, *New Phytol.* **85**:181–189.

Dangeard, P. A., 1900, *Rhizophagus populinus* Dangeard, *Le Botaniste* **7**:285–291.

Daniels, B. A., and Duff, D. M., 1978, Variation in germination and spore morphology among four isolates of *Glomus mosseae, Mycologia* **70**:1261–1267.

Daniels, B. A., and Trappe, J. M., 1979, *Glomus epigaeus* sp. nov., a useful fungus for vesicular-arbuscular mycorrhizal research, *Can. J. Bot.* **57**:539–542.

Dehne, H. W., and Schönbeck, F., 1978, Investigations on the influence of endotrophic mycorrhiza on plant diseases. III. Chitinase activity and ornithine cycle, *Z. Pflanzenkr. Pflanzenschutz* **85**:666–678.

Dimbleby, G. W., 1953, Natural regeneration of pine and birch on the heather moors of northeast Yorkshire, *Forestry* **26**:41–52.

Dominik, T., 1961, Trials on a natural isolation of mycorrhizal fungi from cultivated soils and wasteland in the surroundings of Szczecia, *Pr. Inst. Badaw. Lesn.* **227**:3–37.

Dominik, T., 1963, Investigations on mycorrhizal fungi in moribund stands on former ploughland, *Pr. Inst. Badaw. Lesn.* **257**:4–27.

Dominik, T., 1969, Key to ectotrophic mycorrhizae, *Folia For. Poln. Lesn.* **15**:309–328.

Doncaster, C. C., 1962, A counting dish for nematodes, *Nematologia* **7**:334–336.

Dowding, E. S., 1955, *Endogone* in Canadian rodents, *Mycologia* **47**:51–57.

Dowding, E. S., 1959, Ecology of *Endogone, Trans. Br. Mycol. Soc.* **42**:449–457.

El-Giahmi, A. A., Nicolson, T. H., and Daft M. J., 1976, Endomycorrhizal fungi from Libyan soils, *Trans. Br. Mycol. Soc.* **67**:164–169.

Elmes, R., and Mosse, B., 1980, in: *Rothamsted Experimental Station Report for 1979* (Part 1), p. 188.

Fassi, B., and De Vecchi, E., 1962, Richerche sulle micorrhize ectotrofiche del pino strobo in vivaio. 1: Descrizione di alcune forme pin diffuse in Piemonte, *Allionia* **8**:133–152.

Fassi, B., and Palenzona, M., 1969, Sintesi micorrizica tra "Pinus strobus," "Pseudotsuga douglasii" ed *Endogone lactiflua, Allionia* **15**:105–114.

Fassi, B., Jodice, R., and Van Den Driesche, R., 1972, Matière organique et mycorhizes dans le développement des semis de *Pinus strobus, Allionia* **18**:13–32.

Fitter, A. H., 1977, Influence of mycorrhizal infection on competition for phosphorus and potassium by two grasses, *New Phytol.* **79**:119–125.

Fogel, R., 1979, Mycorrhizae and nutrient cycling in natural forest ecosystems, in: *Abstracts of the 4th North American Conference on Mycorrhizae,* Fort Collins, Colorado.

Foster, R. C., and Marks, G. C., 1967, Observations on the mycorrhizas of forest trees. II. The rhizosphere of *Pinus radiata* D. Don, *Aust. J. Biol. Sci.* **20**:915–926.

Fowells, H. A., and Krauss, R. W., 1959, The inorganic nutrition of loblolly pine and virginia pine with special reference to nitrogen and phosphorus, *For. Sci.* **5**:95–112.

Francke, H. L., 1934, Beiträge zur Kenntniss der Mycorrhiza von *Monotropa hypopitys* L. Analyse und Synthese der Symbiose, *Flora (Jena)* **129**:1–52.

Frank, A. B., 1885, Über die auf Wurzelsymbiose beruhende Ernährung gewisser Baüme durch unterirdische Pilze, *Ber. Dt. Bot. Ges.* **3**:128–145.

Fries, N., 1944, Beobachtungen über die thamniscophage Mykorrhiza einiger Halophyten, *Bot. Not.* **2**:255–264.

Fries, N., 1978, Basidiospore germination in some mycorrhiza-forming Hymenomycetes, *Trans. Br. Mycol. Soc.* **70**:319–324.

Fries, N., 1979, The taxon-specific spore germination reaction in *Leccinum, Trans. Br. Mycol. Soc.* **73**:337–341.

Furlan, V., and Fortin, J. A., 1973, Formation of endomycorrhizae by *Endogone calospora* on *Allium cepa* under three temperature regimes, *Naturaliste Can.* **100**:467–477.

Furlan, V., and Fortin, J. A., 1975, A flotation-bubbling system for collecting Endogonaceae spores from sieved soil, *Nat. Can.* **102**:663–667.

Furlan, V., and Fortin, J. A., 1977, Effects of light intensity on the formation of vesicular-arbuscular endomycorrhizas on *Allium cepa* by *Gigaspora calospora, New Phytol.* **79:**335–340.

Furman, T. E., and Trappe, J. M., 1971, Phylogeny and ecology of mycotrophic achlorophyllous angiosperms, *Q. Rev. Biol.* **46:**219–225.

Garbaye, J., Kabré, A., Le Tacon, F., Mousain, D., and Piou, D., 1979, Fertilisation minérale et fructification des champignons supérieurs en hêtraie, *Ann. Sci. For.* **36:**151–164.

Gaümann, F., Müller, E., Nüesch, J., and Rimpau, R. H., 1961, Über die Wurzelpilze von *Loroglossum hircinum* (L.) Rich., *Phytopath. Z.* **41:**89–96.

Geering, H. R., Hodgson, J. F., and Sdano, C., 1969, Micronutrient cation complexes in soil solution. IV. The chemical state of manganese in soil solution, *Soil Sci. Soc. Amer. Proc.* **33:**81–85.

Gerdemann, J. W., 1955a, Relation of a large soil-borne spore to phycomycetous mycorrhizal-infections, *Mycologia* **47:**619–632.

Gerdemann, J. W., 1955b, Wound healing of hyphae in a phycomycetous mycorrhizal fungus, *Mycologia* **47:**916–918.

Gerdemann, J. W., and Bakshi, B. K., 1976, Endogonaceae of India: Two new species, *Trans. Br. Mycol. Soc.* **66:**340–343.

Gerdemann, J. W., and Nicolson, T. H., 1963, Spores of mycorrhizal *Endogone* species, extracted from soil by wet-sieving and decanting, *Trans. Br. Mycol. Soc.* **46:**235–244.

Gerdemann, J. W., and Trappe, J. M., 1974, The Endogonaceae in the Pacific Northwest, *Mycol. Mem.* **5:**1–76.

Gerdemann, J. W., and Trappe, J. M., 1975, Taxonomy of the Endogonaceae, in: *Endomycorrhizas* (F. F. Sanders, B. Mosse, and P. B. Tinker, eds.), pp. 35–51, Academic Press, New York and London.

Gildon, A., 1981, The Influence of a Vesicular-Arbuscular Fungus on the Host Plant Uptake of Heavy Metals, Ph.D. Thesis, University of London.

Gilligan, C. A., 1979, Modelling rhizosphere infection, *Phytopathology* **69:**782–784.

Gilmore, A. E., 1968, Phycomycetous mycorrhizal organisms collected by open-pot culture methods, *Hilgardia* **39:**87–105.

Gimingham, C. H., 1972, *Ecology of Heathlands,* Chapman and Hall, London.

Giovannetti, M., and Mosse, B., 1980, An evaluation of techniques for measuring vesicular-arbuscular mycorrhizal infection in roots, *New Phytol.* **84:**489–500.

Göbl, F., and Pümpel, B., 1973, Einfluss von "Grünkupfer Linz" auf Pflanzenausbildung, Mykorrhizabesatz sowie Frosthärte von Zirbenjungpflanzen, *Eur. J. For. Pathol.* **3:**242–245.

Godfrey, R. M., 1957, Studies of British species of *Endogone.* II. Fungal parasites, *Trans. Br. Mycol. Soc.* **40:**136–144.

Göttsche, D., 1972, Distribution of tender roots and mycorrhizae in a soil profile of a beech and spruce stand in Solling, *Mitt. Bundesforschungsanst. Forst. Holzwirtsch.* **88:**1–102.

Graw, D., 1979, The influence of soil pH on the efficiency of vesicular-arbuscular mycorrhiza, *New Phytol.* **82:**687–695.

Graw, D., and Rehm, S., 1977, Vesikular-arbuscular Mykorrhiza in den Fruchtträgern von *Arachis hypogea* L., *Z. Acker Pflanzenbau,* **145:**75–78.

Graw, D., Moawad, M., and Rehm, S., 1979, Untersuchungen zur Wirts- und Wirkungsspezifität der VA-Mycorrhiza, *Z. Acker Pflanzenbau* **148:**85–98.

Green, N. E., Graham, S. O., and Schenck, N. C., 1976, The influence of pH on the germination of vesicular-arbuscular mycorrhizal spores, *Mycologia* **68:**929–934.

Greenall, J. M., 1963, The mycorrhizal endophytes of *Griselinia littoralis* (Cornaceae), *N. Z. J. Bot.* **1:**389–400.

Greenwood, D. J., 1967, Studies on the transport of oxygen through the stems and roots of vegetable seedlings, *New Phytol.* **66:**337–347.

Grente, J., and Delmas, J., 1974, *Perspectives pour une trufficulture moderne,* 3rd ed., I.N.R.A. Station de Pathologie Vegetale, Clermont, France, 65 pp.

Grime, J. P., 1979, *Plant Strategies and Vegetation Processes,* John Wiley, Chichester.

Hacskaylo, E., 1961, Influence of cycloheximide on growth of mycorrhizal fungi and on mycorrhizae of pine, *For. Sci.* **7**:376–379.

Hacskaylo, E., and Palmer, J. G., 1957, Effects of several biocides on growth of seedling pines and incidence of mycorrhizae in field plots, *Plant Dis. Rep.* **41**:354–358.

Hacskaylo, E., and Vozzo, J. A., 1965, Effect of temperature on growth and respiration of ectotrophic mycorrhizal fungi, *Mycologia* **57**:748–756.

Hadley, G., 1969, Cellulose as a carbon source for orchid mycorrhiza, *New Phytol.* **68**:933–939.

Hadley, G., 1970, Non-specificity of symbiotic infection in orchid mycorrhiza, *New Phytol.* **69**:1015–1023.

Hadley, G., and Ong, S. H., 1978, Nutritional requirements of orchid endophytes, *New Phytol.* **81**:561–569.

Hadley, G., and Williamson, B., 1972, Features of mycorrhizal infection in some Malayan orchids, *New Phytol.* **71**:1111–1118.

Hall, I. R., 1976, Vesicular mycorrhizas in the orchid *Corybas macranthus, Trans. Br. Mycol. Soc.* **66**:160.

Hall, I. R., 1977, Species and mycorrhizal infections of New Zealand Endogonaceae, *Trans. Br. Mycol. Soc.* **68**:341–356.

Hall, I. R., 1978, Effects of endomycorrhizas on the competitive ability of white clover, *N. Z. J. Agric. Res.* **21**:509–515.

Hall, I. R., and Fish, B. J., 1979, A key to the Endogonaceae, *Trans. Br. Mycol. Soc.* **73**:261–270.

Handley, W. R. C., 1963, Mycorrhizal associations and *Calluna* heathland afforestation, Forestry Comm. Bulletin 36, HMSO London.

Harley, J. L., 1969, *The Biology of Mycorrhiza,* 2nd ed., Leonard Hill, London.

Harris, M. M., and Jurgensen, M. F., 1977, Development of *Salix* and *Populus* mycorrhizae in metallic mine tailings, *Plant Soil* **47**:509–517.

Harvais, G., and Hadley, G., 1967, The development of *Orchis purpurella* in asymbiotic and inoculated cultures, *New Phytol.* **66**:217–230.

Haselwandter, K., 1979, Mycorrhizal status of ericaceous plants in alpine and subalpine areas, *New Phytol.* **83**:427–431.

Haselwandter, K., and Read, D. J., 1979, Mycorrhizal status of high-alpine plant communities in: *Abstracts of the 4th North American Conference on Mycorrhizae,* Fort Collins, Colorado.

Hatch, A. B., 1936, The role of mycorrhizae in afforestation, *J. For.* **34**:22–29.

Hatch, A. B., 1937, The physical basis of mycotrophy in the genus *Pinus, Black Rock For. Bull.* **6**:1–168.

Hay, R. K. M., 1976, The temperature of the soil under a barley crop, *J. Soil Sci.* **27**:121–128.

Hayman, D. S., 1970, *Endogone* spore numbers in soil and vesicular-arbuscular mycorrhiza in wheat as influenced by season and soil treatment, *Trans. Br. Mycol. Soc.* **54**:53–63.

Hayman, D. S., 1974, Plant growth responses to vesicular-arbuscular mycorrhiza. VI. Effect of light and temperature, *New Phytol.* **73**:71–80.

Hayman, D. S., 1975, The occurrence of mycorrhiza in crops as affected by soil fertility in: *Endomycorrhizas* (F. E. Sanders, B. Mosse, and P. B. Tinker, eds.), pp. 495–509, Academic Press, New York and London.

Hayman, D. S., 1978, Mycorrhizal populations of sown pastures and native vegetation in Otago, New Zealand, *N. Z. J. Agric. Res.* **21**:271–276.

Hayman, D. S., and Mosse, B., 1971, Plant growth responses to vesicular-arbuscular mycorrhiza. 1. Growth of *Endogone*-inoculated plants in phosphate-deficient soils, *New Phytol.* **70**:19–27.

Hayman, D. S., and Stovold, G. E., 1979, Spore populations and infectivity of vesicular-arbuscular mycorrhizal fungi in New South Wales, *Aust. J. Bot.* **27**:227–233.

Hayman, D. S., Johnson, A. M., and Ruddlesdin, I., 1975, The influence of phosphate and crop species on *Endogone* spores and vesicular-arbuscular mycorrhiza under field conditions, *Plant Soil* 43:489–495.

Hayman, D. S., Barea, J. M., and Azcón, R., 1976, Vesicular-arbuscular mycorrhiza in Southern Spain: its distribution in crops growing in soils of different fertility, *Phytopathol. Mediterr.* 15:1–6.

Heap, A. J., and Newman, E. I., 1980, Links between roots by hyphae of vesicular-arbuscular mycorrhizas, *New Phytol.* 25:169–171.

Hepper, C. M., 1979, Germination and growth of *Glomus caledonius* spores: The effects of inhibitors and nutrients, *Soil Biol. Biochem.* 11:269–277.

Hepper, C. M., and Mosse, B., 1975, Techniques used to study the interaction between *Endogone* and plant roots, in: *Endomycorrhizas* (F. E. Sanders, B. Mosse, and P. B. Tinker, eds.), pp. 469–484, Academic Press, New York and London.

Hepper, C. M., and Smith, G. A., 1976, Observations on the growth of *Endogone* spores, *Trans. Br. Mycol. Soc.* 66:189–194.

Hijner, J. A., and Arditti, J., 1973, Orchid mycorrhiza: Vitamin production and requirements by the symbionts, *Am. J. Bot.* 60:829–835.

Hirrel, M. C., and Gerdemann, J. W., 1980, Improved growth of onion and bell pepper in saline soils by two vesicular-arbuscular mycorrhizal fungi, *Soil Sci. Soc. Am. J.* 44:654–655.

Hirrel, M. C., Mehravaran, H., and Gerdemann, J. W., 1978, Vesicular-arbuscular mycorrhiza in the Chenopodiaceae and Cruciferae: Do they occur? *Can. J. Bot.* 22:2813–2817.

Ho, I., and Trappe, J. M., 1979, Interaction of a VA-mycorrhizal fungus and a free-living nitrogen fixing bacterium on growth of tall fescue, in: *Abstracts of the 4th North American Conference on Mycorrhizae,* Fort Collins, Colorado.

Hornby, D., 1969, Quantitative estimation of soil-borne inoculum of the take-all fungus (*Ophiobolus graminis* [Sacc.] Sacc.), in: *Proceedings of the Fifth British Insecticides and Fungicides Conference* 1:65–70.

Hornby, D., 1979, Take-all decline: A theorist's paradise, in: *Soil-borne Plant Pathogens* (B. Schippers and W. Gams, eds.), pp. 133–156, Academic Press, London.

Iqbal, S. H., and Qureshi, K. S., 1976, The effect of vesicular-arbuscular mycorrhizal associations on growth of sunflower (*Helianthus annus* L.) under field conditions, *Biologia (Pakistan)* 23:189–196.

Janos, D. P., 1980, Vesicular-arbuscular mycorrhizae affect lowland tropical rain forest plant growth, *Ecology* 61:151–162.

Jarl, K. and Marx, D. H., 1979, Use of encapsulated seeds to inoculate conifers with *Pisolithus tinctorius,* in: *Abstracts of the 4th North American Conference on Mycorrhizae,* Fort Collins, Colorado.

Jasper, P. A., Robson, A. D., and Abbott, L. K., 1979, Phosphorus and the formation of vesicular-arbuscular mycorrhizas, *Soil Biol. Biochem.* 11:501–505.

Johnson, P. M., 1977, Mycorrhizal *Endogone* in a New Zealand forest, *New Phytol.* 78:161–170.

Johnson, P. N., 1976, Effects of soil phosphate level and shade on plant growth and mycorrhizas, *N. Z. J. Bot.* 14:333–340.

Jonsson, L., and Nylund, J. E., 1979, *Favolaschia dybowskyana* (Singer) Singer (Aphyllophorales), a new orchid mycorrhizal fungus from tropical Africa, *New Phytol.* 83:121–128.

Katznelson, H., Rouatt, J. W., and Peterson, E. A., 1962, The rhizosphere effect of mycorrhizal and nonmycorrhizal roots of yellow birch seedlings, *Can. J. Bot.* 40:377–382.

Khan, A. G., 1971, Occurrence of *Endogone* species in West Pakistan soils, *Trans. Br. Mycol. Soc.* 56:217–224.

Khan, A. G., 1972, Mycorrhizae in the Pakistan Ericales, *Pak. J. Bot.* 4:183–194.

Khan, A. G., 1974, The occurrence of mycorrhizas in halophytes, hydrophytes and xerophytes, and of *Endogone* spores in adjacent soils, *J. Gen. Microbiol.* 81:7–14.

Khan, A. G., 1978, Vesicular-arbuscular mycorrhizas in plants colonizing black wastes from bituminous coal mining in the Illawarra region of New South Wales, *New Phytol.* **81**:53–63.

Kigel, J., 1980, Analysis of regrowth patterns and carbohydrate levels in *Lolium multiflorum* Lam., *Ann. Bot.* **45**:91–101.

Koske, R. E., 1975, *Endogone* spores in Australian sand dunes, *Can. J. Bot.* **53**:668–672.

Koske, R. E., 1981a, Multiple germination by spores of *Gigaspora gigantea, Trans. Br. Mycol. Soc.* **76**:328–330.

Koske R. E., 1981b, A preliminary study of interactions between species of vesicular–arbuscular fungi in a sand dune, *Trans. Br. Mycol. Soc.* **76**:411.–416.

Koske, R. E., Sutton, J. C., and Sheppard, B. R., 1975, Ecology of *Endogone* in Lake Huron sand dunes, *Can. J. Bot.* **53**:87–93.

Kruckelmann, H. W., 1975, Effects of fertilizers, soils, soil tillage, and plant species on the frequency of *Endogone* chlamydospores and mycorrhizal infection in arable soils, in: *Endomycorrhizas* (F. E. Sanders, B. Mosse, and P. B. Tinker, eds.), pp. 511–525, Academic Press, New York and London.

Krupa, S. and Fries, N., 1971, Studies on the ectomycorrhizae of pine. I. Production of volatile organic compounds, *Can. J. Bot.* **49**:1425–1431.

Kusano, S., 1911, *Gastrodia elata* and its symbiotic association with *Armillaria mellea, J. Coll. Agr. Imp. Univ. Tokyo* **4**:1–65.

Laiho, O., and Mikola, P., 1964, Studies on the effect of some eradicants on mycorrhizal development in forest nurseries, *Acta For. Fenn.* 77.

Lamb, R. J., 1979, Factors responsible for the distribution of mycorrhizal fungi of *Pinus* in eastern Australia, *Aust. For. Res.* **9**:25–34.

Lambert, D. H., and Cole, H., Jr., 1980, Effects of mycorrhizae on establishment and performance of forage species in mine spoil, *Agron. J.* **72**:257–260.

Lambert, D. H., Cole, H., Jr., and Baker, D. E., 1980, Adaptation of vesicular-arbuscular mycorrhizae to edaphic factors, *New Phytol.* **85**:513–520.

Lamont, B., 1972, The morphology and anatomy of proteoid roots in the genus *Hakea, Aust. J. Bot.* **20**:155–174.

Langlois, C. G., and Fortin, J. A., 1978, Absorption of phosphorus ^{32}P by excised ectomycorrhizae of balsam fir *(Abies balsamea)* from low concentrations of $H_2PO_4^-$, *Nat. Can.* **105**:417–424.

Largent, D. L., Sugihara, N., and Brinitzer, A., 1980a, *Amanita gemmata,* a non-host-specific mycorrhizal fungus of *Arctostaphylos manzanita, Mycologia* **72**:435–439.

Largent, D. L., Sugihara, N., and Wishner, C., 1980b, Occurrence of mycorrhizae on ericaceous and pyrolaceous plants in northern California, *Can. J. Bot.* **58**:2274–2279.

Last, F. T., Pelham, J., Mason, P. A., and Ingleby, K., 1979, Influence of leaves on sporophore production by fungi forming sheathing mycorrhizas with *Betula* spp., *Nature* **280**:168–169.

LeTacon F., and Valdenaire, J. M., 1980, La mycorhization contrôlée en pépinière. Premiers résultats obtenus à la pépinière du Fonds Forestier National de Peyrat le Chateau sur Epicéa *(Picea exelsa)* et Douglas *(Pseudotsuga douglasii), Rev. For. Fr. (Nancy)* **3**:281–293.

Levisohn, I., 1954, Occurrence of ectotrophic and endotrophic mycorrhizas in forest trees, *Forestry* **27**:145–146.

Levisohn, I., 1956, Growth stimulation of forest tree seedlings by the activity of free living mycorrhizal mycelia, *Forestry* **29**:53–59.

Levisohn, I., 1957, Antagonistic effects of *Alternaria tenuis* on certain root-fungi of forest trees, *Nature* **179**:1143–1144.

Levisohn, I., 1965, Mycorrhizal investigations, in: *Experiments on Nutrition Problems in Forest Nurseries* (B. Benzian, ed.), *For.* Comm. *Bull.* **37**:228–235.

Lewis, D. H., 1973, Concepts in fungal nutrition and the origin of biotrophy, *Biol. Rev.* **48**:261–278.

Lewis, D. H., 1975, Comparative aspects of the carbon nutrition of mycorrhizas, in: *Endomycorrhizas* (F. E. Sanders, B. Mosse, and P. B. Tinker, eds.), pp. 119–148, Academic Press, New York and London.

Lindeberg, G., 1948, On the occurrence of polyphenol oxidases in soil-inhabiting Basidiomycetes, *Physiol. Plant.* 1:196–205.

Lindsey, D. L., Cress, W. A., and Aldon, E. F., 1977, The effects of endomycorrhizae on growth of rabbit-brush, four wing saltbush and corn in coal mine spoil material, *USDA Forestry Service Research Note* RM-343.

Loneragan, J. F., 1975, The availability and absorption of trace elements in soil-plant systems and their relation to movement and concentrations of trace elements in plants, in: *Trace Elements in Soil-Plant-Animal Systems* (D. J. D. Nicholas and A. R. Egan, eds.), pp. 109–134, Academic Press, London.

Lutz, R. W., and Sjolund, R. D., 1973, *Monotropa uniflora:* ultrastructural details of its mycorrhizal habit, *Am. J. Bot.* 60:339–345.

Lyr, H., 1963, Zur Frage des Streuabbaues durch ectotrophe Mykorrhizapilze, in: *International Mykorrhizasymposium Weimar 1960,* pp. 123–142, Gustav Fischer Verlag, Jena.

MacFadyen, A., 1975, Some thoughts on the behaviour of ecologists, *J. Anim. Ecol.* 44:351–363.

Marks, G. C., Ditchburne, N., and Foster, R. C., 1968, Quantitative estimates of mycorrhiza populations in radiata pine forests, *Aust. For.* 32:27–38.

Marx, D. H., 1973, Mycorrhizae and feeder root diseases, in: *Ectomycorrhizae* (G. D. Marks and T. T. Kozlowski, eds.), pp. 351–382, Academic Press, New York.

Marx, D. H., 1975, Mycorrhizae and establishment of trees on strip-mined land, *Ohio J. Sci.* 75:288–297.

Marx, D. H., 1977, Tree host range and world distribution of the ectomycorrhizal fungus *Pisolithus tinctorius, Can. J. Microbiol.* 23:217–223.

Marx, D. H., 1980, Ectomycorrhizal fungus inoculations: a tool for improving forestation practices, in: *Tropical Mycorrhiza Research* (P. Mikola, ed.), pp. 13–71, Oxford University Press, Oxford.

Marx, D. H., Bryan, W. C., and Davey, C. B., 1970, Influence of temperature on aseptic synthesis of ectomycorrhiza by *Telephora terrestris* and *Pisolithus tinctorius* on Loblolly Pine, *For. Sci.* 16:424–431.

Marx, D. H., Bryan, W. C., and Cordell, C. F., 1976, Growth and ectomycorrhizal development of pine seedlings in nursery soils infested with the fungal symbiont *Pisolithus tinctorius, For. Sci.* 22:91–100.

Marx, D. H., Bryan, W. C., and Cordell, C. F., 1977a, Survival and growth of pine seedlings with *Pisolithus* ectomycorrhizae after two years on reforestation sites in North Carolina and Florida, *For. Sci.* 23:363–373.

Marx, D. H., Hatch, A. B., and Mendicino, J. F., 1977b, High soil fertility decreases sucrose content and susceptibility of loblolly pine roots to ectomycorrhizal infection by *Pisolithus tinctorius, Can. J. Bot.* 5:1569–1574.

Maser, C., Trappe, J. M., and Nussbaum, R. A., 1978, Fungal–small mammal relationship with emphasis on Oregon coniferous forests, *Ecology* 59:799–809.

Mason, E., 1928, Note on the presence of mycorrhiza in the roots of salt marsh plants, *New Phytol.* 27:193–195.

McIlveen, W. D., and Cole, H., 1976, Spore dispersal of Endogonaceae by worms, ants, wasps and birds, *Can. J. Bot.* 54:1486–1489.

McLennan, E. I., 1935, Non-symbiotic development of seedlings of *Epacris impressa* Labill., *New Phytol.* 34:55–63.

McNabb, R. F. R., 1961, Mycorrhiza in the New Zealand Ericales, *Aust. J. Bot.* 9:57–61.

Medve, R. J., 1979, The mycorrhizae of the Cruciferae, in: *Abstracts of the 4th North American Conference on Mycorrhizae,* Fort Collins, Colorado.

Mejstrik, V., 1965, Study of the development of endotrophic mycorrhiza in the association of *Cladietum marisci*, in: *Plant-Microbe Relationships* (Proceedings of a symposium held in Prague September 24–28, 1963), pp. 283–291, The Publishing House of the Czechoslovak Academy of Sciences, Prague.

Melin, E., 1925, *Untersuchungen über die Bedeutung der Baummykorrhiza. Eine ökologische physiologische Studie,* Gustav Fischer Verlag, Jena.

Melin, E., 1948, Recent advances in the study of tree mycorrhiza, *Trans. Br. Mycol. Soc.* **30:**92–99.

Melin, E., 1953, Physiology of mycorrhizal relations in plants, *Annu. Rev. Plant Physiol.* **4:**325–346.

Menge, J. A., and Grand, L. F., 1978, Effect of fertilization on production of epigeous basidiocarps by mycorrhizal fungi in loblolly pine plantations, *Can. J. Bot.* **36:**2357–2362.

Menge, J. A., Grand, L. F., and Haines, L. W., 1977, The effect of fertilization on growth and mycorrhiza numbers in 11-year-old loblolly pine plantations, *For. Sci.* **23:**37–44.

Menge, J. A., Steirle, D., Bagyaraj, D. J., Johnson, E. L. V., and Leonard, R. T., 1978, Phosphorus concentrations in plants responsible for inhibition of mycorrhizal infection, *New Phytol.* **80:**575–578.

Mertz, S. M., Heithaus, J. J., and Bush, R., 1979, Mass production of axenic spores of the endomycorrhizal fungus *Gigaspora margarita, Trans. Br. Mycol. Soc.* **72:**167–169.

Mikola, P., 1948, On the black mycorrhiza of birch, *Arch. Soc. Zool. Bot. Fenn. Vanamo* **1:**81–85.

Mikola, P., 1953, An experiment on the invasion of mycorrhizal fungi into prairie soil, *Karstenia* (Helsinki) **2:**33–34.

Mikola, P., 1970, Mycorrhizal inoculation in afforestation, *Int. Rev. For. Res.* **3:**123–196.

Mikola, P. and Laiho, O., 1962, Mycorrhizal relations in the raw humus layer of northern spruce forests, *Comm. Inst. For. Finh.* **55:**1–13.

Mikola, P., Hahl, J., and Torniainen, E., 1966, Vertical distribution of mycorrhizae in pine forests with spruce undergrowth, *Ann. Bot. Fennici* **3:**406–409.

Miller, O. K., Jr., 1981, Mycorrhizae, mycorrhizal fungi and fungal biomass in subalpine tundra at Eagle Summit, Alaska, *Holarctic Ecol.* (in press).

Miller, O. K., Jr., and Laursen, G. A., 1978, Ecto- and endomycorrhizae of Arctic plants at Barrow, Alaska, in: *Vegetation and Production Ecology of an Alaskan Arctic Tundra, Ecological Studies 29* (L. L. Tieszen, ed.), pp. 229–237, Springer-Verlag, New York.

Miller, R. M., 1979, Some occurrences of vesicular-arbuscular mycorrhiza in natural and disturbed ecosystems of the Red Desert, *Can. J. Bot.* **57:**619–623.

Mishustin, E. N., 1955, Mycotrophy in trees and its value in silviculture, in: *Mycotrophy in Plants,* pp. 17–34, published for the U.S. Department of Agriculture by the Israel Program for Scientific Translations (1967), Jerusalem.

Mmbaga, M. T., 1975, Studies on Vesicular-Arbuscular Mycorrhiza in Rice (*Oryza sativa* L.), Masters Thesis, University of Dar es Salaam, Tanzania.

Molina, R. J., Trappe, J. M., and Stickler, G. S., 1978, Mycorrhizal fungi associated with *Festuca* in the Western United States and Canada, *Can. J. Bot.* **56:**1691–1695.

Momoh, Z. O., and Gbadegesin, R. A., 1980, Field performance of *Pisolithus tinctorius* as a mycorrhizal fungus of pines in Nigeria, in: *Tropical Mycorrhiza Research* (P. Mikola, ed.), pp. 72–79, Oxford University Press, Oxford.

Montfort, C., and Küsters, E., 1940, Saprophytismus and Photosynthese. 1. Biochemische und physiologische Studien an Humus-Orchideen, *Bot. Arch.* **40:**571–633.

Moorman, T., and Reeves, F. B., 1979, The role of endomycorrhizae in revegetation practices in the semi-arid West. II. A bioassay to determine the effect of land disturbance on endomycorrhizal populations, *Am. J. Bot.* **66:**14–18.

Morley, C. D. and Mosse, B., 1976, Abnormal vesicular-arbuscular mycorrhizal infections in white clover induced by lupin, *Trans. Br. Mycol. Soc.* **67:**510–513.

Morrison, T. M., 1957, Host-endophyte relationships in mycorrhizas of *Pernettya macrostigma, New Phytol.* **56**:247–257.

Moser, M., 1956, Die Bedeutung der Mykorrhiza für Aufforstungen in Hochlagen, *Forstwiss. Centralbl.* **75**:8–18.

Moser, M., 1958, Die künstliche Mykorrhizaimpfung von Forstpflanzen. I., *Forstwiss, Centralbl.* **77**:32–40.

Moser, M., 1959, Die künstliche Mykorrhizaimpfung von Forstpflanzen. III. Die Impfmethodik im Forstgarten, *Forstwiss. Centralbl.* **78**:193–202.

Mosse, B., 1956, Studies on the Endotrophic Mycorrhiza of some Fruit Plants, Ph.D. Thesis, University of London.

Mosse, B., 1959, The regular germination of resting spores and some observations on the growth requirements of an *Endogone* sp. causing vesicular-arbuscular mycorrhiza, *Trans. Br. Mycol. Soc.* **42**:274–286.

Mosse, B., 1970, Honey-coloured, sessile *Endogone* spores, I. Life history, *Arch. Mikrobiol.* **70**:167–175.

Mosse, B., 1972, The influence of soil type and *Endogone* strain on the growth of mycorrhizal plants in phosphate deficient soils, *Rev. Écol. Biol. Sol.* **9**:529–537.

Mosse, B., 1975, Specificity in VA mycorrhizas, in: *Endomycorrhizas* (F. E. Sanders, B. Mosse, and P. B. Tinker, eds.), pp. 469–484, Academic Press, New York and London.

Mosse, B., 1977a, Plant growth responses to vesicular-arbuscular mycorrhiza X. Responses of *Stylosanthes* and maize to inoculation in unsterile soils, *New Phytol.* **78**:277–288.

Mosse, B., 1977b, The role of mycorrhiza in legume nutrition on marginal soils, in: *Exploiting the Legume–Rhizobium Symbiosis in Tropical Agriculture* (J. M. Vincent, A. S. Whitney, and J. Bose, eds.), pp. 175–292, University of Hawaii College of Tropical Agriculture Miscellaneous Publication 145.

Mosse, B., and Bowen, G. D., 1968, The distribution of *Endogone* spores in some Australian and New Zealand soils and in an experimental field soil at Rothamsted, *Trans. Br. Mycol. Soc.* **51**:485–492.

Mosse, B., and Hayman, D. S., 1980, Mycorrhiza in agricultural plants in: *Tropical Mycorrhiza Research* (P. Mikola, ed.), pp. 213–230, Oxford University Press, Oxford.

Mosse, B., and Phillips, J. M., 1971, The influence of phosphate and other nutrients on the development of vesicular-arbuscular mycorrhiza in culture, *J. Gen. Microbiol.* **69**:157–166.

Mosse, B., and Smith, G., 1976, Anatomy of VA infections caused by different endophytes, in: *Rothamsted Experimental Station Report for 1975* (Part 1), p. 280.

Mosse, B., Thompson, J. D., and Smith, S. M., 1978, Development of VA mycorrhiza (E₃ and YV) in plants fed with nutrient solution in sand and nutrient film culture, in: *Rothamsted Annual Report for 1978*, p. 235.

Mosse, B., Warner, A., and Clarke, A. C., 1981a, Plant growth responses to vesicular-arbuscular mycorrhiza, XIII. Spread of an introduced endophyte and residual growth effects of inoculation in the second year, *New Phytol.* (in press).

Mosse, B., Smith, G., Warner, A., and Hepper, C., 1981b, Anatomical studies of VA mycorrhiza (in preparation).

Munns, D. N., and Mosse, B., 1980, Mineral nutrition of legume crops, in: *Advances in Legume Science* (R. J. Summerfield and A. H. Bunting, eds.), pp. 115–125.

Nemec, S., 1979, Growth of *Citrus jambhiri* as affected by single and combined inocula of three *Glomus* species, in: *Abstracts of the 4th North American Conference on Mycorrhizae,* Fort Collins, Colorado.

Nicolson, T. H., 1960, Mycorrhiza in the Gramineae. II. Development in different habitats, particularly sand dunes, *Trans. Br. Mycol. Soc.* **43**:132–145.

Nicolson, T. H., 1975, Evolution of vesicular-arbuscular mycorrhizas in: *Endomycorrhizas* (F. E. Sanders, B. Mosse, and P. B. Tinker, eds.), pp. 25–34, Academic Press, New York and London.

Nicolson, T. H., and Johnston, C., 1979, Mycorrhiza in the Gramineae. III. *Glomus fasciculatus* as the endophyte of pioneer grasses in a maritime sand dune, *Trans. Br. Mycol. Soc.* **72**:261–268.

Nicolson, T. H., and Schenck, N. C., 1979, Endogonaceous mycorrhizal endophytes in Florida, *Mycologia* **71**:178–198.

Norkrans, B., 1950, Studies in growth and cellulolytic enzymes of *Tricholoma* with special reference to mycorrhiza formation, *Symb. Bot. Upsaliensis* **11**:1–126.

Nuesch, J., 1963, Defence reactions in orchid bulbs, in: *Symbiotic Associations* (P. S. Nutman and B. Mosse, eds.), pp. 335–343, Symposia of the Society of General Microbiology 13, Cambridge University Press, Cambridge.

Nye, P. H., and Ramzan, M., 1979, Measurement and mechanism of ion diffusion in soil. X. Prediction of soil acidity gradients in acid-base transfers, *J. Soil Sci.* **30**:43–51.

Ocampo, J. A., and Hayman, D. S., 1981, Influence of plant interactions of vesicular-arbuscular mycorrhizal infections. II. Crop rotations and residual effects of non-host plants, *New Phytol.* **87**:333–343.

Ocampo, J. A., Martin, J., and Hayman, D. S., 1980, Influence of plant interactions on vesicular-arbuscular mycorrhizal infections. 1. Host and non-host plants grown together, *New Phytol.* **84**:27–35.

Oswald, E. T., and Ferchau, H. A., 1968, Bacterial associations of coniferous mycorrhizae, *Plant Soil* **28**:187–192.

Palmer, S. G., 1954, Mycorrhizal Development in *Pinus virginiana* as Influenced by Growth Regulators, Ph.D. Thesis, George Washington University, Washington, D.C.

Parke, J. L., and Linderman, R. G., 1980, Association of vesicular-arbuscular mycorrhizal fungi with the moss *Funaria hygrometrica, Can. J. Bot.* **58**:1898–1904.

Pearson, V., and Read, D. J., 1975a, The biology of mycorrhiza in the Ericaceae. 1. The isolation of the endophyte and the synthesis of mycorrhizas in aseptic culture, *New Phytol.* **72**:371–379.

Pearson, V., and Read, D. J., 1975b, The physiology of the mycorrhizal endophyte of *Calluna vulgaris, Trans. Br. Mycol. Soc.* **64**:1–7.

Persidsky, D. J., and Wilde, S. A., 1960, The effect of biocides on the survival of mycorrhizal fungi, *J. For.* **58**:522–524.

Peterson, T. A., Mueller, W. C., and Englander, L., 1980, Anatomy and ultrastructure of a *Rhododendron* root/fungus association, *Can. J. Bot.* **58**:2421–2433.

Peyronel, B., 1923, Fructification de l'endophyte à arbuscules et à vésicules des mycorrhizes endotrophes, *Bull. Soc. Mycol. Fr.* **39**:119–126.

Pirozynski, K. A., and Malloch, D. W., 1975, The origin of land plants: a matter of mycotrophism, *Biosystems* **6**:153–164.

Ponnamperuma, F. N., 1972, The chemistry of submerged soils, *Adv. Agron.* **24**:29–96.

Porter, W. M., 1979, The "most probable number" method for enumerating infective propagules of vesicular arbuscular mycorrhizal fungi in soil, *Aust. J. Soil Res.* **17**:515–519.

Porter, W. M., Abbott, L. K., and Robson, A. D., 1978, Effect of rate of application of superphosphate on populations of vesicular-arbuscular endophytes, *Aust. J. Exp. Agri. Anim. Husb* **18**:573–578.

Potebnja, N. A., 1952, Die Mykorrhiza-Bildung der Wurzeln einjähriger Eichensämlinge in der Waldsteppe und in der Steppe der UdSSR (Lus.). Untersuchungen über die Mykorrhiza von Holzpflanzen, *Izd. Akad. Nauk SSSR*.

Powell, C. L., 1976a, Development of mycorrhizal infections from *Endogone* spores and infected root segments, *Trans. Br. Mycol. Soc.* **66**:439–445.

Powell, C. L., 1976b, Mycorrhizal fungi stimulate clover growth in New Zealand hill country soils, *Nature (London)* **264**:436–438.

Powell, C. L., 1977a, Mycorrhiza in hill country soils. Growth responses in ryegrass, *N. Z. J. Agric. Res.* **20**:495–502.

Powell, C. L., 1977b, Mycorrhizas in hill country soils. III. Effect of inoculation on clover growth in unsterile soils, *N. Z. J. Agric. Res.* **20**:343–348.

Powell, C. L., 1979, Spread of mycorrhizal fungi through soil, *N. Z. J. Agr. Res.* **22**:335–339.

Powell, C. L., and Daniel, J., 1978a, Mycorrhizal fungi stimulate uptake of soluble and insoluble fertilizer from a phosphate-deficient soil, *New Phytol.* **80**:351–358.

Powell, C. L., and Daniel, J., 1978b, Growth of white clover in undisturbed soils after inoculation with efficient mycorrhizal fungi, *N. Z. J. Agric. Res.* **21**:675–681.

Rabatin, S. C., 1980, The occurrence of the vesicular-arbuscular-mycorrhizal fungus *Glomus tenuis* with moss, *Mycologia* **72**:191–195.

Rabatin, S. C., and Wicklow, D. T., 1979, Plant diversity and vesicular-arbuscular mycorrhizal fungus diversity, in: *Abstracts of the 4th North American Conference on Mycorrhizae,* Fort Collins, Colorado.

Ratnayake, M., Leonard, R. T., and Menge, J. A., 1978, Root exudation in relation to supply of phosphorus and its possible relevance to mycorrhizal formation, *New Phytol.* **81**:543–552.

Rayner, M. C., 1938, The use of soil or humus inocula in nurseries and plantations, *Emp. For. Rev.* **17**:236–243.

Read, D. J., 1974, *Pezizella ericae* sp. nov., the perfect state of a typical mycorrhizal endophyte of Ericaceae, *Trans. Br. Mycol. Soc.* **63**:381–419.

Read, D. J., 1978, The biology of mycorrhiza in heathland ecosystems with special reference to nitrogen nutrition of the Ericaceae, in: *Microbial Ecology* (M. W. Loutit and J. A. R. Miles, eds.), pp. 324–328, Springer-Verlag, Berlin.

Read, D. J., and Armstrong, W., 1972, A relationship between oxygen transport and the formation of the ectotrophic mycorrhizal sheath in conifer seedlings, *New Phytol.* **71**:49–53.

Read, D. J., Koucheki, H. K., and Hodgson, J., 1976, Vesicular-arbuscular mycorrhiza in natural vegetation systems. I. The occurrence of infection, *New Phytol.* **77**:641–653.

Read, D. J., Kianmehr, H., and Malibari, A., 1977, The biology of mycorrhiza in *Helianthemum* Mill., *New Phytol.* **78**:305–312.

Read, D. J., Malibari, A., and Whittingham, J., 1979, Nutrient and water transport between plants connected by mycorrhizal fungi, in: *Abstracts of the 4th North American Conference on Mycorrhizae,* Fort Collins, Colorado.

Redhead, J. F., 1968, Mycorrhizal associations in some Nigerian forest trees, *Trans. Br. Mycol. Soc.* **51**:377–387.

Redhead, J. F., 1974, Aspects of the Biology of Mycorrhizal Associations Occurring on Tree Species in Nigeria, Ph.D. Thesis, University of Ibadan, Nigeria.

Redhead, J. F., 1975, Endotrophic mycorrhizas in Nigeria: Some aspects of the ecology of the endotrophic mycorrhizal association of *Khaya grandifoliola* C.DC. in: *Endomycorrhizas* (F. E. Sanders, B. Mosse, and P. B. Tinker, eds.), pp. 447–459, Academic Press, New York and London.

Redhead, J. F., 1977, Endotrophic mycorrhizas in Nigeria: Species of the Endogonaceae and their distribution, *Trans. Br. Mycol. Soc.* **69**:275–280.

Redmond, D. R., 1955, Rootlets, mycorrhiza and soil temperatures in relation to birch dieback, *Can. J. Bot.* **33**:595–627.

Reeves, F. B., Wagner, D., Moorman, T., and Kiel, J., 1979, The role of endomycorrhizae in revegetation practices in the semi-arid West. I. A comparison of incidence of mycorrhizae in severely disturbed natural environments, *Am. J. Bot.* **66**:6–13.

Reid, C. P. P., and Bowen, G. D., 1979, Effects of soil moisture on VA mycorrhiza formation and root development in *Medicago,* in: *The Soil-Root Interface* (J. L. Harley and R. Scott Russell, eds.), pp. 211–219, Academic Press, London.

Rhodes, L. H., and Gerdemann, J. W., 1975, Phosphate uptake zones of mycorrhizal and non-mycorrhizal onions, *New Phytol.* **75**:555–561.

Riffle, J. W., 1967, An effect of an *Aphelenchoides* species on the growth of mycorrhizal and a pseudomycorrhizal fungus, *Phytopathology* **57**:541–544.

Rives, C. S., Bajwa, M., and Liberta, A. E., 1979, Effect of topsoil storage on VA mycorrhiza populations, in: *Abstracts of the 4th North American Conference on Mycorrhizae,* Fort Collins, Colorado.

Robertson, N. F., 1954, Studies on the mycorrhiza of *Pinus sylvestris* I. *New Phytol.* **53**:253–283.

Robinson, R. K., 1972, The production by roots by *Calluna vulgaris* of a factor inhibitory to growth of some mycorrhizal fungi, *J. Ecol.* **60**:219–224.

Robinson, R. K., 1973, Mycorrhiza in certain Ericaceae native to southern Africa, *J. S. Afr. Bot.* **39**:123–129.

Rose, S., and Daniels, B. A., 1979, *Glomus gerdemannii* sp. nov., *Mycotaxon* **8**:297–301.

Rose, S. L., and Trappe, J. M., 1980, Three new endomycorrhizal *Glomus* spp. associated with actinorrhizal shrubs, *Mycotaxon* **10**:413–420.

Rose, S. L., Daniels, B. A., and Trappe, J. M., 1978, *Glomus gerdemannii* sp. nov., *Mycotaxon* **8**:297–301.

Ross, J. P., 1980, Effect of nontreated field soil on sporulation of vesicular-arbuscular mycorrhizal fungi associated with soybean, *Phytopathology* **70**:1200–1205.

Ross, J. P., and Ruttencutter, R., 1977, Population dynamics of two vesicular-arbuscular endomycorrhizal fungi and the role of hyperparasitic fungi, *Phytopathology* **67**:490–496.

Rothwell, F. M., and Holt, C., 1978, Vesicular-arbuscular mycorrhizae established with *Glomus fasciculatus* spores isolated from the faeces of cricetine mice, *U.S. For. Serv. Res. Note NE* **259** (4pp.).

Rothwell, F. M., and Trappe, J. M., 1979, *Aucalospora bireticulata* sp. nov. *Mycotaxon* **8**:471–475.

Saif, S. R., 1977, The influence of host development on vesicular-arbuscular mycorrhizae and endogonaceous spore population in field-grown vegetable crops. I. Summer-grown crops, *New Phytol.* **79**:341–348.

Saif, S. R., and Iffat, N., 1976, Vesicular-arbuscular mycorrhizae in plants and endogonaceous spores in soils in Northern areas of Pakistan, I. Hunza, Nagan and Gilqit, *Pak. J. Bot.* **8**:163–179.

Saif, S. R., Sheikh, N. A., and Khan, A. G., 1975, Ecology of *Endogone* I. Relationship of *Endogone* spore population with physical soil factors, *Islamabad J. Sci.* **2**:1–5.

Salawu, E. O., and Estey, R. H., 1979, Observations on the relationship between a vesicular-arbuscular fungus, a fungivorous nematode and the growth of soya beans, *Phytoprotection* **60**:99–102.

Samuel, G., 1926, Note on the distribution of mycorrhiza, *Trans. R. Soc. South Aust.* **1**:245–246.

Sanders, F. E., 1975, The effect of foliar-applied phosphate on mycorrhizal infections of onion roots, in: *Endomycorrhizas* (F. E. Sanders, B. Mosse, and P. B. Tinker, eds.), pp. 261–276, Academic Press, New York and London.

Sanders, F. E., and Tinker, P. B., 1973, Phosphate flow into mycorrhizal roots, *Pestic. Sci.* **4**:385–395.

Sanders, F. E., Tinker, P. B., Black, R. L. B., and Palmerley, S. M., 1977, The development of endomycorrhizal root systems. I. Spread of infection and growth-promoting effects with four species of vesicular-arbuscular endophyte, *New Phytol.* **78**:257–268.

Sanford, W. W., 1974, The ecology of orchids, in: *The Orchids* (C. L. Withner, ed.), pp. 1–100, John Wiley & Sons, New York and London.

Schenck, N. G., and Hinson, K., 1971, Endotrophic vesicular-arbuscular mycorrhizae on soybean in Florida, *Mycologia* **63**:672–675.

Schenck, N. C., and Kellam, M. K., 1978, The influence of vesicular-arbuscular mycorrhizae on disease development, in: *Technical Bulletin 798,* University of Florida, pp. 15, Gainesville.

Schenck, N. C., and Kinloch, R. A., 1976, Mycorrhizal fungi colonizing field crops on a newly cleared woodland site, *Proc. Am. Phytopathol. Soc.* **3**:274.

Schenck, N. C., and Nicolson, T. H., 1977, A zoosporic fungus occurring on species of *Gigaspora margarita* and other vesicular-arbuscular mycorrhizal fungi, *Mycologia* **69**:1049–1053.

Schenck, N. C., Graham, S. O., and Green, N. E., 1975, Temperature and light effect on contamination and spore germination of vesicular-arbuscular mycorrhizal fungi, *Mycologia* **67**:1189–1192.

Schmidt, E. L., Biesbrock, J. A., Bohlool, B. B., and Marx, D. H., 1974, Study of mycorrhizae by means of fluorescent antibody, *Can. J. Microbiol.* **20**:137–139.

Schonbeck, F., and Dehne, H. W., 1979, Investigations on the influence of endotrophic mycorrhiza on plant diseases. 4. Fungal parasites on shoots, *Olpidium brassicae* TMV, *Z. Pflanzenkr. Pflanzenschutz* **86**:103–112.

Schonbeck, F., and Spengler, G., 1979, Detection of TMV in mycorrhizal cells of tomato by immunofluorescence, *Phytopathol. Z.* **94**:84–86.

Schrader, R., 1958, Untersuchungen zur Biologie der Erbsenmycorrhiza, *Arch. Mikrobiol.* **32**:81–114.

Schramm, J. R., 1966, Plant colonization studies on black wastes from anthracite mining in Pennsylvania, *Trans. Am. Philos. Soc.* **56**.

Scott Russell, R., 1977, *Plant Root Systems,* McGraw-Hill, London.

Seviour, R. J., Willing, R. R., and Chilvers, G. A., 1973, Basidiocarps associated with ericoid mycorrhizas, *New Phytol.* **72**:381–385.

Shemakhanova, N. M., 1962, Mycotrophy of woody plants, in: *Academy of Sciences of the USSR Institute of Microbiology* (Translated by Israel Program for Scientific Translations 1967), p. 329, Jerusalem.

Shuja, N., Gilani, U., and Khan, A. G., 1971, Mycorrhizal associations in some angiosperm trees around the New University Campus, Lahore, *Pak. J. For.* **21**:367–374.

Sieverding, E., 1979, Einfluss der Bodenfeuchte auf die Effektivität der VA-Mycorrhiza, *Angew. Bot.* **53**:91–98.

Singer, R., 1978, Origins of the deficiency of Amazonian soils—a new approach, *Acta Amazonica* **2**:315–320.

Singer, R., 1979, Observations and role of ectomycorrhiza in the neotropics, in: *Abstracts of the 4th North American Conference on Mycorrhizae,* Fort Collins, Colorado.

Singh, K. G., 1974, Mycorrhiza in the Ericaceae with special reference to *Calluna vulgaris, Sven. Bot. Tidskr.* **68**:1–16.

Skinner, M. F., and Bowen, G. D., 1974, The penetration of soil by mycelial strands of pine mycorrhizas, *Soil. Biol. Biochem.* **6**:57–61.

Slankis, V., 1974, Soil factors influencing formation of mycorrhizae, *Annu. Rev. Phytopathol.* **12**:437–457.

Smith, S. E., and Bowen, G. D., 1979, Soil temperature, mycorrhizal infection and nodulation of *Medicago trunculata* and *Trifolium subterraneum, Soil Biol. Biochem.* **11**:469–473.

Smith, G. W., and Skipper, H. D., 1979, Comparison of methods to extract spores of vesicular-arbuscular mycorrhizal fungi, *Soil Sci. Soc. Am. J.* **43**:722–725.

Søndergaard, M., and Laegaard, S., 1977, Vesicular-arbuscular mycorrhiza in some aquatic vascular plants, *Nature* **268**:232–233.

Sparling, G. P., 1976, Effects of Vesicular-Arbuscular Mycorrhizas on Pennine Grassland Vegetation, Ph.D. Thesis, Leeds University.

Sparling, G. P., and Tinker, P. B., 1975, Mycorrhizas in pennine grasslands, in: *Endomycorrhizas* (F. E. Sanders, B. Mosse, and P. B. Tinker, eds.), pp. 545–560, Academic Press, New York and London.

Sparling, G. P., and Tinker, P. B., 1978a, Mycorrhizal infection in Pennine grassland. I. Levels of infection in the field, *J. Appl. Ecol.* **15**:943–950.

Sparling, G. P., and Tinker, P. B., 1978b, Mycorrhizal infection in Pennine grassland. II. Effects of mycorrhizal infection on the growth of some upland grasses on irradiated soils, *J. Appl. Ecol.* **15**:951–958.

St. John, T., 1979, The problem of non-mycorrhizal trees in an Amazonian rain forest, in: *Abstracts of the 4th North American Conference on Mycorrhizae,* Fort Collins, Colorado.

St. John, T. V., 1980, Root size, root hairs and mycorrhizal infection: a re-examination of Baylis's hypothesis with tropical trees, *New Phytol.* **84**:483–487.

St. John, T., and Machado, A. D., 1978, Effects of depth and management system of a firm soil (Latossolo) on mycorrhizal manifestations, *Acta Amazonica* **8**:139–141.

Stahl, M., 1949, Die Mykorrhiza der Lebermoose mit besonderer Berücksichtigung der thallosen Formen, *Planta* **37**:103–148.

Stark, N., 1977, Direct nutrient cycling by endomycorrhizae, in: *Abstracts of the 3rd North American Conference on Mycorrhizae,* Fort Collins, Colorado.

Stone, E. L., 1949, Some effects of mycorrhizae on the phosphorus nutrition of Monterey pine seedlings, *Soil Sci. Soc. Amer. Proc.* **14**:340–345.

Stribley, D. P., and Read, D. J., 1976, The biology of mycorrhiza in the Ericaceae. VI. The effects of mycorrhizal infection and concentration of ammonium nitrogen on growth of cranberry (*Vaccinium macrocarpon* Ait.) in sand culture, *New Phytol.* **77**:63–72.

Stribley, D. P., Tinker, P. B., and Snellgrove, R. C., 1980, Effect of vesicular-arbuscular mycorrhizal fungi on the relations of plant growth, internal phosphorus concentration and soil phosphate analyses, *J. Soil Sci.* **31**:655–672.

Strzemska, J., 1975, Mycorrhiza in farm crops grown in monoculture, in: *Endomycorrhizas* (F. E. Sanders, B. Mosse, and P. B. Tinker, eds.), pp. 537–543, Academic Press, New York and London.

Sutton, J. C., 1973, Development of vesicular-arbuscular mycorrhizae in crop plants, *Can. J. Bot.* **51**:2487–2493.

Sutton, J. C., and Barron, G. L., 1972, Population dynamics of *Endogone* spores in soil, *Can. J. Bot.* **50**:1909–1914.

Sward, R. J., Hallam, N. D., and Holland, A. A., 1978, *Endogone* spores in a heathland area of Southeastern Australia, *Aust. J. Bot.* **26**:29–43.

Tandy, P. A., 1975, Sporocarpic species of Endogonaceae in Australia, *Aust. J. Bot.* **26**:29–43.

Theodorou, C., 1971, Introduction of mycorrhizal fungi into soil by spore inoculation of seed, *Aust. For.* **35**:23–26.

Thomas, G. W. and Jackson, R. M., 1979, Sheathing mycorrhizas of nursery-grown *Picea sitchensis, Trans. Br. Mycol. Soc.* **73**:117–125.

Tinker, P. B. H., 1976, Effects of vesicular-arbuscular mycorrhizas on higher plants, in: *Symbiosis (Symposia of the Society for Experimental Biology,* No. 29), pp. 325–349, Cambridge University Press, U.K.

Tisdall, J. M., and Oades, J. M., 1979, Stabilization of soil aggregates by the root systems of ryegrass, *Aust. J. Soil. Res.* **17**:429–441.

Tommerup, I. C., and Kidby, D. K., 1979, Preservation of spores of vesicular-arbuscular endophytes by L-drying, *Appl. Environ. Microbiol.* **37**:831–835.

Tranquillini, W., 1964, Photosynthesis and dry matter production of trees at high altitudes, in: *The Formation of Wood in Forest Trees* (M. H. Zimmermann, ed.), pp. 505–518, Maria Moors Labor Foundation for Botanical Research, New York.

Trappe, J. M., 1964, Mycorrhizal hosts and distribution of *Cenococcum graniforme, Lloydia* **27**:100–106.

Trappe, J. M., 1977a, Three new Endogonaceae: *Glomus constrictus, Sclerocystis clavispora,* and *Acaulospora scrobiculata, Mycotaxon* **6**:359–366.

Trappe, J. M., 1977b, Selection of fungi for ectomycorrhizal inoculation in nurseries, *Annu. Rev. Phytopathol.* **15**:203–222.

Trappe, J. M., 1979, *Glomus segmentatus* sp. nov., *Trans. Br. Mycol. Soc.* **73**:362–363.

Trappe, J. M., Stanley, E. A., Benson, N. R., and Duff, D. M., 1973, Mycorrhizal deficiency of apple trees in high arsenic soils, *Hortic. Sci.* **8**:52–53.

Trinick, M. J., 1977, Vesicular-arbuscular infection and soil phosphorus utilization in *Lupinus* spp., *New Phytol.* **78:**297–304.

Van der Plank, J. E., 1975, *Principles of Plant Infection,* Academic Press, New York and London.

Vegh, I., Fabre, E., and Gianinazzi-Pearson, V., 1979, Présence en France de *Pezizella ericae* Read, champignon endomycorrhizogène des Ericacées horticoles, *Phytopath. Z.* **96:**231–243.

Vlasov, A. A., 1952, Methods of estimating the amount of mycorrhizae in seedlings, *Tr. Kompleksn. Nauch. Exsped. Po Voprosam Polezahchitnykh Cesorazvedenii* **11:**231–239.

Voiry, H., 1980, Les ectomycorhizes du chêne et du hêtre. Possibilités d'applications forestières, *Rapport Ecole Nationale Ingénieurs des Travaux des Eaux et Forêts 3ᵉᵐᵉ année Institut National de la Recherche Agronomique—Centre National de Recherches Forestierés,* Nancy, France, 120pp.

Vozzo, J. A., and Hacskaylo, E., 1971, Inoculation of *Pinus caribaea* with ectomycorrhizal fungi in Puerto Rico, *For. Sci.* **17:**239–245.

Vysotskii, G. N., 1902, Mycorrhiza of oak and pine seedlings *Lesoprom Vestn.* **29:**504 (in Russian).

Walker, C., 1979, *Complexipes moniliformis:* A new genus and species tentatively placed in the Endogonaceae, *Mycotaxon* **10:**99–104.

Walker, C., and McNabb, H. S., 1979, Population dynamics of spores of Endogonaceae in the rhizosphere of hybrid poplars in central Iowa, in: *Abstracts of the 4th North American Conference on Mycorrhizae,* Fort Collins, Colorado.

Warcup, J. H., 1959, Studies on Basidiomycetes in soil, *Trans. Br. Mycol. Soc.* **42:**45–52.

Warcup, J. H., 1971, Specificity of mycorrhizal association in some Australian terrestrial orchids, *New Phytol.* **70:**41–46.

Warcup, J. H., 1975, A culturable *Endogone* associated with Eucalypts, in: *Endomycorrhizas* (F. E. Sanders, B. Mosse, and P. B. Tinker, eds.), pp. 53–63, Academic Press, New York and London.

Warcup, J. H., 1980, Ectomycorrhizal associations of Australian indigenous plants, *New Phytol.* **85:**531–535.

Warcup, J. H., and Talbot, P. H. B., 1967, Perfect stages of *Rhizoctonia* associated with orchids, *New Phytol.* **66:**631–641.

Warner, A., and Mosse, B., 1979, Factors affecting the spread of VA endophytes, *Rothamsted Annual Report for 1979,* pp. 187–188.

Warner, A., 1980, Spread of Vesicular-Arbuscular Mycorrhizal Fungi in Soil, Ph.D. Thesis, University of London.

Warner, A., and Mosse, B., 1980, Independent spread of vesicular-arbuscular mycorrhizal fungi in soil, *Trans. Br. Mycol. Soc.* **74:**407–410.

Webster, J., 1976, *Pezizella ericae* is homothallic, *Trans. Br. Mycol. Soc.* **66:**173.

Went, F. W., and Stark, N., 1968, The biological and mechanical role of soil fungi, *Proc. Nat. Acad. Sci. U.S.A.* **60:**497.

Wilcox, H. E. R., Ganmore-Neumann, R., and Wang, C. J. K., 1974, Characteristics of two fungi producing ectendomycorrhizae in *Pinus resinosa, Can. J. Bot.* **52:**2279–2282.

Williams, S. E., Wollum, A. G., and Aldon, E. F., 1974, Growth of *Atriplex canescens* (Pursh) Nutt. improved by formation of vesicular-arbuscular mycorrhizae, *Soil Sci. Soc. Amer. Proc.* **38:**962–965.

Wojciechowska, H., 1960, Studien über den Mykotrophismus der Fichte (*Picea excelsa* Link.) in den nördlichen Verbreitungsgrenzen mit besonderer Berücksichtigung des Mykotrophismus der Pflanzengesellschaften in der Försterei Lipowo, Ober-Försterei Sadlowo bei der Ortschaft Biskupiec Reszelski, *Folia For. Pol. Ser. A* **2:**123–166.

Wojciechowska, H., 1966, A review of studies on mycotrophism of saline soil plants from Wielkopolsko-Kujawska lowland, *Zesz. Nauk. Wyzsz. Szk. Roln. Olsztynie* **22:**385–413 (in Polish).

Woods, F. W., and Brock, K., 1964, Interspecific transfer of Ca^{45} and P^{32} by root systems, *Ecology* **45:**686–889.

Worley, J. F., and Hacskaylo, E., 1959, The effect of available soil moisture on Virginia pine, *For. Sci.* **5:**267–268.

Zak, B., 1974, Ectendomycorrhiza of Pacific madrone, *Trans. Br. Mycol. Soc.* **62:**202–204.

Zak, B., 1976, Pure culture synthesis of bearberry mycorrhizae, *Can. J. Bot.* **54:**1297–1305.

5

Lignin Metabolism and the Carbon Cycle
Polymer Biosynthesis, Biodegradation, and Environmental Recalcitrance

J. G. ZEIKUS

1. Introduction

Lignin metabolism is of utmost importance to the carbon cycle on earth. As a major structural component of photosynthetic biomass, lignin makes up approximately 25% of the dry weight of vascular plants and is second only to cellulose, which accounts for about 50% of the dry weight of cellular carbon stored in the biosphere. However, lignin stores more solar energy and hence accounts for more fuel value in wood than does cellulose because of lignin's unique chemical structure. Lignin is nature's most abundant recalcitrant product. The recalcitrance of lignin in certain environments results in the formation of lignites and coals, which are the major forms of fossilized organic matter on earth. Understanding how lignin is mineralized represents a tremendous challenge to microbiologists because so little is known, and the presence of this polymer in biomass dictates that microorganisms and not higher life forms are the principal agents of organic decomposition in terrestial biological food chains.

Lignin is a natural product that is synthesized by polymerization of phenylpropanoid units in vascular plants. To avoid confusion, the following terminology and definitions will be used in this review when describing lignin and aromatic substrates: *natural lignin,* a water insoluble polymer arising from an

J. G. ZEIKUS ● Department of Bacteriology, University of Wisconsin, Madison, Wisconsin 53706.

enzyme-initiated dehydrogenative polymerization of coumaryl, coniferyl, or sinapyl alcohol; *industrial lignin,* chemically or physically modified wood that is largely aromatic; and *model lignins,* water-soluble aromatic compounds that are synthesized from cinnamyl alcohol derivatives and contain an intermonomer linkage(s) present in natural lignin. This review will emphasize the metabolism (formation and degradation) of natural lignin and will not encompass the metabolism of aromatic compounds in general. Indeed, many previous studies of microbial metabolism of aromatic compounds in which relationships to lignin metabolism have been suggested have not yielded useful information because these substrates (e.g., benzoate, vanillate, α-conidendrin, etc.) bear no direct relationship to the synthesis of lignin or what is known about the degradation of natural lignin by microorganisms.

Several recent reviews (Kirk *et al.,* 1977; Ander and Ericksson, 1978; Crawford and Crawford, 1980) and a book (Kirk *et al.,* 1980) have covered knowledge of the mechanisms of microbial degradation of lignin. The emphasis in these reviews was industrial applications of lignin bioconversions, including biological pulping, chemical and fuel production, food–feed production, prevention of wood products deterioration, and pollution abatement. The present review will build on the mode of lignin metabolism but will focus on the biological and environmental parameters of importance to understanding the relationship between lignin metabolism and the carbon cycle.

2. Biological and Chemical Analysis of Natural Lignin

2.1. Presence and Function in Plants

Lignin is confined to specialized cells of vascular plants that, when living, function in conduction of solutes and in mechanical support (Neish, 1968; Wardrop, 1971). Natural lignin is not detected in algae or mosses, which lack the specialized conductive tissue common to woody plants. Lignin is found in the tracheids and fiber of the cell wall. The presence of lignin in plants imparts rigidity to the cell walls, decreases water permeation across the cell walls of xylem tissue, and impedes microbial destruction of plant tissue.

The major chemical cell-wall components in plant vascular tissue include: a framework component composed of α-cellulose, a matric component composed of linear polysaccharides (i.e., hemicellulose), and an encrusting component composed of lignin (Wardrop, 1971). The organization of these chemical components into the ultrastructural features observed in plant cell walls is complex (Cowling and Kirk, 1976; Côté, 1977) Fig. 1 demonstrates the general ultrastructural features of a model wood cell in schematic form. Crystalline cellulose microfibrils are located in the primary and secondary walls. Lignin is

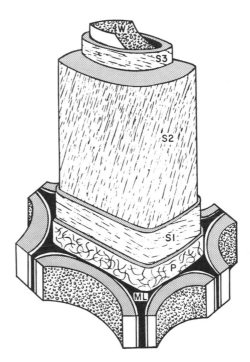

Figure 1. Diagrammatic structural representation of a model wood cell (from Côté, 1977).

found largely in the secondary wall and middle lamella. The linear hemicellulose molecules are found dispersed in cell-wall layers.

2.2. Chemical Structure

Most instructive insights for developing ideas to attack the mode of lignin decomposition by microorganisms are the result of understanding the chemical nature of lignin. Indeed, far more knowledge and excitement will be gained from reading the pioneering manuscript published by Freudenberg and Neish in 1968 than from reading the myriad of papers reported since then which relate ligneous substrate weight loss to microbial decomposition or that concern isolation and characterization of ligninolytic microbes.

Figure 2 represents a schematic structure for lignin in conifer or soft wood and serves to illustrate the chemical and physical complexity and the structural uniqueness of natural lignins. The following properties are important in regard to microbial degradation: (1) lignin has a compact structure that is insoluble in water and is difficult to wet and penetrate by microorganisms; (2) the inter-monomer linkages that account for the structural rigidity of lignin are a variety of carbon-carbon and carbon-oxygen bonds with the beta-aryl ether linkage

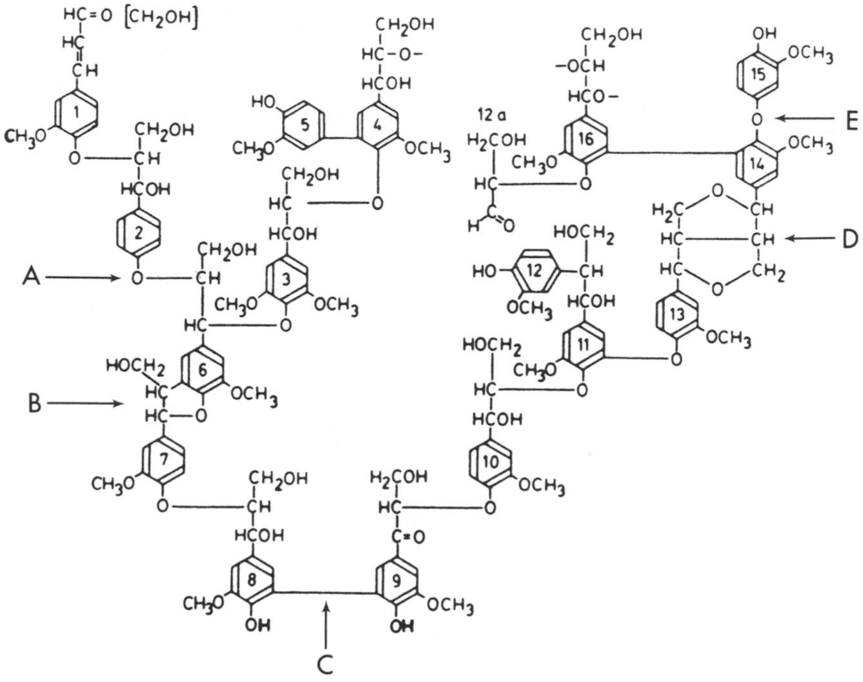

Figure 2. Structural representation of spruce lignin showing the major intermonomer bonds in lignin. The β-aryl ether bond (A) is quantitatively the most significant (from Adler, 1977).

being quantitatively the most significant; and (3) lignins are in essence a form of "natural plastic" (this will be discussed in more detail below) because the intermonomer linkages are not directly hydrolyzable. All natural lignins are synthesized by the dehydrogenative polymerization of p-coumaryl, sinapyl, and/or coniferyl alcohols and contain the kinds of intermonomer linkages shown in Fig. 2. Compounds that do not display these two major criteria are not considered to be lignin (Sarkanen, 1971; Goring, 1971).

Natural lignins are divided into two major classes—guaiacyl lignins and guaiacyl-syringyl lignins (Sarkanen and Hergert, 1971; Higuchi, 1980). Guaiacyl lignins are present in the majority of gymnosperm woods, *Pteridophytea* (i.e., ferns, clubmosses, etc.) and *Cycadales,* whereas guaiacyl-syringyl lignins are found in hardwood species, herbaceous angiosperms, grasses, and other monocotyledons.

It has not been established whether covalent bonds link lignin and the polysaccharides present in plant cell walls (Lai and Sarkanen, 1971). Freudenberg and Neish (1968) suggest that ester linkages between glucuronic and residues of hemicellulose and lignin should be anticipated. Recently, micelle formation based on hydrophobic bond interactions between lignin and carbohydrates has been established (Yaku *et al.*, 1979).

2.3. Synthesis in Plants

The synthesis of lignin in plants can be simplified into two steps, namely, biogenesis of lignin precursors (i.e., cinnamyl alcohol derivatives) and their dehydrogenative polymerization into lignin (Sarkanen, 1971; Gross, 1977). Both steps are highly oxidative and require O_2 directly (i.e., monooxygenase-catalyzed aromatic ring hydroxylations during cinnamyl alcohol synthesis) or indirectly (i.e., formation of H_2O_2 as a reactant for dehydrogenative polymerization). The synthesis of lignin precursors (i.e., monomers) is entirely enzyme mediated and follows the general path: $CO_2 \rightarrow$ carbohydrate $\rightarrow$ shikimic acid $\rightarrow$ phenylpropanoid amino acid $\rightarrow$ cinnamic acid derivatives $\rightarrow$ cinnamyl alcohol derivatives. A variety of plant aromatic compounds are synthesized from cinnamic acids (e.g., flavanoids, coumarins) or from stereospecific coupling of cinnamyl alcohols to lignans (Neish, 1968). However, natural lignin is formed only via the dehydrogenative polymerization of cinnamyl alcohols.

The polymerization process in lignin synthesis differs significantly from that which occurs for all other major biomass polymers. Specifically, lignin is synthesized via a free-radical condensation mechanism and is not under strict enzymatic control (Sarkanen, 1971; Gross, 1977). In other words, there is no polymerase *per se* that is associated with lignin synthesis. Figure 3 depicts the formation of phenoxy radicals by the peroxidase-catalyzed reaction of H_2O_2 and cinnamyl alcohols. Chemical dimerization of these radicals followed by further dehydrogenation and polymerization reactions leads to the polyoligomeric lignin macromolecule. It is still not exactly known how the H_2O_2 is supplied to the plant cell wall, although its origin is thought to be from $O_2/O_2 \cdot ^-$ and reduced pyridine nucleotides in the cytoplasm (Gross, 1977). Because of

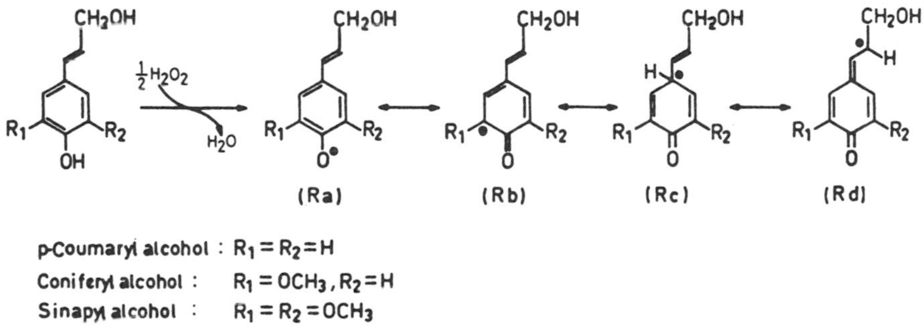

p-Coumaryl alcohol : $R_1 = R_2 = H$
Coniferyl alcohol : $R_1 = OCH_3, R_2 = H$
Sinapyl alcohol : $R_1 = R_2 = OCH_3$

Figure 3. Conversion of cinnamyl alcohols to phenoxy radicals, the immediate precursors for chemical polymerizations of lignin in woody plants (from Gross, 1977).

the chemical-free radical polymerization mechanism, no two natural lignins are exactly alike, even in the same plant species.

2.4. Polymer Evolution

As a biopolymer, lignin is truly a unique macromolecule. The synthesis of lignin obligately requires O_2 for the enzymatic synthesis of phenoxy radical monomers. Thus, lignin evolved when the earth's atmosphere was rich in O_2 as a consequence of photosynthesis. Lignin can be viewed as a "natural plastic" by microbial ecologists because the polymerization of lignin from direct cell precursors is not enzyme mediated, and the polymer contains multiple inter-monomer bonds that are not directly hydrolyzable by enzymes.

The *raison d'etre* for lignin evolution is best described by Neish (1968), who speculated that lignin arose as a consequence of the problems associated with waste-product excretion in plants. Higher plants do not have efficient systems for external excretion and thus depend primarily on chemical transformation of metabolic by-products into volatile, insoluble, or nontoxic compounds. This feature of higher plants is manifested in the presence of a variety of aromatic secondary metabolites (i.e., natural lignins, flavanoids, anthocyanins, coumarins, etc.). Once plants gained the ability to convert phenylpropanoid amino acids to phenoxy radicals, lignin arose, and tracheids and fibers developed in cell walls. When this occurred, lignin was no longer a waste material but rather a requirement for the development of higher plants.

The recognition of lignin as a secondary metabolite by Neish (1968) is most interesting. Lignin differs from primary plant metabolites because it is not essential for growth, although it is beneficial to the organism possessing it. Thus, some plants contain lignin, whereas others do not. Plants, algae, fungi, and bacteria produce a variety of secondary metabolites. Many secondary metabolites of plants, although not lignin in chemical structure, are synthesized in a manner analogous to lignin. For example, *Sphagnum* does not contain lignin as a secondary metabolite but produces large amounts of polyphenolic material via free-radical condensation reactions (Neish, 1968). Likewise, both fungi and bacteria produce aromatic secondary metabolites via phenol oxidase-catalyzed reactions (Arai and Mikami, 1972; Kirk *et al.,* 1975a; Ha-Huy-Kê and Luckner, 1979; Bartsch *et al.,* 1979). It is worth noting that the most actively ligninolytic fungus described to date, *Phanerochaete chrysosporium,* also produces a *p*-benzoquinone as a secondary metabolite (Chen *et al.,* 1977).

In summary, lignin evolved as a recalcitrant molecule as a consequence of its chemical polymerization and shares a mechanism of natural recalcitrance that is common to skins and protective coatings of plant and animal tissue and perhaps to spore coats of aerobic microorganisms.

3. Biological Degradation of Natural Lignin

This section will emphasize the biological decomposition of natural lignins and will not cover in any detail the microbial metabolism of industrial lignins (see Section 4.5) or the general metabolism of aromatic compounds by microbes. Readers are advised to see the reviews of Evans (1977), Dagley (1978a,b), Ander and Ericksson (1978), and Crawford and Crawford (1980) to understand better the general and industrial aspects of microbial aromatic metabolism. Biological degradation of natural lignin is manifested by specific changes in the polymer that include modification of the molecular weight, partial modifications of the chemical structure, and/or complete destruction of the chemical structure.

3.1. Measurement of Polymer Decomposition

Tremendous advances have been made in understanding biological decomposition of lignin during the past five years because of the development and refinement of specific assays for the microbial metabolism of natural lignin. These assays are based on the biological conversion of [^{14}C]lignins to $^{14}CO_2$. Other studies that employed weight loss, phenolic color indicators, loss of fractions resistant to acid hydrolysis, and/or growth on chemically ill-defined lignaceous substrates as indicators of microbial lignin metabolism are not reviewed here because of their questionable relevance because of the lack of a definitive bioassay. The specific bioassays for lignin decomposition involve the use of ^{14}C-labeled lignins that are obtained either from extraction of plants that were fed ^{14}C-precursors or from the synthesis of polymers that were formed via the enzymatic dehydrogenation of [^{14}C]cinnamyl alcohols. The procedures for preparation of these radioactive natural lignins are based on previous chemical studies of lignin metabolism in plants and are expertly reviewed by Freudenberg and Neish, (1968), Neish (1968), and Sarkanen (1971).

Sorensen (1963) and Mayaudon and Batistic (1970) first developed bioassays for lignin degradation based on analysis of microbial decomposition of [^{14}C]lignins prepared from plants that were fed radioactive precursors. This technique was further developed and refined by Haider and Trojanowski (1975), Crawford and Crawford (1976), and Crawford et al. (1977a,b). The greatest shortcoming of using ^{14}C-natural lignins prepared from extractive isolation of radioactive plants in a bioassay is that ^{14}C-plant components other than lignin may also be present in the test substrate. For example, [^{14}C]Ferulic acid used for preparation of ^{14}C-maize lignin by Haider and Trojanowski (1975) and [^{14}C]phenylalanine employed for production of ^{14}C-wood lignins by Crawford et al. (1977a,b) are precursors for lignin as well as other plant components (e.g., flavonoids, coumarins, lignans, alkaloids, and proteins). The extractive procedures employed by these investigators remove the soluble ^{14}C-aromatic fractions in plant material; however, the amounts of other [^{14}C]nonlignin-insoluble components present were not chemically determined. In this regard, it is worth noting that the insoluble fraction left after hot ethanol extraction of ^{14}C-plants that were labeled by infusion of [^{14}C]phenylalanine during growth contained a considerable amount (ca 20%) of denatured protein (Neish, 1968).

Biological assays for lignin decomposition that employ synthetic ^{14}C-labeled lignin (Kirk *et al.*, 1975b; Haider and Trojanowski, 1975) can afford an unequivocal method for monitoring biological lignin decomposition provided that the chemical properties of the synthetic lignins used are well characterized (Kirk *et al.*, 1977). Kirk *et al.* (1975b) described the synthesis of lignins that were labeled with ^{14}C in either the ring, propane side chain, or methoxyl moieties of lignin. These [^{14}C]lignins were chemically well characterized and shown to contain the same intermonomer linkages found in wood lignins and to have an average molecular weight of greater than 1400. The synthesis of ring-labeled [^{14}C]lignin via the procedures used by Kirk *et al.* (1975b) is schematically represented as three separate steps in Fig. 4. The first step involves the chemical conversion of commercial U[^{14}C]phenol to 2(2-methoxy phenoxy)-5 nitrobenzophenone, which is hydrolyzed to guaiacol and then formylated to vanillin. The ring-labeled vanillin is condensed with ethyl hydrogen malonate to form ethyl ferulate, which is reduced by lithium aluminum hydride to yield coniferyl alcohol, an immediate lignin precursor. The second step of the synthesis involves the horseradish peroxidase-catalyzed formation of phenoxy radicals from ring-labeled coniferyl alcohol. Lignin is produced in the last step by the carefully controlled chemical polymerization of phenoxy radicals. Synthetic lignins formed by these procedures are often called dehydrogenative polymerizates (DHPs). The close chemical identity between synthetic [^{14}C]lignins (or [^{14}C]-DHPs) and lignins prepared from wood (Kirk *et al.*, 1975b) and the demonstration that a model wood lignin (Kirk *et al.*, 1977), milled wood lignin (Wayman and Obiaga, 1974), and [^{14}C]-DHPs form a common metabolic pool (i.e., wood lignins dilute the specific activity of [^{14}C]-DHPs in assays of biological decomposition) suggest that these synthetic lignins are representative of lignins in plants. It is of tremendous importance that dehydrogenative polymerizates of cinnamyl alcohols used for lignin biodegradation assays be chemically characterized to insure a high-molecular-weight product (i.e., >1400 molecular weight) and to verify that they contain the intermonomer linkages present in plant lignins. Failure to do this only raises questions as to whether the synthetic product is representative of lignin or is only a lignin model compound that is of low molecular weight or that lacks important lignin intermonomer bonds. [^{14}C]Lignins formed by dehydrogenative polymerization or by incorporation of radioactive tracers in plants are both considered natural lignins because the formation of active monomers (i.e., phenoxy radicals) is bio-

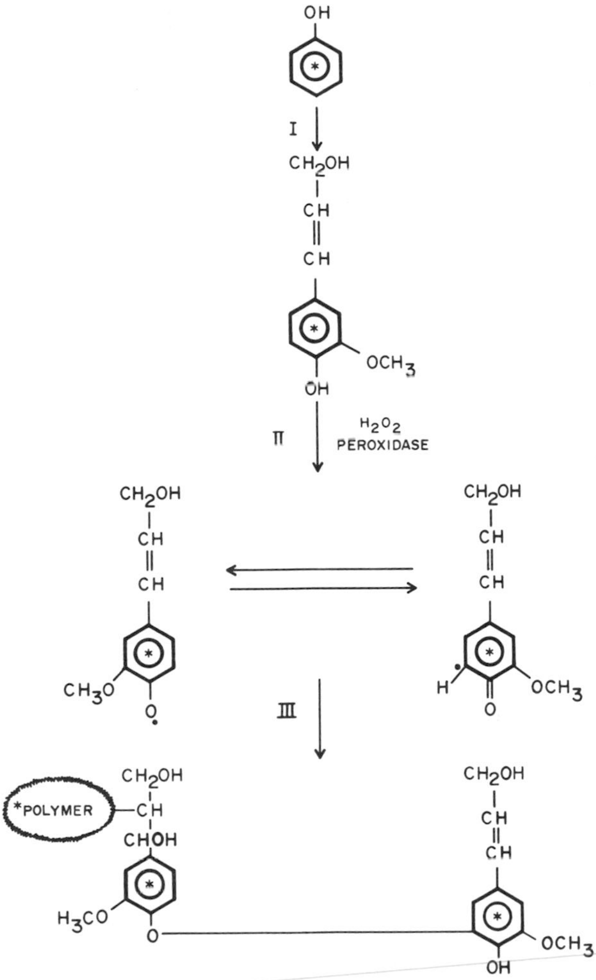

Figure 4. Schematic illustration of the main steps in the chemical synthesis of ^{14}C-ring-labeled dehydrogenative polymerizates (i.e., natural lignins or DHPs).

logically mediated (i.e., via enzyme catalysis) whereas the polymerization step is synthetically mediated (i.e., via chemical catalysis). "Natural" is used here in the broadest sense of the word and is employed to denote what is to be expected of lignins in nature. Unfortunately, because of nonspecific incorporation of ^{14}C-precursors and the chemical extraction procedures used for isolation of ^{14}C-labeled plant lignins and the numerous chemical steps employed in synthesis of [^{14}C]-DHPs, these substrates are not natural lignins *sensu strictu.*

^{14}C-Labeled plant lignins and [^{14}C]-DHPs have been used effectively to study lignin metabolism in natural materials or in laboratory cultures (see Sections 3.2 and 4.3). By and large, biological ligninolytic activity is monitored by quantification of the amount of substrate degraded to CO_2 (i.e., percent [^{14}C]lignin converted to $^{14}CO_2$). Crawford and Crawford (1980) have suggested that the sum of [^{14}C]lignin converted to $^{14}CO_2$ and water-soluble ^{14}C is more indicative of ligninolytic activity. This seems reasonable unless ^{14}C-plant lignins are used that contain high amounts of [^{14}C]nonlignin insolubles (e.g., protein). The use of the [^{14}C]lignins described above for analysis of biological ligninolytic activity requires establishment of a minimum value for decomposition (i.e., percent [^{14}C]lignin converted to $^{14}CO_2$ or $^{14}CO_2$ plus water-soluble ^{14}C) in terms of ascribing the significance of lignin decomposition by a given organism or a mixed biological population. In general, if greater than 3% decomposition of chemically characterized [^{14}C]-DHPs or greater than $\sim$10% degradation of ^{14}C-plant lignins is observed in biological experiments, then it should be assumed that the organism(s) significantly modified the chemical structure of lignin.

3.2. Polymer-Degrading Organisms

3.2.1. Eucaryotic Microbes

Among all organisms examined to date, only eucaryotic microbes have unequivocally been shown to destroy completely the chemical structural features of natural lignins. Nonetheless, the effects of ligninolytic microorganisms, including those of eucaryotic species, on modification of the lignin molecule are quite varied (see Table I). The use of two different methods for analysis of structural-chemical changes in lignin has demonstrated how fungal metabolic activity modifies and/or destroys the integrity of the molecule. These methods are based on quantification of ligninolytic activities of individual species with specifically labeled [^{14}C]lignins (i.e., lignin labeled in the methoxyl, aryl, or alkyl moieties) or comparative chemical-structural analysis of untreated and fungal decayed wood. The ligneous activities of the white-, brown-, and soft-rot fungi have been best characterized because the effects of these eucaryotic species on woods are often visibly discernible.

White-rot fungi are the most actively ligninolytic organisms yet characterized. White-rot fungi can also completely degrade all major chemical components of intact wood into CO_2 and H_2O (Kirk and Highley, 1973). *Pleurotus ostreatus, Phanerochaete chrysosporium,* and *Coriolus versicolor* were shown to decompose [^{14}C]-DHPs labeled in either the ring, methoxyl, or side chain moieties to $^{14}CO_2$ by Haider and Trojanowski (1975, 1980) and Kirk *et al.* (1975b). Kirk and Highley (1973) demonstrated by chemical analysis of decayed wood that significant lignin decomposition (i.e., $>$45%) by *Ganod-*

Table I. Representative Organisms that Degrade and/or Modify Natural Lignin[a]

Organism	Effect on lignin		
	Modify molecular size or weight[b]	Partial chemical modification[c]	Complete chemical decomposition[d]
I. Eucaryotic microbes			
White-rot fungi			
Pleurotus ostreatus	+	+	+
Phanerochaete chrysosporium	+	+	+
Coriolus versicolor	+	+	+
Polyporus anceps			
Brown-rot fungi			
Lenzites trabea	+	+	−
Gloeophyllum trabeum	+	+	−
Poria cocos	+	+	−
Lentinus lepideus	+	ND	−
Poria monticola	+	ND	−
Soft-rot fungi			
Thielavia terrestris	+	+	ND
Preussia fleishhakii	+	+	ND
Chaetomium piluliferum	+	+	ND
Others			
Fusarium solani	+	+	ND
II. Procaryotic microbes			
Actinomycetes			
Streptomyces badius	+	+	ND
Nocardia autotrophica	+	+	ND
Others			
Bacillus megaterium	+	+	ND
Azotobacter species	+	+	−
Pseudomonas species	+	+	ND
III. Lignocellulose-destroying animals			
Steer	+	−	−
Millipede	+	−	−
Termite	+	+	ND

[a]Compiled from references reviewed in the text. Abbreviations +, occurs; −, not demonstrated; ND, not determined.
[b]As determined by a physical or biochemical action on the polymer that decreases size or molecular weight.
[c]Determined by chemical analysis of decayed wood or ligninolytic bioassays with ^{14}C-natural lignins.
[d]As determined by detailed chemical–structural analysis of decayed lignocellulose.

erma applanatum and *C. versicolor* was associated with decomposition of cellulose and hemicellulose. The extent ($>50\%$ conversion of ring-labeled DHP to CO_2) and rate ($>15\%$ conversion of ring-labeled DHP to CO_2 in 24 hr) or lignin metabolism in *P. chrysosporium* flask cultures surpasses that of other described microorganisms (Kirk *et al.*, 1978; Fenn and Kirk, 1979). The taxonomic status of ligninolytic white-rot fungi is often confused in the literature (Kirk *et al.*, 1977; Ander and Ericksson, 1978). Most notably, *Sporotrichum pulverulentum* is synonomous with *P. chrysosporium* and should be properly recognized by the later species name (Burdsall and Eslyn, 1974). Similarly, *C. versicolor* is often designated by other generic names (e.g., *Trametes, Polyporus, Polystictus*).

Brown-rot fungi extensively degrade polysaccharides and cause limited chemical structural modification of lignin during growth on wood (Kirk and Highley, 1973). Detailed chemical analysis of wood degraded by *Lenzites trabea* demonstrated that brown-rot decomposition was largely oxidative, and that demethylation of both phenolic and nonphenolic units in wood was the major degradative reaction (Kirk, 1975). Decomposition studies with differentially labeled [^{14}C]-DHPs supported the finding that demethyoxylation was the major ligninolytic activity of the brown-rot fungi *Gloeophyllum trabeum* and *Poria coros* (Kirk *et al.*, 1975b).

Soft-rot fungi degrade all major wood chemical components and appear to decompose hardwood lignins more effectively than those in softwoods. Chemical analysis of woods decomposed by six soft-rot species indicated significant weight loss of lignin (as measured by the "sulfuric acid" method) in red alder and balsam poplar but not in white pine (Eslyn *et al.*, 1975). Haider and Trojanowski (1975, 1980) demonstrated with differentially labeled lignins that soft-rot species *Preussia fleshhikii* and *Chaetomium piluliferum* decomposed ring-labeled [^{14}C]-DHP ($\sim$3.0% conversion to $^{14}CO_2$ in 25 days) and that demethoxylation was more significant ($\sim$8% conversion of [^{14}C]methoxyl-DHP to $^{14}CO_2$ in 25 days) than aryl or alkyl metabolism. These results (Haider and Trojanowski, 1975) may reflect minimal levels of ligninolytic activity in brown-rot fungi because, unlike that reported for white-rot species (Keyser *et al.*, 1978; Fenn and Kirk, 1979), optimized culture parameters for ligninolytic activity in brown-rot fungi are not described.

Eucaryotic microorganisms other than the three major groups of wood-rotting fungi have not been well examined for ligninolytic activity. Most notably, the mold *Fusarium solani* has been reported to grow actively on and significantly degrade DHPs (Iwahara, 1980). This species also catabolizes the model lignin dehydrodiconiferyl alcohol by a novel pathway (Ohta *et al.*, 1979). It will be of interest to examine the ability of other fungal species to degrade ^{14}C-natural lignin, especially those that catabolize soluble aromatic compounds (Cain, 1980) or degrade industrial lignins (Drew and Kadam, 1979).

3.2.2. Procaryotic Microbes

Bacteria have been suggested to possess ligninolytic activity (see Sorensen, 1962; Sundman, 1964; Cartwright and Haldom, 1973), but because of nondefinitive bioassay methods, these studies are subject to criticism. Hackett (1975) demonstrated that both procaryotic and eucaryotic organisms were actively ligninolytic in soils; this was done by comparing decomposition rates of differentially labeled [^{14}C]-DHPs in the presence of specific antibiotics. Hackett further showed that an *Azotobacter* species isolated from soil decomposed both ring-labeled ($\sim$7% conversion to CO_2 in 6 days) and methoxyl-labeled ($\sim$4% conversion to CO_2 in 6 days) lignins when grown on soluble aromatic compounds as the growth substrates. However, these rates were considerably lower than those observed for a mixed bacterial culture enriched from the same environment.

Haider *et al.* (1978) demonstrated that *Nocardia autotrophica* cultures were dramatically more ligninolytic (31% total conversion of alkyl-, aryl-, and methoxyl-lignin to CO_2 in 15 days) than other *Norcardia* or *Pseudomonas* species. These two genera are of particular interest because of their ability to catabolize a variety of aromatic compounds, including model lignins that contain the β-aryl-ether intermonomer bond present in natural lignins (Crawford *et al.*, 1973a, 1975; Kawakami, 1980).

Robinson and Crawford (1978) showed that *Bacillus megaterium* cultures decomposed 12% of ^{14}C-spruce lignin and 9% of [^{14}C]aryl-DHP to ^{14}CO$_2$ in 20 days. Crawford and Sutherland (1979) and Crawford and Crawford (1980) reported that three *Streptomyces* species decomposed ^{14}C-fir lignin to ^{14}CO$_2$. *Streptomyces badius* displayed the greatest ligninolytic activity, and cultures converted $\sim$13% of ^{14}C-wood lignin to ^{14}CO$_2$ and $\sim$16% of this substrate to water-soluble ^{14}C in 42 days. The ability of bacteria like *S. badius* to decompose lignin is especially interesting because actinomycetes share many physiological features in common with fungi. These common growth properties include a hyphal-mycelium growth phase and a complex secondary metabolism phase with spore formation and active metabolite excretion. Sutherland *et al.* (1979) have demonstrated, by weight loss and scanning electron microscopy, that *Streptomyces flavovirens* decays intact cell walls of wood tissue. It will be of interest to learn what specific chemical modifications are associated with this type of bacterial wood-decay process.

3.2.3. Animals

The role animals play in biospheric lignin decomposition should not be ignored despite the fact that their only well documented effect on the polymer appears to involve lowering the molecular weight as a consequence of mechanical disruption by species which feed on lignocellulosic biomass (see Table I). By and large, insects, soil invertebrates, and ruminants contribute greatly to the physical depolymerization of lignin in biomass and increase the digestibility of lignocellulose to microorganisms associated with their own gastrointestinal tracts and in their environment. It should be noted that microorganisms can biochemically degrade lignin, whereas animals appear to physically lower the molecular weight. The digestion of lignin by some higher termites has been demonstrated by Butler and Buckerfield (1979), but the mechanisms involved were not investigated. Detailed studies of the gut structure and physiology in soil-feeding termites (J. M. Anderson, personal communication) have revealed a unique digestive system which is compatible with lignin degradation. The hind gut contains both aerobic and anaerobic bacteria and an anterior paunch in which plant materials are pretreated under highly alkaline conditions (pH 10–11). However, lower termites fed a wood diet did not significantly demethoxylate or deplete the lignin fraction but rather effectively decomposed cellulose and hemicellulose (Esenther and Kirk, 1975). Differentially labeled [^{14}C]-DHPs were not metabolized to $^{14}CH_4$ by steer rumen contents (Hackett et al., 1977) or millipedes (Neuhauser et al., 1978). Organic decomposition in the gastrointestinal tracts of many animals is essentially an anaerobic process, and anaerobic microorganisms have been shown to decompose a wide variety of soluble aromatic compounds but not natural lignin (Zeikus, 1980b).

3.3. Physiology and Biochemistry of Polymer Metabolism in White-Rot Fungi

White-rot fungi at present provide a model system for learning how natural lignin is metabolized by microorganisms because of their well demonstrated ability to completely destroy lignin in plant biomass. Most detailed physiological studies on lignin metabolism by white-rot fungi have been limited to the thermotolerant fungus P. chrysosporium. These studies were conducted with shallow batch cultures in a chemically defined medium, using the conversion of [^{14}C]-DHPs to $^{14}CO_2$ as a measurement of ligninolytic activity (Kirk et al., 1976; Kirk et al., 1978; Keyser et al., 1978).

The development of a defined culture medium for white-rot fungi that employed carbohydrate-free lignin from wood (i.e., milled wood lignin) or synthetic DHPs as energy source led to the remarkable finding that P. chrysosporium could not grow on lignin alone but required, in addition, a readily metabolizable organic co-substrate such as cellulose or glucose (Kirk et al., 1976). This seemed extraordinary in regard to lignin's eminence as an organic molecule in the biosphere. Later studies revealed that (1) wood cellulose as a growth substitute stimulated ligninolytic activity of P. chrysosporium more than glycerol or succinate; and (2) vanillate did not support growth or lignin decomposition (Kirk et al., 1978). During batch culture growth of P. chrysosporium on glucose and DHP, approximately 20 times more co-substrate was consumed than lignin.

Recent studies by Drew and Kadam (1979) demonstrated that white-rot fungi and Aspergillus fumigatus required a co-substrate for growth and decomposition of industrial Kraft lignin. The terms cometabolism (Horvath,

1972) and *cooxidation* (Perry, 1979) have been used recently to describe microbial oxidations of substances (usually hydrocarbons) in which substrate metabolism alone will not allow organismal growth. At first glance, the co-substrate requirement for lignin metabolism by white-rot fungi would seem to suggest that lignin decomposition is a cooxidation process. However, this terminology does not appear directly applicable (see discussion below) because lignin is completely degraded by *P. chrysosporium* to H_2O and CO_2 without the accumulation of toxic metabolites and by nonproliferating (i.e., stationary phase) cultures.

Manipulation of several batch culture parameters revealed that a complex set of requirements was needed to optimize lignin decomposition by *P. chrysosporium* (Kirk *et al.*, 1978; Fenn and Kirk, 1979). Culture agitation suppressed ligninolytic activity. Lignin decomposition was optimal at pH values between 4.0–4.5, but certain pH-effective buffers were inhibitory (e.g., *o*-phthalate). Nitrate, ammonium, or amino acids were sources of nutrient nitrogen for growth of *P. chrysosporium,* but high nitrogen concentrations in culture medium decreased ligninolytic activity. Most importantly to the biochemical mechanism(s) of lignin metabolism by white-rot fungi, high O_2 partial pressures were required for optimal ligninolytic activity (see Fig. 5). Growth was not influenced by O_2 levels, but lignin was not significantly decomposed at 5%

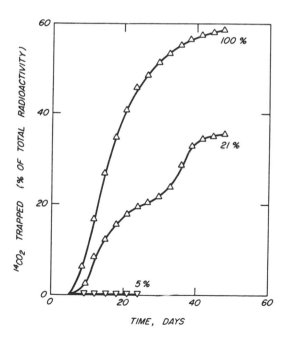

Figure 5. The relation of O_2 partial pressure to ligninolytic activity of *Phanerochaete chrysosporium* in batch cultures (from Kirk *et al.*, 1977).

O_2 and ligninolytic activity was 2- to 3-fold greater at 100% O_2 than in air (21% O_2). The physiological requirement for high O_2 levels for lignin metabolism of *P. chrysosporium* correlates well with the finding that lignins in wood attacked by white-rot fungi contain more oxygen than in the corresponding sound wood (Chang *et al.*, 1980) and with the conclusion that lignin decomposition in general is a highly oxidative process.

The inhibition of lignin decomposition by high levels of nutrient nitrogen in *P. chrysosporium* cultures led to examination of the relationships between organismal growth phase, co-substrate and nitrogen consumption, and ligninolytic activity (see Fig. 6). Most notably, lignin was decomposed by stationary phase (i.e., nongrowing) cells, and in the presence of co-substrate but the absence of a nitrogen source (Keyser *et al.*, 1978). The addition of ammonium to cultures immediately prior to complete nitrogen depletion delayed the appearance of ligninolytic activity. This indicated that nitrogen starvation initiated fungal secondary metabolism and the associated degradation of lignin. It should be noted that the same levels of ligninolytic activity were displayed irrespective of the initial concentration of lignin in the culture medium (i.e., from no lignin to large amounts present). Weinstein *et al.* (1980) demonstrated that degradation of soluble model lignins, which contain β-aryl-ether bonds, by *P. chrysosporium* required a co-substrate, and these model lignins were also

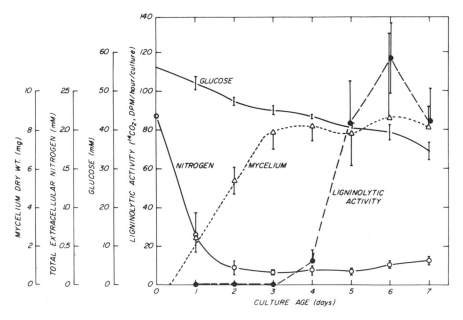

Figure 6. Relationship between growth, nitrogen depletion, and ligninolytic activity in *Phanerochaete chrysosporium* batch cultures (from Keyser *et al.*, 1978).

metabolized in the stationary growth phase in response to nitrogen starvation. As a consequence of secondary metabolism, lignin degradation by white-rot fungi can not be considered properly as cooxidation (Perry, 1979). Nonetheless, secondary metabolism of fungal resting cells is a well-documented phenomenon that is associated in part with nitrogen-induced starvation (Bu'Lock *et al.*, 1974; Bu'Lock, 1975). *Phanerochaete chrysosporium* also produces and degrades veratryl alcohol during secondary metabolism (Lundquist and Kirk, 1978).

The new physiological discoveries concerning lignin metabolism in *P. chrysosporium* raise questions about the mechanism(s) of macromolecular depolymerization and the *raison d'etre* for microbial evolution of ligninolytic activity. In general, evolution of metabolic function in the carbon cycle can be viewed in the following sequence: chemical-synthetic (i.e., nonenzyme mediated) substrate formation → biological–chemical substrate degradation → biological degradative metabolite accumulation → biological resynthesis from metabolites. Lignin has an evolutionary relation unique among major polymers of the biomass involved in the carbon cycle; namely, lignin evolved as a secondary metabolite with its molecular polymerization catalyzed by a chemical-synthetic process. This feature raises the following question—do lignin-metabolizing microorganisms depolymerize and degrade lignin by enzyme-catalyzed reactions alone or are chemical-degradative steps also involved? This question remains unanswered. The evolution of biological lignin synthesis and degradation in the carbon cycle as a secondary metabolic process is thought-provoking. The terms *secondary metabolism* and *primary metabolism* in cell physiology are often confusing in regard to definition. Primary metabolism is defined here as the coupling of cellular metabolic activities to growth, whereas secondary metabolism is the coupling of cellular metabolic activities to survival. The survival value of ligninolytic activity to a wood-rotting fungus is perhaps best addressed by anonymous forest pathologists who say "fungi degrade lignin to eat the cellulose in wood." This statement does make sense in view of the physiochemical properties of wood and the physiology of white-rot fungal metabolism. Lignin in essence physically surrounds the cellulose in wood and makes cellulose, an energy source for growth of fungi, inaccessible to catabolic enzymes. Thus, the removal of lignin during fungal secondary metabolism enables catabolic enzymes to function during growth. It is hard to imagine that white-rot fungi do not gain energy by degrading lignin to CO_2 during secondary metabolism. It is speculated here that energy is gained from lignin degradation but at a low rate that is sufficient for survival but not for growth. Perhaps this ultimately reflects the lack of a specific enzyme mechanism for depolymerization of lignin.

Determination of the biochemical pathway and characterization of the key enzymes in fungal lignin decomposition are inherently difficult tasks because lower enzyme activities are to be expected from a secondary metabolic process than from the catabolism of a specific energy source for growth (e.g., cellulose decomposition). The biochemical pathway and the key polymer-degrading enzymes for lignin decomposition in white-rot fungi are not known; thus, the biochemistry of polymer metabolism is at present open to speculation. Nonetheless, analysis of fungal-decayed lignins in wood provides data that must be incorporated in any proposed biochemical pathway(s), and character-ization of enzymes active on model lignins suggests a role for these activities in lignin decomposition. These data were expertly reviewed by Kirk *et al.* (1977), Ander and Eriksson (1978), and Chang *et al.* (1980) and will be briefly summarized here.

Chemical analysis of fungal white-rotted wood lignins has suggested inter-esting and novel biochemical degradative mechanisms (Kirk and Chang, 1974, 1975; Chang *et al.*, 1980). Most importantly, aromatic moieties of lignin are oxidatively cleaved while part of a polymeric structure. This implicated extra-cellular cleavage of aromatic rings, an unprecedented reaction for oxygenases. Other degradative reactions observed included oxidation of α-carboninol and propanoid side chains. White-rot fungi appear to attack the exposed surfaces of the lignin polymer. Demethylations of guaiacyl and syringyl moieties are the initial degradative reactions followed by ring cleavage and depolymerization. Notably, the highly oxidative degradation of lignin proceeds without the for-mation of significant amounts of identifiable, informative low-molecular-weight degradation products. In conclusion, the physiological studies and chemical analysis of degraded woods imply a unique decomposition process for lignin by white-rot fungi analogous to a biologically-mediated polymer erosion rather than a catabolic decomposition based on enzymes with high substrate affinity and specific activity.

A variety of enzymes have been suggested as important to lignin decom-position. Oxygenases that cleave aromatic rings are undoubtedly important in lignin decomposition (Kirk and Lorenz, 1973; Wood *et al.*, 1977; Kirk *et al.*, 1977; Dagley, 1978a,b; Buswell *et al.*, 1979a; Buswell and Eriksson, 1979), but white-rot fungal oxygenases have not been purified and shown to be active on lignin. Phenol oxidases, including laccases, tyrosinases, and peroxidases, may be important in lignin decomposition (Kirk *et al.*, 1977; Ander and Eriksson, 1978). The formation of free radicals from lignin by action of phenol oxidases and the subsequent coupling of the radicals may induce cleavage of bonds between the aromatic ring and the propane side chain (Ander and Eriksson, 1978). Most notably, a *P. chrysosporium* mutant which lacked phenol oxidase did not degrade lignin, whereas a revertant was ligninolytic (Ander and Eriks-son, 1976). Purified laccase from white-rot fungi is active on milled wood lignin and model lignins, forming quinones and methanol (Ishihara, 1980). However, the specificity of phenol oxidases toward lignin *per se* is still challenged largely because many nonligninolytic fungi have these activities (Froehner and Eriks-son, 1974a,b; Bartsch *et al.*, 1979). Most recently, a new enzyme activity, cellobiose:quinone oxidoreductase, has been suggested as important to both cellulose and lignin decomposition in white-rot fungi (Westermark and Eriksson, 1974b; Hiroi and Eriksson, 1976). Ander and Eriksson (1978) concluded that this enzyme is of importance, although not entirely necessary, in the degradation of lignin by white-rot fungi. It will be especially important to learn if cellobiose:quinone oxidoreductase is formed during the linear or stationary phase of growth because this will reveal the catabolic and/or secondary metabolic function of the enzyme.

Undoubtedly, the missing link in understanding the biochemistry of lignin decomposition is what accounts for specific depolymerization. The kinds of enzymes that are active on the major C—C and C—O intermonomer bonds in lignin are not known or even well hypothesized. The possibility still has not been eliminated that depolymerization of the nonhydrolyzable bonds in intact lignin is effected by specific chemical products of white-rot fungi. In view of the high O_2 requirement for lignin decomposition by *P. chrysosporium*, examination of the possible fungal generation of superoxide radicals (i.e., $O_2 \cdot ^-$) as oxidative agents in lignin metabolism is intriguing. Biological generation of superoxide radicals is important in a variety of secondary metabolic processes, especially the wounding of plant tissue and fruit ripening (Halliwell, 1979).

4. Lignin Decomposition and the Environment

4.1. General Abundance and Recalcitrance of Lignin

The importance of lignin in the carbon cycle is most dramatically revealed by examination of the abundance of biospheric carbon that contains or is derived from lignin and by analysis of the influence of lignin on organic mineralization by biological food chains (see Table II). According to recent estimates

Table II. Abundance and Degradation of Carbon in the Biosphere

Net primary production
  Land: 63×10^{15} g C/yr
  Oceans: 45×10^{15} g C/yr
Total plant biomass
  99.8% is on land
  85% is in forests
Wood
  35- to 45-yr turnover time
  25% lignin
Annual secondary production
  Oceans: >37% consumption of plant biomass
  Land: <7% consumption of plant biomass

(Whittaker and Likens, 1973; Reiners, 1973; Bolin *et al.*, 1979), the amount of organic carbon on earth that contains or is derived from lignin includes (in grams of C): living and dead phytomass, 900×10^{15}; peats, 900×10^{15}; humus, 2800×10^{15}; and, coals–lignites, $>5000 \times 10^{15}$. Net primary production (i.e., photosynthesis) on land (63×10^{15} g C/year) is only 1.4-fold higher than in oceans (45×10^{15} g C/year); however, greater than 99.8% of the total biosphere phytomass is continental (Whittaker and Likens, 1973; Bolin *et al.*, 1979). Most notably, 85% of all land biota is contained in forests. The reason endogenous phytomass persists on land and not in oceans is because lignification of land plant tissue results in natural product recalcitrance. Wood, the net product of forest photosynthesis, has a calculated turnover time of 35–45 years (Bolin *et al.*, 1979) because it contains approximately 25% lignin by weight, which physically limits biological mineralization. Thus, a noticeable influence of ligneous phytomass on biological trophic food chains is seen by comparison of secondary production in marine and land ecosystems. Animals consume 37% of endogenous phytomass in oceans but less than 7% on land because of lignification. The preeminence of microbial metabolism of organic matter in the carbon cycle on earth is thus in part attributable to lignin in biomass. Furthermore, the recalcitrance of lignin to both microbial and animal metabolism is associated with the formation of humus, peats, lignites, and coals.

Microorganisms are challenged to decompose lignin in the environment as a consequence of plant detritus formation via annual litter fall in nature and as a consequence of man's manipulation of the carbon cycle via generation, concentration, and deposition of ligneous residues from biomass. Ligneous litterfall can amount to nearly 80% of net primary production (Reiners, 1973). In the United States, biomass-based industries (i.e., forestry and agriculture) produce approximately 10^9 tons of primary biomass (i.e., crops and wood) and generate another 10^9 tons of biomass-derived residues (i.e., secondary and tertiary biomass) per year (Zeikus, 1980a). Secondary biomass includes crop (e.g., corn stover) and forestry (e.g., branches) residues that have low economic value and are generally deposited and decomposed in the environment without pollution-associated problems. Tertiary biomass (i.e., wastes) generated by agriculture (e.g., animal manures) and forestry (e.g., paper mill sludges) industries generally require pretreatment to alleviate pollution prior to environmental deposition at high concentrations.

The general recalcitrance of lignin and the importance of lignin decomposition in the environment and the carbon cycle has received some recognition (Nickerson, 1971; Alexander, 1975; Dagley, 1975, 1978a,b; Crawford *et al.*, 1977b; Faber, 1979). In light of the knowledge presented in this review, the basis for lignin's recalcitrance in the environment agrees with some general mechanisms of molecular recalcitrance postulated by Alexander (1973); namely, lignin is slowly metabolized because it is an insufficient energy source for growth, it is not soluble in water, and the intermonomer bonds of lignin are not readily accessible to cleavage. In addition, the absence of lignin decomposition in anaerobic environments appears related to the lack of O_2, an essential nutrient for ligninolytic microbes. Two generalizations proposed by Alexander (1973), which are probably applicable to chemically synthesized substrates, may not be related to lignin's recalcitrance—i.e., chemicals that fail to induce the enzyme(s) needed for their breakdown are not degraded, and compounds are not degraded for which there exists no enzyme(s) for catalyzing an initial reaction of the degradative pathway. More detailed studies are required to prove these points, but lignins, as natural products formed by a chemical polymerization process, may be decomposed by microbes in spite of these suggested restrictions. Ultimately, this is probably a reflection of the general infallibility of microbial degradative attack on natural products.

In relation to lignin biodegradation in the environment, it is suggested that microbial ecologists consider lignin as a biological plastic. Indeed, past experiences in synthesis of DHPs (see Kirk et al., 1975b) revealed that improper control of chemical reaction parameters resulted in a plastic product which, unlike natural lignins, was not readily soluble in dimethylformamide. Haraguchi and Hatakeyama (1980) have demonstrated similarities between the degradative pathways of synthetic model polystyrenes and lignins. Most notably, detailed structural analysis of humic substances revealed the widespread occurrence of polymethylene chain structures similar to those present in high-molecular-weight synthetic polyethylenes (D. Grant, 1977).

Substrates that are readily decomposed by microbes (e.g., glucose) are not significantly modified by chemical and physical perturbations in the environment. The persistence of lignin in the environment dictates that microbial as well as chemical and physical conditions are important effectors of its degradation. Lignins are probably structurally modified by such factors as light, pH, O_2, in the environment. The effects of light on photodecomposition of certain aromatic compounds are well documented (e.g., Jethwa et al., 1979). Lignins are generally acid stable but are solubilized under alkaline conditions. Chemical oxidation of lignins in air should be expected. The most notable environmental parameter that influences microbial decomposition of lignins is the absence of O_2. Peats, lignites, and coals resulted in part from lignin deposition in anaerobic environments (Swain, 1970). Peats are considered to contain organic carbon that originates from incomplete decomposition of plants in water-saturated, anaerobic conditions. Recently, Hayatsu et al. (1979) have demonstrated a relationship between lignins and coal. These investigators characterized the phenolic acid components of coal degradation products and showed that lignin-like polymers (i.e., ether-linked aromatic systems) were incorporated into the macromolecules of lignite and bituminous coals. The lignites and bituminous coals examined were more highly cross-linked and aromatized than natural lignins. Hackett et al. (1977) demonstrated that ^{14}C synthetic lignins were not degraded in anaerobic environments, including rumen contents and aquatic sediments, in the presence or absence of nitrate and sulfate. Further studies of Hackett (1975) showed that water saturation of ligninolytic soils inhibits lignin decomposition, a situation that may be paralleled in peats.

A molecular basis for the recalcitrance of lignin to biological decomposition in anaerobic environments has been observed (J. G. Zeikus, A. Wellstein, and T. K. Kirk, unpublished data). [^{14}C]-DHPs were chemically depolymerized into high- and low- (<1000 mol. wt.) molecular-weight fractions, and their decomposition in anaerobic sediments was compared to a variety of ^{14}C-ring-labeled aromatic substrates. Low-molecular-weight [^{14}C]-DHP was significantly decomposed (>5% conversion to $^{14}CO_2$ and $^{14}CH_4$ in 30 days) but not as readily as guiacyl-glycerol-β-(o-methoxyphenyl) ether (a model lignin), phenol, or vanillate (>30% decomposition in 30 days). These data suggest that O_2 is required for the biologically mediated depolymerization of lignin but not for the microbial decomposition of model lignins that contain ether linkages or soluble aromatic compounds that contain an oxygen function in the molecule. The biochemistry of anaerobic aromatic catabolism and the environmental fate of aromatic compounds in anaerobic environments have recently been reviewed (Evans, 1977; Zeikus, 1980b).

4.2. Biodegradation of Lignin in Nature

Detailed understanding of the microbial ecology of lignin decomposition in different ecosystem types is lacking at present. Studies on the decomposition of natural lignins in terrestial (especially in relation to humus formation) and aquatic habitats have been initiated. Biomass mineralization studies that relate to lignin or polyphenolic metabolism in peat land have not been detailed. This is an especially challenging and significant environment to examine organic mineralization because one-half of the total dry organic matter of all soil types is present in global peats (Bolin *et al.*, 1979).

Hackett *et al.* (1977) demonstrated that the biodegradation of ^{14}C-alkyl-labeled lignin varied greatly in different soil types and horizons. Amounts of [^{14}C]aryl lignin mineralized in Wisconsin soils were significantly correlated with organic carbon, organic nitrogen, nitrate-N, exchangeable calcium, and exchangeable potassium. Lignin decomposition in the tropical soils examined did not show dramatic differences from ligninolytic activity in temperate soils. Lignin mineralization was slow in all soils examined (>1% decomposition/day). Further studies of Hackett (1975) explored the microbial ecology of lignin decomposition in the Oa1 horizon of an actively ligninolytic, poorly drained Adrian muck soil. *In vitro* incubations revealed a temperature range for lignin decomposition characteristic of microbial organic decomposition in temperate environments (optimum near 30°C). Lignins labeled in the propane side chains

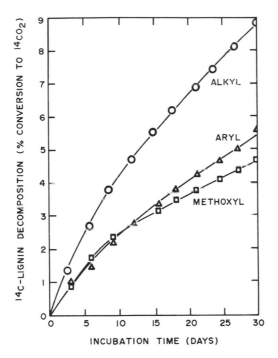

Figure 7. Biodegradation of differentially labeled ^{14}C-DHPs in the Oa1 horizon of an Adrian muck soil (from Hackett, 1975).

were mineralized at significantly faster rates than those labeled in the ring or methoxyl moieties (see Fig. 7). The addition of readily metabolizable carbon (i.e., glucose, xylose, or cellulose) decreased lignin decomposition. The influence of additions of specific antibiotics on the rates of lignin mineralization suggested that both procaryotic and eucaryotic populations were actively ligninolytic in this soil. These findings and the isolation of bacteria which caused limited structural modifications of [^{14}C]lignins (see Section 3.2.2) indicated that microbes other than wood-rot fungi and including bacteria were active agents of lignin mineralization in soil. Crawford *et al.* (1977a) used the detection of ^{14}CO$_2$ from ^{14}C-wood lignin to estimate the most probable number of ligninolytic bacteria in soils and reported higher numbers of lignin degraders in marsh soil ($>2 \times 10^6$/g) than in forest soil (6.2×10^4/g).

Humus is a generic term used to describe refractory organic materials resulting from microbiological decay of plant and animal tissue in soil. The humic acid fraction of humus is regarded as a polyphenol and originates from degradation and condensation reactions associated with the decomposition of amino acids, lignin, and other phenolic polymers such as plant tannins or microbial melanins (Christman and Oglesby, 1971). Tannin decomposition in nature has not been well examined, but *Penicillium adametzi* grows on low-molecular-weight tannins as sole carbon sources (W. D. Grant, 1976). Nelson *et al.* (1979) reported that incorporating ^{14}C-labeled insoluble microbial cell walls or cytoplasmic contents into phenolic or hydroxybenzoic polymers decreased organic mineralization of microbial cells in soil by 31–56%. These studies support the concept of soil humus formation by interaction of cellular components with the products of microbial synthesis. Martin and Haider (1979) compared the degradation of differentially ^{14}C-labeled DHPs, fungal melanins, and phenols in soils. They concluded that lignin, phenols, and melanins are relatively resistant to decomposition, that the greatest rate of carbon loss during decomposition occurs early, and that rates of mineralization are not significantly influenced by addition of readily available organic residues. The relation of these new data on polyphenolic metabolism to humus formation in soil is the topic of a recent review (Martin and Haider, 1980).

Detailed, meaningful studies have not been reported on microbial decomposition of lignin in aquatic ecosystems (see Fenchel and Jorgenson, 1977). Hackett *et al.* (1977) noted that [^{14}C]lignins were slowly mineralized during aerobic *in vitro* incubations of dimictic lake sediments but not during anaerobic incubations. Crawford *et al.* (1977a) noted that the most probable number of ligninolytic organisms was considerably lower in lake water (3.3 × 10/ml) than in soils (2 × 10^6/g). Godshalk and Wetzel (1978) reported that the lignin fraction of aquatic angiosperms was the portion of detritus most resistant to decomposition. The fate of lignin in freshwater lakes is dynamic and deserves detailed study. A major organic fraction of lake sediments is lignin and humus materials. In dimictic lakes that stratify, lignin inputs may only degrade during turnover, and this could lead to gradual carbon accumulation and perhaps play a role in eutrophication. In meromictic lakes, lignin inputs without subsequent mineralization (because of constant anaerobic conditions) may lead to diagenesis of acid bogs.

4.3. Biodegradation of Industrial and Agricultural Waste Lignins

Ligneous wastes generated by agriculture and pulp and paper industries pose several pollution problems, and these wastes should be effectively biodegraded prior to environmental deposition at high concentrations. Major forms of biomass-derived wastes in the United States include (in metric tons): pulp and paper mill waste, 100; feedlot manure, 237; and, urban refuse, 150 (Zeikus, 1980a). The exact chemical composition and amount of lignin in these wastes vary considerably. Pulp and paper mill waste primarily contains chemically modified lignin (Sarkanen and Hergert, 1971). Feedlot manures contain lignocellulosics, but microbial cells are also major constituents (Zeikus, 1980a,b). Cellulose is the major chemical constituent of urban waste (Wilke,

1975). The lignin content of all these biomass-derived wastes limits their biodegradation.

Industrial lignins in pulp and paper mill effluents are virtually unaltered by conventional pollution-abatement processes. Consequently, rivers that receive these effluents are characteristically dark colored and contain large amounts of sedimented lignins. Industrial lignins, as chemical substrates, differ drastically from natural lignins and in general have lower molecular weights, greater solubilities, and altered intermonomer linkages as compared to natural lignins (Sarkanen and Hergert, 1971). Fortunately, the modifications in the chemical structure of industrial lignins generally increase their biodegradability. In the United States, the alkaline sulfate process (i.e., Kraft) produces greater than four times the amount of ligneous wastes than does the sulfite pulping process (Sarkanen and Hergert, 1971). Commercial lignins and lignosulfonates contain a variety of organic chemical constituents (e.g., polysaccharides, sugars, alcohols, aldehydes, etc.) and can vary considerably in physical structure (Glennie, 1971; Marton, 1971). ^{14}C-Industrial lignins have been prepared as model substrates for biodegradation studies by pulping [^{14}C]-DHPs or ^{14}C-wood lignins (Lundquist et al., 1977; Crawford and Crawford, 1980). Perhaps pulping of uniformly labeled ^{14}C-wood would be a more representative substrate for biodegradation studies than those described to date.

A variety of aerobic microorganisms, including bacteria and fungi, grow on commercial lignosulfonates as sole sources of carbon and energy (Selin and Sundman, 1971; Pandila, 1973; Ban and Glauser-Soljan, 1979). Selin and Sundman (1971) demonstrated that mixed bacterial populations utilized only low-molecular-weight fractions (<1000 mol. wt.) as energy sources for growth and polymerized higher-molecular-weight fractions of lignosulfonates. Mixed, anaerobic bacterial populations that include Desulfovibrio grow on spent sulfite liquor and transform lignosulfonates to lignins and H_2S (Jurgensen and Patton, 1979).

White-rot fungi actively degrade lignosulfonates (Selin et al., 1975; Hiroi et al., 1976; Lundquist et al., 1977). Selin et. al. (1975) demonstrated that white-rot fungi grow on calcium lignosulfonate fractions with a molecular weight of 1350 as sole source of carbon and energy, but not on fractions of higher molecular weight. Most notably, growth of white-rot fungi on lignosulfonate is associated with induction of phenoloxidase that causes polymerization of lignosulfonates (Hiroi et al., 1976; Räihä and Sundman, 1975). This has also been observed with other fungi (Huttermann et al., 1977).

Despite the importance of Kraft lignins, the biodegradation of such lignins has not received much attention. Both bacteria and fungi degrade Kraft lignins (Tansey et al., 1977; Lundquist et al., 1977; Forney and Reddy, 1979). Drew and Kadam (1979) demonstrated that white-rot fungi and A. fumigatus degraded Kraft lignin but required a co-substrate (e.g., cellulose) for growth.

Interestingly, a mixed bacterial culture grows on a high-molecular-weight fraction (1500 mol. wt.) of Kraft lignin as sole energy source (Forney and Reddy, 1979).

New pollution-abatement technology is aimed at total destruction of industrial lignins by microbiological and/or chemical means. Promising bioconversions have been suggested that employ white-rot fungi, and include decolorization of pulp mill effluents and single cell protein production using industrial lignins as growth substrates (Kirk *et al.*, 1980). Stern and Gasner (1974) demonstrated that pretreatment of Kraft lignins by ozonation renders the waste fermentable as a growth substrate for *Aspergillus niger, Rhizopus stolonifer, Penicillium chrysogenum,* and *Candida utilis.*

Urban refuse and feedlot manure are produced in larger volumes than pulp and paper mill wastes. These ligneous waste pollutants are receiving considerable attention as starting substrates for chemical or fuel production via anaerobic microbial fermentations. Most notably, newsprint from garbage is fermented to ethanol and manures are fermented to methane (Wilke, 1975; Zeikus, 1980a). Alternatively, the ligneous components of manures and garbage can be converted to compost (Walters, 1974) or can be pretreated by microorganisms to increase their digestibility as animal feeds (Kaneshiro, 1977; Rosenberg, 1979). Lignin in these wastes greatly limits their bioconvertibilities, especially in anaerobic processes. A variety of pretreatments have been examined to increase the fermentability of lignocellulosic wastes by anaerobic microorganisms, and chemical rather than biological treatments are most successful (Decker and Richards, 1973; Millet *et al.*, 1975; Zeikus, 1980a). For example, alkaline–heat treatment of lignocellulosics in garbage or manures greatly enhances anaerobic fermentability to methane (McCarty *et al.*, 1976). This treatment destroys the physical structure of lignocellulose and also converts lignin to low-molecular-weight aromatic compounds that are fermentable substrates for mixed-culture methanogenesis (Healy and Young, 1979).

5. Summary

The relationship of lignin to the carbon cycle in nature is very dynamic (Fig. 8). The majority of organic carbon on earth is in the form of lignocellulosic biomass synthesized from CO_2 and in products of degradation of the lignocellulosic biomass (humus, peats, lignites, and coals). Lignocellulose is slowly mineralized in nature because lignin, a biological plastic, provides an effective barrier to impede microbial degradation of the cellulosic component by catabolic enzymes. In aerobic environments, the lignin polymer is biologically eroded by nonspecific microbial enzymes to degradative products that are

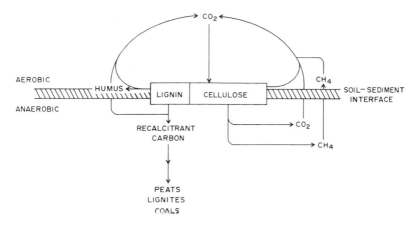

Figure 8. The importance of lignocellulose in the carbon cycle.

converted to CO_2 or complexed with other biochemical constituents in soil to form humus. In anaerobic environments, lignin and humus are recalcitrant (owing to an oxidative requirement for effective depolymerization and solubilization), accumulate, and contribute to diagenesis of peats, lignites, and coals.

ACKNOWLEDGMENTS. This research was supported by the College of Agricultural and Life Sciences, University of Wisconsin, and in part by grant PFR-7910084 from the National Science Foundation. This article is dedicated to T. K. Kirk, who helped pioneer meaningful studies on the microbial degradation and modification of lignin.

References

Adler, T., 1977, Lignin chemistry—past, present and future, *Wood Sci. Technol.* **11:**169–218.

Alexander, M., 1973, Nonbiodegradable and other recalcitrant molecules, *Biotechnol. Bioeng.* **15:**611–647.

Alexander, M., 1975, Environmental and microbiological problems arising from recalcitrant molecules, *Microb. Ecol.* **2:**17–27.

Ander, P., and Eriksson, K. -E., 1976, The importance of phenol oxidase activity in lignin degradation by the white-rot fungus *Sporotrichum pulverulentum, Arch. Microbiol.* **109:**1–8.

Ander, P., and Eriksson, K. -E., 1978, Lignin degradation and utilization by microorganisms, in: *Progress in Industrial Microbiology,* Vol. 14 (M. J. Bull, ed.) pp. 1–38 Elsevier, Amsterdam.

Arai, T., and Mikami, Y., 1972, Chromogenicity of *Streptomyces, Appl. Microbiol.* **23:**402–406.

Ban, S, and Glauser-Soljan, M., 1979, Rapid biodegradation of calcium lignosulfonate by means of a mixed culture of microorganisms, *Biotechnol. Bioeng.* **21**:1917–1928.

Bartsch, E., Lerbs, W., and Luckner, M., 1979, Phenol oxidase activity and pigment synthesis in conidiospores of *Penicillium cyclopium, Z. Allg. Mikrobiol.* **19**:75–82.

Bolin, B., Degens, E. T., Kempe, S., and Ketner, P., 1979, *The Global Carbon Cycle,* John Wiley and Sons, New York.

Bu'Lock, J. D., 1975, Secondary metabolism in fungi and its relationship to growth and development, in: *The Filamentous Fungi* (J. E. Smith and D. R. Berry, eds.), *Industrial Mycology,* Vol. 1, pp. 33–58, Wiley, New York.

Bu'Lock, J. D., Detroy, R. W., Hostalek, A., and Munum-Al-Shakarcki, A., 1974, Regulation of secondary biosynthesis in *Gibberella fujikuroi, Trans. Br. Mycol. Soc.* **62**:377–389.

Burdsall, H. H., and Eslyn, W. E., 1974, A new *Phanerachaete* with a *chrysosporium* state, *Mycotaxon* **1**:123–133.

Buswell, J. A., and Eriksson, K.-E., 1979, Aromatic ring cleavage by the white-rot fungus *Sporotrichum pulverulentum, FEBS Lett.* **104**:258–260.

Buswell, J. A., Ander, P., Pettersson, B., and Eriksson, K.-E., 1979a, Oxidative decarboxylation of vanillic acid by *Sporotrichum pulverulentum. FEBS Lett.* **103**:98–101.

Buswell, J. A., Hamp, S., and Eriksson, K.-E., 1979b, Intracellular quinone reduction in *Sporotricum pulverulentum* by a NAD(P)H: Quinone osidoreductase. *FEBS Lett.* **108**:229–232.

Butler, J. H. A., and Buckerfield, J., 1979, Digestion of lignin by termites, *Soil Biol. Biochem.* **11**:507–513.

Cain, R., 1980, The uptake and catabolism of lignin-related aromatic compounds and their regulation in microorganisms, in: *Lignin Biodegradation: Microbiology, Chemistry and Potential Applications* (T. K. Kirk, T. Higuchi, and H. M. Chang, eds.) pp. 21–60, CRC Press, Boca Raton, Florida.

Cartwright, N. J., and Haldom, K. S., 1973, Enzymic lignin, its release and utilization by bacteria, *Microbios* **8**:7–14.

Chang, H. M., Chen, C.-L., and Kirk, T. K., 1980, Chemistry of lignin degraded by white-rot fungi, in: *Lignin Biodegradation: Microbiology, Chemistry and Potential Applications* (T. K. Kirk, T. Higuchi, and H. M. Chang, eds.), pp. 215–230, CRC Press, Inc., Boca Raton, Florida.

Chen, C.-L., Chang, H.-M., and Kirk, T. K., 1977, Betulachrysoquinone Hemeketal: Ap-Benzoquinone Hemiketal macrocyclic compound produced by *Phanerochaete chrysosporium, Phytochemistry* **16**:1983–1985.

Christman, R. F., and Oglesby, R. T., 1971, Lignins: Microbiological degradation and the formation of humus, in: *Lignins: Occurrence, Formation, Structure and Reactions* (K. V. Sarkanen and C. H. Ludwig, eds.), pp. 769–793, Wiley-Interscience, New York.

Côté, W. A., 1977, Wood ultrastructure in relation to chemical composition, in: *The Structure, Biosynthesis, and Degradation of Wood* (F. A. Loewus and V. C. Runeckles, eds.), *Recent Advances in Phytochemistry,* Vol. 11, pp. 1–44, Plenum Press, New York.

Cowling, E. B., and Kirk, T. K., 1976, Properties of cellulose and lignocellulosic materials as substrates for enzymatic conversion processes, *Biotechnol. Bioeng. Symp.* **6**:95–123.

Crawford, D. L., and Crawford, R. L., 1976, Microbial degradation of lignocellulose: The lignin component, *Appl. Environ. Microbiol.* **31**:714–717.

Crawford, D. L., and Crawford, R. L., 1980, Microbial degradation of lignin, *Enzyme Microbiol. Technol.* **2**:11–22.

Crawford, D. L., and Sutherland, J. B., 1979, The role of actinomycetes in the decomposition of lignocellulose, in: *Developments in Industrial Microbiology,* Vol. 20 (L. A. Underkofler, ed.), pp. 143–152, Proceedings of Thirty-Fifth General Meeting of the Society for Industrial Microbiology, Arlington, Virginia.

Crawford, R. L., Kirk, T. K., Harkin, J. M., and McCoy, E., 1973a, Bacterial cleavage of an aryl glycerol-β-aryl ether bond, *Appl. Microbiol.* **25**:322–324.

Crawford, R. L., McCoy, E., Harkin, J. M., Kirk, T. K., and Obst, J. R., 1973b, Degradation of methoxylated benzoic acids by a *Nocardia* from a lignin-rich environment: Significance of lignin degradation and effect of chloro substitutes, *Appl. Microbiol.* **26**:176–184.

Crawford, R. L., Kirk, T. K., and McCoy, E., 1975, Dissimilation of the lignin model compound veratryl-glycerol-β-(o-methoxyphenyl) ether by *Pseudomonas acidovorans:* Initial transformations, *Can. J. Microbiol.* **21**:577–579.

Crawford, D. L., Crawford, R. L., and Pometto, A. L., III, 1977a, Preparation of specifically labeled ^{14}C-(lignin)- and ^{14}C-(cellulose)-lignocelluloses and their decomposition by the microflora of soil, *Appl. Environ. Microbiol.* **33**:1247–1251.

Crawford, R. L., Crawford, D. L., Olafsson, C., Wikstrom, L., and Wood, J. M., 1977b, Biodegradation of natural and man-made recalcitrant compounds with particular reference to lignin, *J. Agric. Food Chem.* **25**:704–708.

Dagley, S., 1975, Microbial degradation of organic compounds in the biosphere, *Am. Sci.* **63**:681–689.

Dagley, S., 1978a, Determinants of biodegradability, *Q. Rev. Biophys. II* **4**:577–602.

Dagley, S., 1978b, Microbial catabolism, the carbon cycle and environmental pollution, *Naturwissenschaften* **65**:85–89.

Dekker, R. F. H., and Richards, G. N., 1973, Effect of delignification of the *in vitro* rumen digestion of polysaccharides of Bagasse, *J. Sci. Food Agric.* **24**:375–379.

Drew, S. W., and Kadam, K. L., 1979, Lignin metabolism by *Aspergillus fumigatus* and white-rot fungi, in: *Developments in Industrial Microbiology,* Vol. 20 (L. A. Underkofler, ed.), pp. 153–164, Proceedings of the Thirty-Fifth General Meeting of the Society for Industrial Microbiology, Arlington, Virginia.

Esenther, G. R., and Kirk, T. K., 1974, Catabolism of aspen sapwood in *Reticulitermes flavipes* (Isoptera: Rhinotermitidae), *Ann. Entomol. Soc. Am.* **67**:989–991.

Eslyn, W. E., Kirk, T. K., and Effland, M. J., 1975, Changes in the chemical composition of wood caused by six soft-rot fungi, *Phytopathology* **65**:473–476.

Evans, N. C., 1977, Biochemistry of the bacterial catabolism of aromatic compounds in anaerobic environments, *Nature* **270**:17–22.

Faber, M. D., 1979, Microbial degradation of recalcitrant compounds and synthetic aromatic polymers, *Enzyme Microbiol. Technol.* **1**:226–232.

Fenchel, T. M., and Jorgenson, B. B., 1977, Detritus food chains of aquatic ecosystems: The role of bacteria, in: *Advances in Microbial Ecology,* Vol. 1 (M. Alexander, ed.), pp. 1–58, Plenum Press, New York.

Fenn, P., and Kirk, T. K., 1979, Ligninolytic system of *Phanerochaete chrysosporium:* Inhibition by o-phthalate, *Arch. Microbiol.* **134**:307–309.

Forney, L. J., and Reddy, C. A., 1979, Bacterial degradation of Kraft lignin, in: *Developments in Industrial Microbiology,* Vol. 20 (L. A. Underkofler, ed.), pp. 138–142, Proceedings of the Thirty-Fifth General Meeting of the Society for Industrial Microbiology, Arlington, Virginia.

Freudenberg, K., and A. C. Neish, 1968, The constitution and biosynthesis of lignin, in: *Constitution and Biosynthesis of Lignin,* pp. 1–168, Springer-Verlag, New York.

Froehner, S. C., and Eriksson, K.-E., 1974a, Induction of *Neurospora crassa* Laccase with protein synthesis inhibitors, *J. Bacteriol.* **1**:450–457.

Froehner, S. C., and Eriksson, K.-E., 1974b, Purification and properties of *Neurospora crassa* Laccase, *J. Bacteriol.* **120**:458–465.

Glennie, D. W., 1971, Lignins: Reactions in sulfite pulping, in: *Lignins: Occurrence, Formation, Structure and Reactions* (K. V. Sarkanen and C. H. Ludwig, eds.), pp. 597–638, Wiley-Interscience, New York.

Godshalk, G. L., and Wetzel, R. G., 1978, Decomposition of aquatic angiosperms. II. Particulate components, *Aquat. Bot.* **5**:301–327.

Goring, D. A. I., 1971, Polymer properties of lignin and lignin derivatives, in: *Lignins: Occurrence, Formation, Structure and Reactions* (K. V. Sarkanen and C. H. Ludwig, eds.), pp. 695–768, Wiley-Interscience, New York.

Grant, D., 1977, Chemical structure of humic substances, *Nature* **270**:709–710.

Grant, W. D., 1976, Microbial degradation of condensed tannins, *Science* **193**:1137–1139.

Gross, G. G., 1977, Biosynthesis of lignin and related monomers, in: *The Structure, Biosynthesis, and Degradation of Wood* (F. A. Loewus and V. C. Runeckles, eds.), pp. 141–184, *Recent Advances in Phytochemistry*, Vol. 11, Plenum Press, New York.

Hackett, W. F., 1975, Microbial Degradation of Synthetic ^{14}C-Lignins in Natural Environments, M. S. Thesis, University of Wisconsin, Madison.

Hackett, W. F., Connors, W. J., Kirk, T. K., and Zeikus, J. G., 1977, Microbial decomposition of synthetic ^{14}C-labeled lignins in nature: Lignin biodegradation in a variety of natural materials. *Appl. Environ. Microbiol.* **33**:43–51.

Ha-Huy-Kê, and Luckner, M., 1979, Structure and function of the conidiospore pigments of *Penicillium cyclopium, Z. Allg. Mikrobiol.* **2**:117–122.

Haider, K., and Trojanowski, J., 1975, Decomposition of specifically ^{14}C-labeled phenols and dehydropolymers of coniferyl alcohol as models for lignin degradation by soft and white rot fungi, *Arch. Microbiol.* **105**:33–41.

Haider, K., and Trojanowski, J., 1980, A comparison of the degradation of ^{14}C-labeled DHP and cornstalk lignins by micro- and macrofungi and by bacteria, in: *Lignin Biodegradation: Microbiology, Chemistry and Potential Applications* (T. K. Kirk, T. Higuchi, and H. M. Chang, eds.) pp. 111–134, CRC Press, Boca Raton, Florida.

Haider, K., Trojanowski, J., and Sundman, V., 1978, Screening for lignin degrading bacteria by means of ^{14}C-labeled lignins, *Arch. Microbiol.* **119**:103–106.

Halliwell, B., 1979, Oxygen-free radicals in living system: Dangerous but useful? in: *Strategies of Microbial Life in Extreme Environments* (M. Shilo, ed.), pp. 195–221, Dahlem Konferenzen, Berlin.

Haraguchi, T., and Hatakeyama, H., 1980, Biodegradation of lignin related polystyrenes, in: *Lignin Biodegradation: Microbiology, Chemistry and Potential Applications* (T. K. Kirk, T. Higuchi, and H. M. Chang, eds.), pp. 147–160, CRC Press, Boca Raton, Florida.

Hayatsu, R., Winans, R. E., McBeth, R. L., Scott, R. G., Moore, L. P., and Studier, M. H., 1979, Lignin-like polymers in coals, *Nature* **278**:41–43.

Healy, J. B., and Young, L. Y., 1979, Anaerobic biodegradation of eleven aromatic compounds to methane, *Appl. Environ. Microbiol.* **38**:84–89.

Higuchi, T., 1980, Lignin structure and morphological distribution in plant cell walls, in: *Lignin Biodegradation: Microbiology, Chemistry and Potential Applications* (T. K. Kirk, T. Higuchi and H. M. Chang, eds.), pp. 1–20, CRC Press, Boca Raton, Florida.

Hiroi, T., and Eriksson, K.-E., 1976, Microbiological degradation of lignin. Part 1. Influence of cellulose on the degradation of lignins by the white-rot fungus *Pleurotus ostreatus, Sven. Papperstidn.* **5**:157–161.

Hiroi, T., Eriksson, K.-E., and Stenlund, B., 1976. Microbiological degradation of lignin. Part 2. Influence of cellulose upon the degradation of calcium lignosulfonate of various molecular sizes by the white-rot fungus *Pleurotus ostreatus, Sven. Papperstidn.* **5**:162–166.

Horvath, R. S., 1972, Microbial co-metabolism and the degradation of organic compounds in nature. *Bacteriol. Rev.* **36**:146–155.

Huttermann, V.-A., Gebauer, M., Volger, C., and Rosger, C., 1977, Polymerisation ured abbau von natrium—lignin sulfonat durch *Fomes annosus* (Fr.), *Cooke. Holzforschung* **31**:83–89.

Ishihara, T., 1980, The role of laccase in lignin biodegradation, in: *Lignin Biodegradation:*

Microbiology, Chemistry, and Potential Applications (T. K. Kirk, T. Higuchi, and H. M. Chang, eds.) pp. 17–32, CRC Press, Boca Raton, Florida.

Iwahara, S., 1980, Microbial degradation of DHP, in: *Lignin Biodegradation: Microbiology, Chemistry and Potential Applications* (T. K. Kirk, T. Higuchi and H. M. Chang, eds.) pp. 151–170, CRC Press, Boca Raton, Florida.

Jethwa, S. A., Stanford, J. B., and Sugden, J. K, 1979, Light stability of vanillin solutions in ethanol, *Drug Dev. Ind. Pharmacy* **5**:79–85.

Jurgensen, M. F., and Patton, J. T., 1979, Bioremoval of lignosulphonates from sulphite pulp mill effluents, *Process Biochem.* **14**.

Kaneshiro, T., 1977, Lignocellulosic agricultural wastes degraded by *Pleurotus ostreatus*, in: *Developments in Industrial Microbiology*, Vol. 18, pp. 591–597, Society for Industrial Microbiology, Arlington, Virginia.

Kawakami, M., 1980, Metabolism of lignin related compounds by several Pseudomonas, in: *Lignin Biodegradation: Microbiology, Chemistry and Potential Applications* (T. K. Kirk, T. Higuchi and H. M. Chang, eds.), pp. 103–126, CRC Press, Boca Raton, Florida.

Keyser, P., Kirk, T. K., and Zeikus, J. G., 1978, Lignionolytic enzyme system of *Phanerochaete chrysosporium*: Synthesized in the absence of lignin in response to nitrogen starvation, *J. Bacteriol.* **135**:790–797.

Kirk, T. K., 1975, Effects of a brown-rot fungus, *Lenzites trabea*, on lignin in spruce wood, *Holzforschung* **29**:99–107.

Kirk, T. K., and Chang, H.-M., 1974, Decomposition of lignin by white-rot fungi. I. Isolation of heavily degraded lignins from decayed spruce, *Holzforschung* **28**:217–222.

Kirk, T. K., and Chang, H.-M., 1975, Decomposition of lignin by white-rot fungi. II. Characterization of heavily degraded lignins from decayed spruce, *Holzforschung* **29**:56–64.

Kirk, T. K., and Highley, T. L., 1973, Quantitative changes in structural components of conifer woods during decay by white- and brown-rot fungi, *Phytopathology* **63**:1338–1342.

Kirk, T. K., and Lorenz, L. F., 1973, Methoxyhydroquinone, an intermediate of vanillate catabolism by *Polyporus dichrous*, *Appl. Microbiol.* **26**:173–175.

Kirk, T. K., Lorenz, L. F., and Larsen, M. J., 1975a, Partial characterization of a phenolic pigment from sporocarps of *Phellinus igniarius*, *Phytochemistry* **14**:281–284.

Kirk, T. K., Connors, W. J., Bleam, R. D., Hackett, W. F., and Zeikus, J. G., 1975b, Preparation and microbial decomposition of synthetic [^{14}C] lignins, *Proc. Nat Acad. Sci.* **72**:2515–2519.

Kirk, T. K., Connors, W. J., and Zeikus, J. G., 1976, Requirement for a growth substrate during lignin decomposition by two wood-rotting fungi, *Appl. Environ. Microbiol.* **32**:192–194.

Kirk, T. K., Connors, W. J., and Zeikus, J. G., 1977, Advances in understanding the microbiological degradation of lignin, in: *The Structure, Biosynthesis, and Degradation of Wood* (F. A. Loewus and V. C. Runeckles, eds.), *Recent Advances in Phytochemistry*, Vol. 11, pp. 369–394, Plenum Press, New York.

Kirk, T. K., Schultz, E. M., Connors, W. J., Lorenz, L. F., and Zeikus, J. G., 1978, Influence of culture parameters on lignin metabolism by *Phanerochaete chrysosporium*, *Arch. Microbiol.* **117**:277–285.

Kirk, T. K., Higuchi, T., and Chang, H. M., 1980, *Lignin Biodegradation: Microbiology, Chemistry and Potential Applications*, CRC Press, Boca Raton, Florida.

Lai, Y. Z., and Sarkenen, K. V., 1971, Isolation and structural studies, in: *Lignins: Occurrence, Formation, Structure and Reactions* (K. V. Sarkanen and C. H. Ludwig, eds.), pp. 165–240, Wiley-Interscience, New York.

Lundquist, K., and Kirk, T. K., 1978, *De novo* synthesis and decomposition of veratryl alcohol by a lignin-degrading Basidiomycete, *Phytochemistry* **17**:1676.

Lundquist, K., Kirk, T. K., and Connors, W. J., 1977, Fungal degradation of Kraft lignin and lignin sulfonates prepared from synthetic ^{14}C-lignins, *Arch. Microbiol.* **112**:291–296.

Martin, J. P., and Haider, K., 1979, Biodegradation of ^{14}C-labeled model and cornstalk lignins, phenols, model phenolase humic polymers, and fungal melanins as influenced by a readily available carbon source and soil, *Appl. Environ. Microbiol.* **38**:283–289.

Martin, J. P., and Haider, K., 1980, Microbial degradation and stabilization of ^{14}C-labeled lignins, phenols and phenolic polymers in relation to soil humus formation, in: *Lignin Biodegradation: Microbiology, Chemistry and Potential Applications* (T. K. Kirk, T. Higuchi and H. M. Chang, eds.), pp. 77–100, CRC Press Inc., Boca Raton, Florida.

Marton, J., 1971, Lignins: reactions in alkaline pulping, in: *Lignins: Occurrence, Formation, Structure and Reactions* (K. V. Sarkanen and C. H. Ludwig, eds.), pp. 639–694, Wiley-Interscience, New York.

Mayaudon, J., and Batistic, L., 1970, Degradation biologique de la lignine ^{14}C dans le sol, *Ann. Inst. Pasteur* **118**:191–198.

McCarty, P. L., Young, L. Y., Stuckey, D. C., and Healy, J. B., 1976, Heat treatment for increasing methane yields from organic materials, in: *Microbial Energy Conversion* (H. G. Schlegel and J. Barnea, eds.), pp. 179–200. Erich Goltze K. G., Gottingen.

Millet, M. A., Baker, A. J., and Satter, L. D., 1975, Pretreatments to enhance chemical, enzymatic and microbiological attack of cellulose materials, in: *Cellulose as a Chemical and Energy Resource* (C. R. Wilke, ed.), Biotechnology and Bioengineering Symposium No. 5, pp. 193–220, John Wiley & Sons, New York.

Neish, A. C., 1968, Monomeric intermediates in the biosynthesis of lignin, in: *Constitution and Biosynthesis of Lignin*, pp. 169–238, Springer-Verlag, New York.

Nelson, D. W., Martin, J. D., and Ervin, J. O., 1979, Decomposition of microbial cells and components in soil and their stabilization through complexing with model humic acid-type phenolic polymers, *Soil Sci. Soc. Am. J.* **43**:84–88.

Neuhauser, E. F., Hartenstein, R., and Connors, W. J., 1978, Soil invertebrates and the degradation of vanillin, cinnamic acid, and lignins, *Soil Biol. Biochem.* **10**:431–435.

Nickerson, W. J., 1971, Decomposition of naturally-occurring organic polymers, in: *Organic Compounds in Aquatic Environments* (S. D. Faust and J. V. Hunter, eds.), pp. 599–609, Marcel Dekker, New York.

Ohta, M., Higuchi, T., and Iwahara, S., 1979, Microbial degradation of dehydrodiconiferyl alcohol, a lignin substructure model, *Arch. Microbiol.* **121**:23–28.

Pandila, M. M., 1973, Microorganisms associated with microbiological degradation of lignosulphonates: A review of literature, *Pulp Pap. Mag. Can.* **74**:T78–T82.

Perry, J. J., 1979, Microbial cooxidations involving hydrocarbons, *Microbiol. Rev.* **43**:59–72.

Räihä, M., and Sundman, V., 1975, Characterization of lignosulfonate-induced phenol oxidase activity in the atypical white-rot fungus *Polyporus dichrous, Arch. Microbiol.* **105**:73–76.

Reiners, W. A., 1973, Terrestrial detritus and the carbon cycle, in: *Carbon and the Biosphere* (G. M. Woodswell and E. V. Pecan, eds.), pp. 303–327, *Proceedings of the 24th Brookhaven Symposium in Biology*, Technical Information Center, Office of Information Services, U.S. Atomic Energy Commission, Springfield, Virginia.

Robinson, L. E., and Crawford, R. L., 1978, Degradation of ^{14}C-labeled lignins by *Bacillus megaterium, FEMS Microbiol. Lett.* **4**:301–302.

Rosenberg, S. L., 1979, Physiological studies of lignocellulose degradation by the thermotolerant mold, *Chrysosporium pruinosum,* in: *Developments in Industrial Microbiology,* Vol. 20 (L. A. Underkofler, ed.), Proceedings of the 35th General Meeting of the Society for Industrial Microbiology, Arlington, Virginia.

Sarkanen, K. V., 1971, Lignins: Precursors and their polymerization, in: *Lignins: Occurrence, Formation, Structure and Reactions* (K. V. Sarkanen and C. H. Ludwig, eds.), pp. 95–164, Wiley-Interscience, New York.

Sarkanen, K. V., and Hergert, H. L., 1971, Lignins: Classification and distribution, in: *Lignins: Occurrence, Formation, Structure and Reactions* (K. V. Sarkanen and C. H. Ludwig, eds.), Wiley-Interscience, New York.

Selin, J.-F., and Sundman, V., 1971, Microbial actions on lignosulfonates, *Finska Kemists. MEDD.* **80N:**11–19.

Selin, J.-F., Sundman, V., and Räihä, M., 1975, Utilization of polymerization of lignosulfonates by wood-rotting fungi, *Arch. Mikrobiol.* **103:**63–70.

Sorensen, H., 1962, Decomposition of lignin by soil bacteria and complex formation between autoxidized lignin and organic nitrogen compounts, *J. Gen. Microbiol.* **27:**21–34.

Sorenson, H., 1963, Studies on the decomposition of ^{14}C-labeled barley straw in soil, *Soil Sci.* **9:**45–51.

Stern, A. M., and Gasner, L. L., 1974, Degradation of lignin by combined chemical and biological treatment, *Biotechnol. Bioeng.* **16:**789–805.

Sundman, V., 1964, A description of some ligninolytic soil bacteria and their ability to oxidize simple phenolic compounds, *J. Gen. Microbiol.* **36:**171–183.

Sutherland, J. B., Blanchette, R. A., Crawford, D. L., and Pometto, A. L., III, 1979, Breakdown of Douglas fir phloem by a lignocellulose-degrading *Streptomyces, Curr. Microbiol.* **2:**123–126.

Swain, F. M., 1970, *Non-marine Organic Geochemistry,* Cambridge Earth Science Series, Cambridge University Press, Cambridge.

Tansey, M. R., Murrmann, D. N., Behnke, B. K., and Behnke, E. R., 1977, Enrichment, isolation and assay of growth of thermophilic and thermotolerant fungi in lignin-containing media, *Mycologia* **69:**463–476.

Walters, A. H., 1974, Microbial biodeterioration of materials: Relevance to waste recycling, *Chem. Ind. N.Y.* **4:**365–372.

Wardrop, A. B., 1971, Lignins: Occurrence and formation in plants, in: *Lignins: Occurrence, Formation, Structure and Reactions* (K. V. Sarkanen and C. H. Ludwig, eds.), pp. 14–42, Wiley-Interscience, New York.

Wayman, M., and Obiaga, T. I., 1974, Molecular weights of milled-wood lignins, *Tappi* **57:**123–126.

Weinstein, D. A., Krisnangkura, K., Mayfield, M., and Gold, M. H., 1980, Metabolism of radiolabeled β-guanacyl ether-linked lignin dimeric compounds by *Phanerochaete chrysosporium, Appl. Environ. Microbiol.* **39:**535–540.

Westermark, U., and Eriksson, K.-E., 1974a, Carbohydrate-dependent enzymic quinone reduction during lignin degradation, *Acta Chem. Scand.* **28:**204–208.

Westermark, U., and Eriksson, K.-E., 1974b, Cellobiose: Quinone oxidoreductase, a new wood-degrading enzyme from white-rot fungi, *Acta Chem. Scand.* **28:**209–214.

Whittaker, R. H., and Likens, G. E., 1973, Carbon in the biota, in: *Carbon and the Biosphere* (G. M. Woodwell and E. V. Pecan, eds.), *Proceedings of 24th Brookhaven Symp. in Biology,* Technical Information Center, Office of Information Services, U.S. Atomic Energy Commission, Springfield, Virginia.

Wilke, C. R., 1975, *Cellulose as a Chemical and Energy Resource,* Biotechnology and Bioengineering Symposium #5, John Wiley & Sons, New York.

Wood, J. M., Crawford, R. L., Munck, E., Zimmerman, R., Lipscomb, J. D., Stephens, R. S., Bromley, J. W., Que, L., Jr., Howard, J. B., and Orme-Johnson, W. H., 1977, Structure and function of dioxygenases. One approach to lignin degradation, *J. Agric. Food Chem.* **25:**698–704.

Yaku, F., Tsuji, S., and Koshijima, T., 1979, Lignin carbohydrate complex. Part III. Formation of micelles in the aqueous solution of acidic lignin carbohydrate complex, *Holzforschung* **33:**54–59.

Zeikus, J. G., 1980a, Chemical and fuel production by anaerobic bacteria, *Annu. Rev. Microbiol.*

Zeikus, J. G., 1980b, Fate of lignin and related aromatic substrates in anaerobic environments, in: *Lignin Biodegradation: Microbiology, Chemistry, and Applications* (T. K. Kirk, T. Higuchi and H. M. Chang, eds.), pp. 101–110, CRC Press, Boca Raton, Florida.

Index